RV Electrical & Appliance Systems

Why They Don't Work and How They Could Be Fixed

Thomas Childs

ISBN: 978-0-692-12657-8

All photographs & drawings are originals done by Thomas Childs, except painting on p. 46, done by my mother, Frances Childs.

(actual fortune of author)

Safety Warning/Disclaimer

This book makes every effort to provide accurate and adequate safety information, and tries to steer people away from dangerous RV devices and practices. Still, all devices have their individual idiosyncrasies. The author is not familiar with individual brand names nor specific models of gadgets, therefore any RV users/customizers should always consult manufacturers regarding safety concerns, and should not hold the author/printer/ distributor of this book liable for either injuries or poor performance. (Stated more simply, use this book as a *guide* or starting point for information, but not as a *final authority*.) Here are the more important and relevant RV safety rules:

1) Lead-acid batteries often produce hydrogen gas which can cause batteries or poorly ventilated compartments containing batteries to explode if ignited. Hydrogen may be ignited by open flames of appliances (including pilot lights), lit cigarettes, or sparks made by switches turning appliances on/off, or sparks made within appliances themselves. Lead-acid batteries have sulfuric acid, which can be propelled in any direction during an explosion, and if landing on a nearby person, can cause severe skin burns or even blindness. Always wear eye protection when around batteries. Also, have water handy to quickly wash acid off people in the event of an accident. Get immediate medical help if acid burns do occur. Other non-lead-acid types of batteries have their own safety concerns, which are beyond the scope of this book. Consult manufacturers and other authors.

2) Carbon monoxide is a deadly gas, which can accumulate indoors (especially in small RVs) and kill people. It is made as a byproduct of combustion of propane, gasoline, oil, and other flammable stuff. Read and understand the information and heed the warnings on carbon monoxide, contained in Chapter 4: "Propane."

3) Do not take, spread or use flammable solvents (especially gasoline, which is extremely flammable) indoors, in open or unsealed containers, especially when any propane appliances or candles or cigarettes are burning, or when some appliance or switch might spark and thus ignite these. Ignited solvents can cause fire, explosion or both. This can cause severe injury or death to anyone caught within such.

4) Unburned propane or butane, mixed with air in an enclosed RV space, is poisonous and explosive. Read and heed warning information in Chapter 4: "Propane."

5) Always have a working carbon monoxide and propane gas detector inside your RV. Also have a working smoke detector. To ensure quick escape, be sure the exit door isn't locked from the inside.

6) High voltage electricity can electrocute people. Even 12 Volt low voltage devices may transform the lower voltage to higher voltages.

7) When designing or modifying an appliance, YOU TAKE COMPLETE RESPONSIBILITY for safety. Do your own safety testing.

<u>If I Were Emperor of the World...</u>

Many things annoy me. My annoyance levels range from minor irritation through all out rage. Usually I am powerless to change the things in the world that I do not like, so to cope, I sometimes fantasize that I have absolute power over all other humans, as "Emperor of the World." I then dream up punishments that I would bestow upon people whom I do not like. I know, I know, it is very unlikely that I will ever be "Emperor of the World." The logistics of actually achieving such a title approach impossibility, and even if I somehow did manage to claim it, I'd be overthrown or assassinated quite quickly. But let me dream a little! I've got to have some fun!

Since I have a lot of interest in RVs, and not much love for the people who design and make them, if I were Emperor of the World, imposing at least some form of torture upon these people would be very high upon my "to do" list. No, I wouldn't go psycho--- nobody would go to the guillotine and nobody would be drawn and quartered (whatever that is). But I do have something in mind for them that they definitely would not like: *I would make them live in their own RVs for awhile!* I would issue an executive non-reviewable order requiring each RV designer and manufacturer to live in one of their RVs for three continuous months, as actual RVers, in wilderness areas without municipal electricity (can't plug in--- ever). All persons refusing such an order would be allowed as much time as they needed to rethink their positions and decide to do the task (but they would have to spend all their rethinking time in the dungeon).

Once they began their live-in assignments, the fun would begin. I wouldn't want to just find out what happens to them from afar; I would park my RV near them, and enjoy observing their follies personally. On cold nights their furnaces would stop heating. I would laugh and they would cry. (My heart would be as cold to their suffering as the temperatures on the insides of their RVs!) Their lights, water faucets and refrigerators would all stop working (they would be forced to throw the food from their fridges away). I would laugh and they would cry. When they parked in the sun on hot days, indoor temperatures would start climbing. They would try to turn on their generators to power up their air conditioners (and everything else), but those generators just wouldn't start, so they would have no air conditioning. Then, temperatures inside their RVs would skyrocket. They would then be forced to sit outside all day, at which point, I would laugh and they would cry. Then, I might walk on over to them and ask, "Do you think it is possible that there might actually be a technical literal hell? A version of which is right here on Earth? Inside of your RVs?" They would be at the depths of despair.

I would then show them a picture of my mom. "Look at this woman. I watched all this happen to her--- and it was not fun. Now I get to watch this happen to you--- and it *is* fun! You took her money and drove her to tears. You must be very proud of those 'F's you got in school. What did my mother ever do bad to you? If this situation was good enough for her, why is it somehow not good enough for you?"

Before leaving, I would say: "Cheer up. RVing is about seeing wonderful things that you've never seen before. Today, I will show you something wonderful that you have never seen before." Instead of showing them some breathtaking natural feature, I would toss them a book about the basics of electricity, and offer to let them keep it.

For the really hideous designers, I'd impose an extra punishment for dessert. On a cold night, I'd invite them into my camper for coffee/tea and a light snack (attendance required). They would stay for a half an hour, all the time of which I would talk incessantly about how well my lights worked, how reliably my furnace worked, and how much I really looked forward to sleeping warm as toast that night. Then, I would kick them out and force them to sleep in their own cold dark campers. Yes, I am evil. It is probably a good thing that I am not Emperor of the World.

What if somebody made an RV with an electrical & appliance system that actually worked? My Emperor's decree would still stand. But if that system worked, living in the RV would be a pleasure, not be a punishment. That would be alright with me. My Emperor decree wouldn't actually punish or reward anybody, it would simply make them punish or reward themselves in direct proportion to their own RV design integrity. It would be beautiful cosmic justice. Maybe I am not so evil after all.

But alas! The fantasy fades. What was I thinking? A world in which RV designers bear the consequences of their own designs? How silly! RV designers decide how RVs are made. RV users have no say. RV designers & manufacturers simply get away with whatever they do. And it will probably stay that way for quite some time. They laugh and we cry. Like I said, usually I am powerless to change the things in the world that I do not like.

I am not Emperor of the World. I am just a dreamer with lots of ideas, who reads books, studies, likes scientific mechanical and electrical concepts and camping. I cannot coerce anyone into doing anything, and maybe that's just as well. I must try persuasion instead. The best I can do is write a book and hope that somebody reads it.

RV Electrical & Appliance Systems

Why they don't work and
How they could be fixed

Thomas Childs

Table of Contents

Introduction

Who am I?	14
Why did I write this book?	14
How did my RV days get started?	15
What experience inspired this book?	22
Who is likely to benefit from this book?	23
Who is likely to dislike this book?	24
What does the term "electrical & appliance system" mean?	24
What topics does this book cover that others don't?	25
What does this book *not* do?	25
How do I differ from many other RV author/users?	26
How did I obtain my knowledge?	29
Tips On Reading This Book	30

Chapter 1
Electricity and RVs

The Role of Electricity in RVs 32
Plug In or Don't Plug In? 32
The Case for Going Off Grid 33
The Case for Grid Dependency 44
Bringing Municipal Electricity Into Campgrounds 47
Should RVs Be Designed to Function Without Municipal Electricity? 51

Electricity Basics 53
What is Electricity? 53
Voltage 53
Current and Charge 55
Power and Energy 58
Energy-Charge Relationship 60
Power-Current Relationship 63
Voltage-Current-Resistance Relationship (Ohm's Law) 64
Summaries of Terms & Formulas 66

Components of RV Electrical Systems 67
Two RV "House" Electrical Systems
The 120 Volts AC System 67
The 12 Volts DC System 69
AC Verses DC 70
Power Converters (AC to DC) 72
Power Inverters (DC to AC) 75
Two Operation Modes: "Generator On" and "Generator Off" 75
Batteries
Battery Basics 75
Problems With Batteries 78
Battery Charging Systems
The "Charge As You Drive" System 82
The 120 Volt AC Charging System 85
Generators
What Can They do? Are They the Answer? 86
Generators Make Noise! 88
Physical Factors *(88)*
Social Aspects *(90)*
Political Aspects *(91)*
Easing Tensions *(92)*
My Personal Feelings *(93)*
Some Disregard the Feelings of Others *(93)*
How I Deal With Generator Users *(94)*

Chapter Conclusion: No Easy Answers 95

Chapter 2
RV Electrical Systems Are Defective!
(Problems & Solutions)

An Analogy: The "Stupercar" 96

The Defects:

- #1: Insufficient Batteries 98
- #2: Inadequate Chargers 100
  - (Efficiency of 3 Amps Charging by Generator) *(104)*
- #3: Generators that Start From "Dead" Batteries 108
- #4: Energy Hogging Appliances 109
- #5: The "Octopus Effect" 111
  - (20 Things That Can Shut an RV Electrical System Down) *(112)*

The Defects Working Together 113

What Happens When Existing RV Electrical Systems are Used 114

The "Official Plan" For RV Electricity 118

Proposed Solutions to the Electrical Defects:

- #1: Have More Batteries 121
- #2: (Standing Systems) Charge With More Current 125
  - (Moving Systems) Have More Batteries, Cut Electrical Waste 135
- #3: Start Generators From Other Energy Sources 136
- #4: Use Energy Saving Appliances 139
- #5: A Computer to Monitor the System 142

Chapter Conclusion: The Biggest Defect of Them All 146

Chapter 3
Making Electricity & Charging Batteries

Harvesting Electricity From Nature 148
Where Does Electricity Come From? 148
How Can Electricity Be Made? 148
From What Could Electricity Be Produced Locally at RVs? 149

Battery Charging / Electricity Making Strategies 150
I.) Alternators (Charge As You Drive): 150
Getting Electricity From the "Car" Into the "House" *(150)*
Battery Swapping *(152)*
Battery Paralleling *(153)*
Plugging the "House" Into the "Car" *(156)*
How I (the author) Learned to
Charge Batteries (while driving) 162
II.) Solar Panels:
What Are Solar Panels? How do They Work? 169
Advantages of Solar panels 169
Disadvantages of Solar Panels 170
Differences Between Solar Panels and Generators 171
Practical Considerations for RV Solar Panel Use 172
Technical Points 174
III.) Wind Turbines:
What Are Wind Turbines? How do They Work? 176
Advantages of Wind Turbines 176
Disadvantages of Wind Turbines 177
Practical Considerations for RV Wind Turbine Use 177
Wind Verses Sun 178
Finding Wind 179
IV.) Thermoelectric Generators:
What Are Thermoelectric Generators? How do They Work? 182
Technical Points 187
Thermoelectric Generators Verses Solar Panels 188
My Own Proposal for an RV Thermoelectric Generator 188
Advantages of Thermoelectric Generators 190
Disadvantages of Thermoelectric Generators 191
V.) Fuel Cells:
What Are Fuel Cells? How do They Work? 193
How Do Fuel Cells Differ From Batteries? 193
Advantages of Fuel Cells 195
Disadvantages of Fuel Cells 196
VI.) Generators Powering 120 Volts AC Chargers:
Generators Versus Other Electricity-Production Methods 198
VII.) Charging From Municipal Electricity: 199

Supercapacitors 201
What are Supercapacitors? 201
Differences Between Batteries and Supercapacitors 202
Energy Storage and Voltage Drop 204
Supercapacitor Aided Battery Charging 206
The case for a supercapacitor aided system (206)
The case against a supercapacitor aided system (208)

Comparing Battery Charging / Electricity Making Strategies: 209
Advantages/Disadvantages Summery 209
Efficiencies 212
Costs per KiloWatt Hour 213
Ideal Battery Charging 215
Which Electricity-Making Strategies Work Well Together? 216

Chapter Conclusion: The Simplest Might Be the Best 217

Chapter 4
Propane

A Natural Fuel Choice 218
Electricity Versus Propane 218

Properties of Propane 218
Detecting Propane 219
Combustion Byproducts of Propane 221
Carbon monoxide *(222)*
Carbon dioxide *(225)*

Trailer Propane Tanks 227
A Defect in Many Motorhome Propane Systems 230
Proposed Solutions to the Motorhome Defect *(231)*

Chapter Conclusion: Choose Propane for Heating 233

Chapter 5
Appliances

Introduction 235

Do You Speak RV?

General Criteria for RV Appliances

The Appliances 236

Lights 236

Types of Lights (236)

I. Incandescent (236), II. Fluorescent (237), III. LED (238), IV. Gas Mantle (240)

Which Light Types are Best Suited For RVs? 243

History of Artificial Lighting 250

Electricity Verses Fire--- Which is Better? 252

Portable Lights 258

Best Overall RV Lighting Plan 259

How Much Electricity Could Be Saved Using Efficient Lighting? 259

Furnaces 261

Furnace Characteristics 261

Types of Furnaces 262

I. Passive Convection, II. Central Radiant, III. Forced-Air, IV. Hot Water/Steam, V. Catalytic, VI. Direct Discharge, VII. Electric Space Heaters, VIII. Heat Pumps

Which Furnace Types are Best Suited For RVs? 264

Best Overall RV Heating Plan 273

How Much Electricity Could be Saved Using Efficient Heating? 274

Furnace Stories:

"My Stove is Also My Furnace" *(275)*

"Some Like it Cold" *(275)*

"I Just Can't Figure Out…" *(277)*

"My Old Reliable Furnace" *(277)*

The 99 Night Field Test 279

Progress--- But In What Direction? 283

Final Thought on Furnaces 284

Stoves & Ovens 285

Types of Stoves/Ovens 285

Indoor Gas Stoves, Outdoor Fold-Down Stoves, Outdoor Portable Stoves, Conventional Ovens, Microwave Ovens, Charcoal Grills

If you don't have an oven… 290

Alternatives to Cooking with Electricity 292

Refrigerators 294
Types of Refrigerators/Coolers 294
I. Ice Boxes & Coolers, II. Compression,
III. Ammonia Absorption, IV. Thermoelectric
Which Refrigerator Types are Best Suited for RVs? 295
Best Overall Refrigeration Plan 303

Air Conditioners 304
Active Air Conditioning (uses power sources) 304
Types of Air Conditioners 305
I. Compression, II. Evaporative,
III. Ammonia Absorption, IV. Solar Powered Fans
Which Air Conditioner Types are Best Suited for RVs? 306
Passive Air Conditioning (no power sources) 313
Passive Air Conditioning Variables 313
I. Reflection (313), II. Insulation (317),
III. Ventilation (322), IV. The Greenhouse Effect (323),
V. Heat Storage Within Objects (327),
VI. The Chimney Effect (328)
Best Overall Comfort-Cooling Plan 330

Water Systems 331
Water Pumps
Types of Water Pumps 336
I. Demand Pumps, II. Compressed Air, III. Hybrids,
IV. Gravity, V. Manual Pumps, VI. Municipal Water
Which Water Pump Types are Best Suited For RVs? 337
Hot Water Heaters
Types of Hot Water Heaters 354
I. Pilot Light, II. DSI., III. Electric, IV. Tankless,
V. MotorAid, VI. Combined With Furnaces
Which Hot Water Heater Types are Best Suited For RVs? 355
Toilets
Types of Toilets 360
I. Folding Toilets, Bags & Jugs, II. DryFlush,
III. Composting, IV. RV Water Toilets,
V. Small Portable Chemical Toilets
Which Toilet Types are Best Suited For RVs? 361
Washing Machines & Dryers 369

TVs, Stereos & Other Electronics 371
120 Volts AC Appliances in RVs *(371)*,
TVs *(372)*, Stereos *(374)*, Computers *(374)*, Printers *(375)*,
Small electronics with rechargeable batteries *(375)*,
Electronics then and now… *(376)*

Things that Support Appliances 377
Switches
Master Disconnects 377
Timer Switches 377
Connectors
Trailer Connectors 378
Aluminum is a poor metal choice for electrical connectors 378
Appliance & wire connections 381

Testing Appliance Currents 382

Chapter Conclusion: Choices Exist But One Must Look 386

Conclusion

Building Better RVs
A New Plan 387
Dealing With Existing Building Codes 388
What Would a Thomas Childs RV Be Like? 389
The Appliances *(389)*
The Batteries *(392)*
The Charging System *(392)*
Other (Non-Electrical) Requirements *(394)*
Testing RVs 395
The Three Month Field Test *(396)*
Responding to Customer Feedback / Warranties 398

The RV World is Loaded With Problems
Why Are RVs Built to Lower Standards than Other Things? 399
Why are RV Electrical & Appliance Systems Particularly Bad? 399
Other Problems 400
Is There Hope? 400

What Can We All Do?
What Can New RV Buyers Do? 401
What Can Existing RV Owners Do? 401
What Can RV Manufacturers Do? 401
Let's Not Allow Off-Grid Camping To Die 402
Let's Communicate 402

Some Final Philosophical Thoughts
The RV as a Metaphor For Waste in America 403
The Future of RVing (Can RVs Go "Green?") 404
What Have I Learned From My RVing Days? 405
Camping as a Metaphor For Life 406

Index 407

Introduction

Who am I?

Hello. My name is Thomas Childs. I have many interests in life. These include photography, playing the piano, teaching, clean renewable energy and having dogs. Of most notable interest for readers of this book, I am an avid RV user and lover of wilderness. I also love to write, which is a good characteristic for anyone to have, who has information that might be of interest to others.

There is nothing quite like the RV lifestyle. Nothing matches the feeling you get from waking up in your own bed, 1000+ miles away from your regular "home," packing up, driving several hundred miles to still another part of the country you've never seen, then still sleeping in your own bed that night (and every night). Of course, the leftovers from that great pot of soup you've recently made are still in your refrigerator, waiting to satisfy you the next time you get hungry. The can opener, the scissors, the peeler, the dishes, and all your other worldly camping possessions are still in the same spots of your drawers and cupboards, even though you may have relocated hundreds of miles away since putting them there.

I've had and seen a lot of really great things right in my own camping "yards": saguaro cactuses, redwood trees, sequoia trees, a petrified forest (which looked like jewelry), lava-rock beds, herds of elk and bison, snow covered mountains (which were indigo just before sunrise, then suddenly turned pink when sunlight first hit them), stunning dessert rock formations, babbling brooks, riverfronts, lakefronts (some with half a dozen distinct color bands in the morning, others with pure blues and greens), alligators and cypress trees in the Everglades swamp, geothermal features with color combinations I didn't even know existed. I love experiencing this country in an RV.

But there is something about RVing that I don't like. For many people, the RV experience gets ruined by poorly designed and mismatched RV equipment. It also often gets inadvertently ruined by the RV users themselves, who though not stupid, are often unaware as to what equipment they should have and how they should use it.

Why did I write this book?

I wrote this book because over the years, I have gotten really good at living in an RV on primitive public lands without electricity, and feel I should share this knowledge with others who may be interested in doing the same. Whether or not I am world class at RV living off the grid is not for me to judge, but I will say this: While out and about, I meet many other RV users who struggle to stay comfortable, but I no longer meet people who have their system of gadgets and procedures working as well as mine.

<u>How did My RV Days Get Started</u>?

In 1985, while in my early twenties, I was recovering from a psychiatric episode and a bad college experience. (I was without worthy goals.) I had been down in the dumps for months. My mind started cleaning house. It was time to run away and not look back. I quit going to college, quit trying to get a college degree, quit my psychiatrist and quit taking all anti-depressant drugs. Then I won a case for some disability benefits from the government. It wasn't much money, but it was enough to get by.

One night shortly thereafter, at about 4:00 in the morning, I had a wild idea as to how I should spend this new found "wealth" to jump start my life: buy an old dilapidated camping trailer, and a car to tow it, fix the trailer up, and use the trailer as my home while traveling, visiting & photographing the beautiful National Parks and Monuments of the United States. Try as I might, I couldn't think of any reason not to do this (other than maybe I should live my life solely to please other people--- but that's just rubbish). So I went for it. Hard work, patience and a lot of swearing turned this dream into a reality. After six months, the initial repairs were done making it ready for use, and after about five years of cycles of trailer tearing apart, rebuilding, then camping, I had rebuilt the trailer to a like-new state. After that, I just maintained it, fixing things as they broke.

To date, I have visited over 40 national treasures while camping in 32 different states for a combined total camping time length of over three years (if all trips occurred end to end).

I still have the trailer and I still use it today in 2017.

I have been free and sober from anti-depressant drugs since 1984. I don't miss them.

Warning: If you want to get off of antidepressant drugs, you should seek medical help. There are some doctors who are willing to help you do this. Trying to do it yourself or quitting "cold turkey" can result in suicidal behavior. Generally, the way to get off is to gradually lower the dosage to zero over a period of several (maybe six) months.

I never did get a college degree, though I did go back to college "buffet style."

I took a number of science/technology classes.
College can be a great experience but only if one goes for the right reasons.

1985

above: junky trailer sitting in the driveway of the lady who sold it to me
below: brought it home

<u>What was I thinking</u>? I accepted that trailer about as much as an organ recipient accepts a mismatched kidney. But it was mine. I did this. Oh well…

Time to get to work…

above: experimenting with the plumbing system
below: making & painting cupboard doors

above: cutting holes in new countertop (for stove & sink)
below: making & gluing together kitchen table

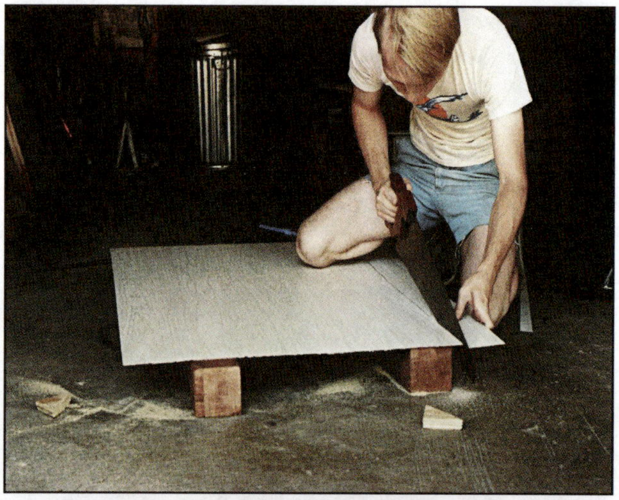

"You made your bed, now sleep in it."

Well, I *made* my bed. And I *do sleep* in it.
It is really quite comfortable.

Folks, this was 1985. I don't look like a middle school student anymore.

this page: the floor

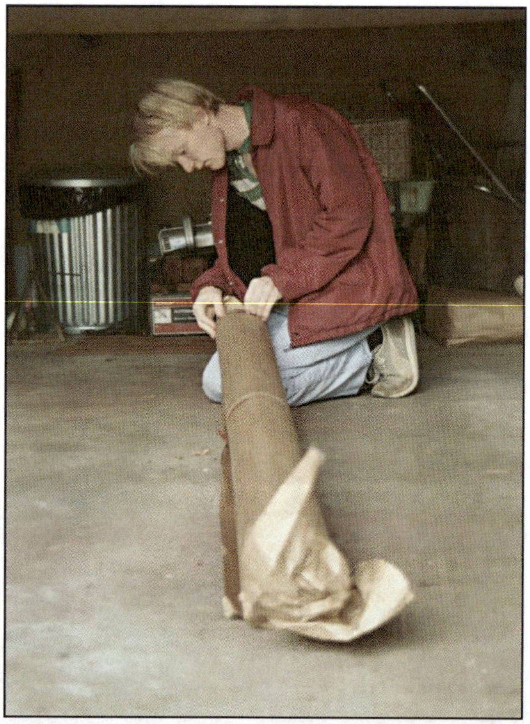

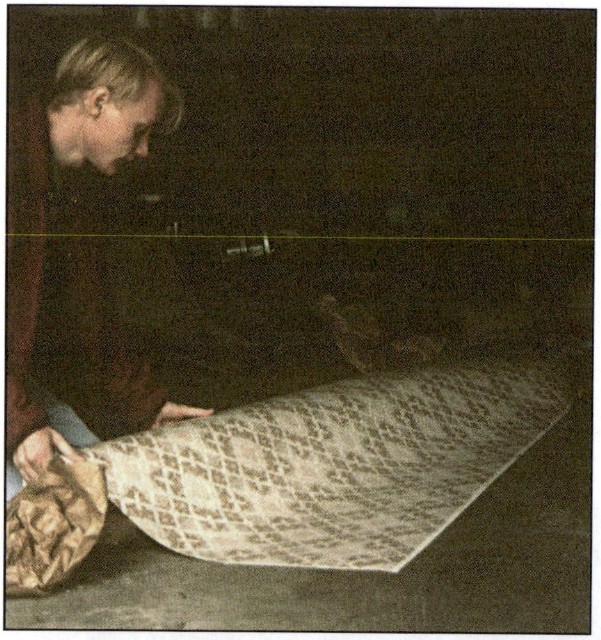

It took quite a bit of nerve to cut this floor, since I only had one chance to do it right. The taped together construction paper pattern I made in the trailer was my cutting guide. This cut (*left*) turned out to be a perfect fit. This floor is still in my trailer in 2017 and still looks quite good.

1989

below: putting on new siding

What experience inspired this book?

Twenty years later, in 2005, my mother joined the RV camping equation. After having been widowed for three years, Mom told me that she was interested in visiting some National Parks while touring this country in a motorhome. "Great!" I said. "I'll help you pick one out. We'll caravan together. You camp in yours and I'll camp in mine. Motorhomes are designed to function well and keep people comfortable in wilderness." Talk about words that have come back to haunt me! Boy, I could not have been more wrong. That last statement will probably go down as one of the top five stupidest remarks I've made in my entire life.

That summer, after buying a used motorhome, we drove it from her home in Florida to my home in New York, then went as far west as Colorado before parking in a National Forest. There, it happened. Everything in her motorhome started to shut down just a few days after being parked. She shivered in the cold at night, as her furnace shut down. She baked in the day, when the sun hit her poorly insulated roof, turning her RV into a "hot box" (her words). Her refrigerator stopped working. Water stopped coming out of her faucets. Her lights stopped working. Her generator wouldn't start. Her battery charger wouldn't charge. We even had trouble filling her propane tank. I tried as best I could to get her system working. I figured, if anybody was capable of getting that system working, I should have been, because after all, I'd been an RV user for twenty years, and had long since solved the equipment problems unique to RV living. (My trailer parked in the campsite next to her functioned just fine during this ordeal.) But no matter how I tried, I could not get her system working, except for very brief time periods. Early one morning, after spending the night shivering in the cold, Mom knocked on my trailer door: "I want to go home!" So I put my van and trailer into storage in Colorado, then helped her drive her motorhome back to her house in Florida. After we got there, I took a plane back to Colorado, reunited with my van and trailer, then finished my trip alone, camping for a little over two months. Returning to Florida marked the end of Mom's "living-in-an-RV" days. Although we did get some use out of her motorhome for weekend camping in Florida, the only camping she got out of it that was long-term & far-away was: *one day* visiting *one National Park.*

Although I had been aware of some of the RV design problems prior to this trip, from conversations with other RV users, I had been largely in the dark as to how bad typical RV electrical & appliance systems really are. After this trip, I developed a keen interest in them and began studying them. I read technical RV gadgetry data in websites. I read other books on RV life-off-the-grid. I read books on electricity in general. As I began to understand these RV systems, three things became clear: First, although Mom's motorhome was twenty years old, these failures were not due to the age or wear-and-tear of the gadgetry, these were actually design failures (problems with the plan itself). Second, Mom's problems were just the tip of the iceburg. Nearly all RVs have the same design problems as her RV did. And third, as I began crunching some numbers, comparing overall electricity consumption rates with battery storage capacities and charger charging rates, I came to a conclusion: It is an understatement to say that RV electrical & appliance systems *don't* work; it is more correct to say that these systems *can't* work.

above (2017 photos): Mom's RV decaying in yard several years after her ill-fated trip. RVs are not the only things that decay, so too do people's dreams.

So in 2008, I began writing this book to blow the whistle on some really lousy engineering and take some revenge against some really lousy engineers. But I also wrote this book to help people to redesign and modify their RVs to get those RVs working right. (People must be willing to invest some time, money and effort in order to do this.) I've been writing the book for over nine years now (2017). I don't know why I can't seem to complete anything within a reasonable time frame. I don't know if I will ever finish this book.

Who is likely to benefit from this book?

Anyone who actively uses or is just considering using an RV in places without municipal electricity is likely to benefit from this book. RV owners and customizers, willing to make modifications, will probably love this book, especially if they are willing to experiment and use less-well-known equipment. Though not intended for this purpose, I think that people with permanent remote stationary houses or vacation cottages or tiny houses, not connected to municipal electricity, will also benefit from this book, because many of the same principles that apply to off-grid RVs also apply to off-grid houses and cottages (and tiny houses on wheels are technically RVs anyway). Although this book harshly criticizes current RV designs, some RV designers, who are genuinely interested in improving off-grid performance of their RVs, may actually like this book. Perhaps Park Rangers and Forest Rangers, who have a lot of contact with campers, will also find this book helpful for giving technical advice to campers or just understanding camper stories.

Who is likely to dislike this book?

RV designers and manufacturers, who want to continue making and selling stuff with the same old dysfunctional designs, while hoping that nobody notices, will hate this book. RV designers and manufacturers, who think that they are "too good" to take any advice from RV users, will hate this book. RV designer and manufacturers, who want to cut corners to boost profits, will hate this book. Some private RV Park owners may dislike this book, as they may feel that camping in wilderness instead of at their RV parks may encroach upon their profits. Persons who just want to learn how to use their existing RV systems without making modifications, will not find this book helpful. Finally, RV users, who always stay in places with municipal electricity, are unlikely to find this book helpful.

What does the term "electrical & appliance system" mean?

Many authors use the term "electrical system" to mean the sum total of all related components having anything to do with electricity. Any complete "electrical system" can be divided into three branches: *making* electricity, *storing* electricity and *using* electricity. The appliances that use electricity are often overlooked by other authors (but not me), so instead of using the term "electrical system," I use "electrical & appliance system," to indicate that I also consider the (branch of) appliances that use electricity. My definition goes further. Some appliances can be designed to use no electricity at all. Any non-electric appliance, used instead of one which normally consumes electricity, is also a part of an "electrical & appliance system," because non-electric alternatives directly effect such a system by lightening its electrical load. (By this definition, I consider the entire propane system to be part of the "electrical & appliance system.) Finally, anything that effects the efficiency of appliances (such as insulation in walls, which effect both furnaces and air conditioners), is also a part of an electrical & appliance system, because these too effect electrical loads.

What topics does this book cover that others don't?

Making electricity (for battery charging)

1) One of the best methods of making electricity is to charge batteries while driving a detached car/truck. This is the cheapest, easiest (both to install and use), and possibly the oldest of all the methods, and it makes lots of electricity. Yet, other prominent books on RV electricity don't even mention it. This book covers it in detail.

2) One of the best methods of charging batteries for RVs parked in wilderness may be with thermoelectric generators. Yet I've seen no other books on RV electricity that propose their use nor even mention them--- I don't think these authors even know what they are. What are thermoelectric generators? Why might they work well in RVs? Find out in Chapter 3.

All three branches of RV electricity (making, storing and using it)

Many books on RV and 12 Volt electricity systems focus on only two of the branches of electricity (making and storing it). They go into enormous detail about electricity making and batteries, only to allow the system to be neutered with the most energy-wasting appliances imaginable. This book covers electricity-using in detail (emphasizing energy-conserving appliances). Other authors focus entirely on one branch, the batteries, going into elaborate detail on how to charge them, while erroneously assuming that RV users are going to find electrical outlets all over the place to plug the battery chargers into. (This approach leads to finely detailed information that is completely useless in the wilderness.) This book considers not only battery chargers, but where the electricity (or energy) needed to run those chargers is supposed to come from.

The good and the bad

Some authors are just too optimistic about certain equipment (such as solar panels, which don't work in the shade). I like to see both the good and the bad. While some gadgets work almost universally and others fail universally, most fall somewhere between those extremes. Most gadgets work well under certain conditions but fail under others. I like to note the differing conditions under which a given piece of equipment is likely to either work or fail.

What does this book *not* do?

This book does not consider specific models nor specific manufacturers of RVs and RV components nor quality of equipment that's out there. Information is only given for general design types, including which designs are or aren't likely to work, and under which conditions those designs are or aren't likely to work. I would need a sizable staff and budget to independently test and report on specific brands & models of RVs and their components. I have neither of these.

.

How do I differ from many other RV author/users?

Many authors are far too friendly toward the RV industry and their designs. I'm not friendly toward that industry. I'm not afraid to call a "cow plop" a "cow plop."

Many other RV author/users plug into municipal electricity once in awhile. I camp exclusively off-grid.

Much RV literature is written by people who use RVs *as they are*. (They teach others how to do the same.) I write about RVs as they *could be* or *should be*.

Many people buy new RVs, which gradually get old. I bought an RV old, which gradually got new.

Many people keep their RVs only a few years before trading them in. I've kept mine for over three decades.

Many RV users spend big bucks on their RVs. I built mine on a shoestring budget.

Elaborating a bit…

Calling a "cow plop" a "cow plop"

Some other authors may be receiving heavy financial backing from the RV industry. Others may be getting funding from private RV parks. This can influence author judgments. Readers may think they are getting objective reviews, when in fact, they are getting cleverly disguised commercials. Other authors may just want to be a bit too nice and positive. I am free of all of these vices. I give an independent point of view.

Camping exclusively off-grid

You may be wondering: "Does this guy cheat a little bit? Does he really *never* plug into electricity when he camps?" Well, everybody cheats a little, I suppose--- even Amish people cheat a little on their religious/lifestyle choice of foregoing electricity, by sometimes using flashlights and radios. Consider this: 1) Multiple times, I have gone three continuous months without plugging my trailer into electricity even once. [Twice, I went over four continuous months without plugging in.] 2) I almost always stay at places without electricity. When I do stay at a park that has both electric and non-electric sites [not that often], I almost always choose a non-electric site. I choose an electric site only when that's the only type of site available. 3) Once, while working as a photography director at a children's summer camp, I brought my trailer to serve as my residence. [I was the only "kid" at camp who brought his own house.] Although electricity was within reach, for the entire two and a half months that I worked there, I never bothered plugging in. 4) The only substantial "cheating" I do is at Laundromats. While my cloths are washing & drying, I often can be found sitting at a table with my laptop plugged into the wall, writing a book about RV electricity.

I believe that the only way to truly break free of the municipal electricity habit, is to never have it. If I plugged into municipal electricity even sometimes, I might begin to like it and become dependent upon it (or at least more comfortable with it). By never having it, I've forced myself to learn how to make and use my electricity, and to use it sparingly, and find other ways of accomplishing tasks normally done with electricity. I can therefore be equally comfortable anywhere.

RVs as they *could be* or *should be*

When evaluating designs, I care not one iota about what "everybody else does" nor if "that's the way it's been done a long time." I just want to know *if it works*. Look, if tomorrow all RV manufacturers decided to do away with round wheels and replace them with square wheels, some people might actually "go with the flow" and write books on how to get the most of those square-wheeled RVs, and how to keep the corners of those wheels nice and sharp, but I wouldn't. I would tell people to lose those square wheels and replace them with round ones ASAP. I have no desire to learn how to use, recommend, nor teach others how to use dysfunctional equipment. If you just want to learn how to use your existing RV equipment *as it is*, throw this book away and find one written by some author other.

Buying old and becoming new

Starting with an old trailer and having it gradually become new had advantages. I needed only a minimal investment for the initial trailer purchase ($750--- believe it or not). I gradually added money to it for raw repair materials, *as needed*, *when I felt like doing the work* and *when I had money to buy stuff*. My trailer then *improved* with the passage of time. This is in sharp contrast to the experiences of most, who put a lot of money up front for their initial RV purchases. They often must borrow that money, which locks them into payments. Then, their RVs *deteriorate* with the passage of time. They often encounter further financial burdens, when their RVs need repairs far more frequently and that are far more costly than anticipated, possibly putting their RVs outside their workable budgets. Then they try to sell their RVs, only to discover that RVs depreciate extremely rapidly. I've had to deal with none of these problems.

Keeping an RV for three decades

Suppose two middle-aged guys wrote books on marriage. The first was married five different times, each for a few years. The second was married only once but for thirty years. The five-wife guy would know a lot more about "what's out there in the playing field" or "reeling them in," but would lack knowledge as to how to "make it last." The one-wife guy would lack knowledge as to "what's out there in the playing field" or "reeling them in" (because he only needed to do this once), but would know a lot more about how to "make it last." Whose perspective would be better? I dare not say. I think they both would be experts, just in different areas. I've had a single RV for 30+ years. I know how to keep an RV alive a long time. I've also had a lot of time to can-tinker and experiment, to get the right stuff in the right places and get everything in my RV working as it should. I don't think this would have been possible, had I owned multiple RVs, each for a few years. Just want you to know where I stand.

A shoestring budget

I bought my initial trailer "as is" for $750 in 1985. Over the next few months, I put about another $750 into it in the form of repair materials. I did the labor myself, at a cost of $0, so I saved big. The initial car to tow it was $1100 plus about $600 in professionally done repairs. Over a six year period, the total cost of the trailer repair & customizing materials was just below $6000 (including the trailer itself), which was about the price of a new trailer at that time. Over the years, I have kept the trailer up. It probably averages just a few hundred dollars per year in maintenance. (In 2014, I installed a new refrigerator, costing about $1000 for everything. This was atypical. Most years, the maintenance is probably less than $200.) The biggest overall expense is for replacing and maintaining the tow vehicles, the last of which I paid $6100 for in 2005. I estimate that adjusted for inflation to 2017 prices, an equivalent trailer could be built for about $12,000. If a used tow vehicle were added, the entire package could be obtained for under $20,000.

I believe that RVing and National Parks visitation should be for everyone who wants the experience, not just for rich people. I am living proof that a person of modest means can successfully RV. For the first fifteen years that I did this, my only income was SSI. (Since then, I've gotten Social Security Disability, which is somewhat more.) I've been on Medicaid, even sometimes on food stamps, yet still was able to do this. But if you're richer than I, keep reading. All the tricks that I've learned over the years could work just as well in expensive RVs as they have in my simple one. (In fact, many of my suggestions are geared toward higher cost RVs.) It bothers me when someone with an RV, costing six times what mine did, parks next to me, and their equipment doesn't hold up nearly as well nor as long as mine does. I would like it if luxury RVs could someday function as well as my shoestring budget RV does.

How did I obtain my knowledge?

I do not attribute my RV knowledge or success to superhuman intelligence (which I don't believe I possess). Instead, I mastered the art of RV living off the grid primarily by doing a lot of it. There is a saying: "practice makes perfect." Of course, this is untrue; there is no such thing as perfection. However, practice does make people very good at things. If you play the piano a lot, you just might get good at it. If you ice skate a lot, you might get good at it. If you drive a car a lot, you might get good at it. RV living off the grid is a skill, like all of these--- if you do a lot of it, you just might get good at it. It also helps to be observant, willing to experiment, willing to listen to others, do some internet research and have good intuition.

Elaborating a bit...

Willing to experiment: Besides physically reconstructing much of my RV, I have personally selected and installed all my appliances. I have modified many of them to make them work better.

Listening to others: Much of my knowledge comes from conversations with other RVers. They have such good insights, that it would be a shame if nobody wrote their thoughts down, and shared them as public knowledge with other RVers everywhere.

Internet research: I get a lot of really good information from tuning onto internet blogs posted by other RV users. I also like to sift through RV dealers' parts catalogues, to see what's available, and to get and analyze technical specifications (like consumption rates) that often go missed by others until they actually buy something, put it in their RV, then find out that it doesn't perform the way they'd hoped.

Intuition: I know the realities of RV off-the-grid living well enough such that I can make good educated guesses as to what designs will or won't work and the conditions under which they will or won't work. (Gadgets are effected primarily by weather, but other variables also effect performance.)

One more attribute that gives me good RV knowledge: When most children grow up and first move out on their own away from mommy and daddy, they usually go live in an apartment or a college dorm (or maybe military housing). Not me. The first time I ventured away from my parents, I lived in my RV trailer. RV living therefore became very natural for me. Sometimes I feel more "at home" in my trailer than I do in my regular traditional house.

Tips On Reading This Book

Yes, I know, it's a long book. I suppose I could have written a shorter book, but RV users suffer far more often from having too little written information than they do from having too much information.

If you have limited time...

If you can only read one chapter, read Chapter 2, the "Defects" chapter. It is the backbone of this book.

If you have just a bit more time, read the first and last sections of Chapter 1 as well. "The Role of Electricity in RVs" is easy non-technical reading, dealing with the philosophical and social/political trends & issues of RV life on or off the grid. "Components of RV Electrical Systems" gives you a basic overview of typical RV systems unmodified.

If you have still more time, read the sections: "I. Alternators (Charge As You Drive)" in Chapter 3 and "Lights" and "Furnaces" in Chapter 5. Charging batteries as you drive is a great way to charge batteries and may even be able to produce all the electricity you need. Lights and furnaces are two of the most energy wasting components of typical RVs, so converting to energy saving counterparts can save an overall system.

The rest of this book you may (or may not) then choose to read on an as-needed basis. For sure, use the table of contents as your guide. (For example, if you don't have and don't plan on getting a wind turbine, you probably don't need to read the subsection on "Finding Wind.") Bear in mind though that alternative equipment, that you may not have considered, might hold the solutions to some of your specific problems, so if and when you do have time, it might be a good idea to try to read most of this book.

A Bit About the Math...

Don't get hung up on the "Electricity Basics" section in Chapter 1. If you do start reading it and become overwhelmed, skip it for the time being. Although my conclusions are all supported and documented by math, they are also explained in plain English, so I think you could understand most of them without looking at the equations. After you've read most of this book, glanced at the equations and begun to see the need for the supporting math, you may want to go back and read the "Electricity Basics" section, then once again look at the equations to see if you can understand them.

Most of the math in this book is just multiplication and division, with only a hint of algebra. (So, if you can understand that $4 \times 5 = 20$, $20 / 5 = 4$ and $20 / 4 = 5$, and have a calculator, you can understand most of the equations.) Note though that you need to get the units of measurement right. Getting Volts, Amps, Watts, Joules, or that which they describe (Voltage, Current, Power, Energy) mixed up, produces erroneous results. (It would be like describing the distance between San Francisco and San Diego in kilograms, minutes, pounds per square inch or degrees Fahrenheit. Acceptable units of measurement for this distance are miles or kilometers--- these other measurement units make no sense at all.) At the end of the "Electricity Basics" section is a "Summaries of Terms & Formulas" page. If you can understand what's on that page, you are in really good shape for understanding the math in this book.

A Bit About the Photography and Illustrations...

This is not a completely illustrated book; it doesn't even come close to being that. I have however included *some* photographs and drawings. The decision to (or not to) include each photograph was made based upon a combination of what's important, what's available for me to photograph, the difficulty involved in taking and Photoshopping a good photo, and the time involved in the process. Look, I've been a one-man-show for awhile. I wrote all the text, did all the drawings, took all the photos, fixed the photos up in Photoshop, laid the document out and designed the covers all by myself. Trying to put out a completely illustrated book would add years to the production time and result in a forever-unfinished book stuck in my laptop that nobody could ever read. Consider the photos to be just a bonus.

Chapter 1
Electricity and RVs

The Role of Electricity in RVs

Plug in or Don't Plug In?

There are two strategies for meeting electrical needs of parked occupied RVs. The first is the "plug it in" method, also known as *grid dependency*. With this strategy, users find RV parks with electrical outlets and plug their entire RVs in there. The second is the "take it, make it or learn to live without it" method, also known as *going off grid*. With this electricity strategy, users "take it" to their campsites (in batteries), "make it" (using generators, solar panels or other methods) or "learn to live without it," (by using non-electric devices, such as candles instead of electric lights, or by conserving). Which strategy is best?

Some Camping Terms

hookup--- Any connection between an RV and an external source or drain of something (including cable TV).

full hookups camping--- Camping at sites that have complete hookups to electricity, water and sewers.

electric only camping--- Camping at sites that have hookups to electricity but not water nor sewers.

dry camping--- Camping at sites/places that have no hookups at all. The campground itself may or may not have water hydrants. (Hydrants are not available for hookup though; water must be carried to RVs.)

> ***dry campground***--- Campground without drinking water anywhere on the premises.
>
> Note: Although the terms "dry" and "wet" usually denote the presence or absence of water in the general English language, in RV lingo, "dry camping" usually means camping without municipal electricity. This is probably because electricity seems to be far more important than water for RV users.

boondocking *(also.* ***dispersed camping)***--- An extreme form of dry camping: camping in places where absolutely nothing is provided. (No water, electricity, showers, toilets, dump station, picnic tables, fireplaces, trash collection nor even clearly defined campsites or boundaries within which to park.)

RV sites--- Campsites with full hookups (electricity, water and sewer).

tent sites--- 1) Campsites where only tents are allowed. (They may be walk-in sites or accessible by vehicle.)
2) Campsites without electricity where both tents and small RVs are allowed.

non-electric sites--- Campsites without electricity. Some people also call these *tent sites*, even though RVs are also allowed to occupy them. I know it's confusing, but not all people agree on the meaning of terms.

The Case For Going Off Grid

Three places that I visited:
(I'll tell you about them, then I'll make my points.)

1) Lake Powell National Recreational Area (Arizona, visited October 2005): I found this place by following a map logo and some road signs. I unhitched and camped on a beautiful lakeshore for about a week.

Lake Powell from about a quarter of a mile away
(Notice those tiny bright objects by beach. Those are RVs--- one of them is mine.)

From my windows and front yard, I had a stunning view of the pinkish white rock formations protruding from and across that lake. I sat and watched boats float and move about. One day, when temperatures were warm enough, I went for a swim, just a few steps from my front door. The lake itself often had pure saturated colors, probably because it is a desert lake. In the desert, there are usually few clouds in the sky, so with little variation in sky color to reflect, lakes can take on pure colors. The lake exhibited different colors during the mornings and evenings, presumably due to different sun angles. Sometimes it was turquoise, sometimes it was pure deep blue, though I no longer remember which times of day revealed which colors. The camping area itself had few facilities; just chemical outhouses. There were no designated sites; people just found places on the beach to park. Water, dump station and trash dumpsters were located on the way into the area, which although not close, was not inconvenient, since there was only one road leading to the area, so campers had to pass by these facilities on their way in and out. What was the price of this lovely spot? $8/night, or $4/night with Golden Age or Access (disability) card. (I paid $4/night, because I have a Golden Access card.) On my way to this camping spot, I passed by a more developed RV park, also within the Lake Powell National Recreation Area. I think it was run by a private concession. (Sometimes private concessions are issued special permits to develop and operate services, including campgrounds, within the borders of National Park Service or National Forest Service lands.) This RV park had full hookups to electricity, water and sewer, as well as shower facilities. From this vantage point, one still had a good view of the desert but not the lake, because it was several miles away. What was the price for these non-lakefront sites? $29/ night, with no discounts for senior or disabled persons.

2) Yellowstone National Park and vicinity (visited September 2009): I camped in a National Forest campground just east of the park (photos below). My site was huge and located close to the bank of a river. It had a great view of that river and the rock formations on the other side. There was a rustic wooden fence along the dropoff (presumably to prevent people from accidentally wandering off the small cliff and getting hurt), which really added to the aesthetic beauty of the spot. The campground was simple with no hookups. It did have drinkable hand pumped well water, outhouses and garbage dumpsters. The price for this spot was $15/night or $7.50/night with Golden Age/Access card. Inside Yellowstone National Park, I came across a full- hookups campground, Fishing Bridge RV Park, run by a private concession. The campground had nice civilized goodies, including showers and laundry facilities, which could also be used by non-campers passing through, for modest fees. But, as far as beauty is concerned, this place was pretty bland. There were trees all around, but this was nothing special, as trees were common in the area. The place certainly was not riverfront. What was the price for one of these sites? About $39/night (including taxes), with no discounts for senior or disabled persons. Want to camp there on a whim? Good luck trying to get a site there. When I visited, this campground was full to capacity.

3) Grand Canyon National Park (South Rim, visited October 2009): Just south of the southwest entrance, in the small town of Tusayan, is a private RV park. Campsites there have full hookups, and the general grounds have showers, a propane filling station and even a small camp store. I have to admit, the place was clean, well kept and pleasant looking, and the staff was very friendly. The price was $25/night. This might be a nice place for families with small children, who like lots of accommodations, who don't have lots of time, and for whom money is not much of an issue. But would you believe that less than three miles away from this RV park, you can camp in the nearby Kaibab National Forest for FREE (no facilities)?

Well you can! And I did! *(My campsite above.)* It's totally legal (for occupations of up to 14 days). You could also try camping at the Mather Campground within Grand Canyon National Park. But forget this notion, if you require electrical outlets. Despite being in one of the most visited parks in the world, this campground doesn't have them. Also, you can't stay at Mather campground on a whim, unless you arrive off season. This campground is by *reservation only*.

Have I described places that are unusual? Hardly. These places reflect the *norm*, not the *exceptions*. I could go on and on with numerous other examples from my personal travels, but these three places are enough to demonstrate the fundamental reasons why camping off grid is often so preferable to grid dependency.

Reasons For Camping Off-Grid:

1) Saving Money: Considerable savings can be realistically achieved, by cutting the dependency to the grid. The Lake Powell example typifies the extreme price differences between primitive and full service camping. There, the price difference between $29/night and $4/night (with G.A. discount) was $25/night. (The cost of a site at the RV park with hookups was more than *seven times* the discounted cost for the site without hookups by the beach.) In the Yellowstone example, the price difference between $39/night and $7.50/night (with G.A. discount) was $31.50/night. (The cost of a site with hookups was more than *five times* the discounted cost of a huge riverfront site without hookups.) In the Grand Canyon example, the price difference between the private RV park and the FREE Kaibab National Forest was $25/night. (The cost of a site with hookups is *infinity* times the cost of the free primitive site.) Even without Golden Age/Access discounts, the savings would still be substantial: $21/night, $24/night and $25/night for each of these three places respectively. For the purposes of calculation, let's assume an average figure of savings by camping off grid to be about $25/night. This may not sound like much, to people who just camp on occasional weekends, but consider what kind of savings could be achieved, for a single cross country trip, lasting a considerable amount of time. Then, consider how much money could be saved from several extended cross country trips. A savings of $25/night, amounts to $750/month. A single three month trip would result in a savings of a whopping $2250. That money could be used to purchase special equipment, necessary to make an RV functional without municipal electricity. Or that money could be used for other good stuff, like fuel, food, etcetera. And consider another fact. Some people just don't have the huge amounts of money needed to stay in RV parks with full hookups on a regular basis. For these people, the difference between primitive camping and full service camping may be the difference between *being able* to take a trip and *not being able* to take a trip. I figure the amount of money I saved by staying exclusively in primitive campgrounds, instead of RV parks, during my two longest trips combined (about nine months), saved me enough money to buy the van that I currently use to tow my trailer, which was $6100 + tax. This kind of savings is serious money.

2) Attaining Greater Aesthetic Beauty: Would you like to camp by a beach? By the bank of a river? By yourself on the edge of a mesa, with a beautiful view of a mountain range, visible from your own front window? Well, you can. I can tell you where. Oh, that's right--- you can't--- at least, not if you are dependent upon municipal electricity, because these places don't have it. Back in the 1970s, when my parents did a lot of family camping with us kids, my father observed: "The really beautiful campsites almost never have electricity. The campsites that have electricity are usually pretty plain." Although my father was generalizing about State Parks in our home State of New York long ago, my observations of camping opportunities on Federal lands, all over the place (especially out west), in the present day, have been the same. The RV world seems be backwards, when it comes to temporary rented camping real estate and scenic beauty. With regular permanently privately owned real estate, beachfront property commands a premium. A house on the beach sells for quite a bit more than a similar house located a mile or two away from the beach. People sometimes buy beachfront

houses just for the lots, tear down those houses, then build newer houses to their liking. In the RV world, the Lake Powell example is typical, in which the mile-away-from-the-beach property rents for many times what the beachfront property rents for. I find this pretty bizarre.

Views of my FREE campsite by Lake Mead.
This was double beachfront property (beaches in both front and back yards).

3) <u>Having Greater Choices of Places to Camp</u>: In the western United States, there are vast stretches of public lands. There are many places to camp on these lands. Although some have electricity, the vast majority do not. There are probably three reasons for this: 1) *Many of these camping places are rather remote.* Out west, distances between populated areas can be considerable. Stringing electric lines the long distances to remote places would be cost prohibitive. Many campgrounds could never recuperate the money needed in order to electrify them, no matter how high their fees got. 2) *Many camping areas only get lightly used.* Remember, unoccupied campsites bring in $0, regardless of what prices are posted at the entrance. In order for lightly used camping areas to even exist, prices must be low, so facilities need to be minimal. Some of these places don't even have water. 3) *Many places to camp aren't even in campgrounds.* National Forests and BLM lands have vast opportunities for siteless no-facilities (no electric) free (dispersed) camping. Dispersed camping may be the only option in certain places during off-season, when developed campgrounds are closed. Dispersed camping never closes, although RV camping in certain areas may be inadvisable or impossible, if roads become impassable due to snow, or mud.

Overall, the camping places that don't have electricity greatly outnumber the camping places that do have electricity.

top: panoramic view from "side yard" tongue (that small white object is another RV) at
above: my FREE campsite on the edge of a mesa near Canyonlands N.P. in Utah

above: pencil drawing I made of LaSalle Mountains as viewed from my Utah camp kitchen window (parallel to door)

right: Sun about to rise from behind Utah fins early AM, as seen and photographed from my Utah camp front door. I woke up several mornings just to see this--- I didn't wake up intentionally--- something just seemed to be calling me.

4) Not Having to Deal With "No Vacancy" Too Often: Would you like to camp at a heavily visited area during peak tourist season (summertime)? If you require a campsite with an electrical hookup, you almost certainly will need a reservation. Good luck trying to get one! In some heavily visited places, it is difficult even to *make* a reservation. (You almost need a *reservation* to *make* a reservation!) Many places allow reservations to be made no sooner than specific time lengths before the actual camping dates, and these reservations often get booked within only minutes of becoming available. What if you can't get a reservation? Too bad, cancel your plans. Suppose you get lucky and do get a reservation. It will clearly define the day you are expected to arrive and the day you will need to leave. This certainly can be reassuring to know that there will be a place for you--- that is, if everything works out on schedule. But what about the unexpected? What if you travel a considerable distance and have vehicle problems on the way out? Your reservation date won't change, just because your vehicle goes to the road shoulder, with something smoking under the hood, and you have to spend a couple of days "camping" at some automotive repair station. (Yes, this happens.) Worse yet, in many campgrounds, if you don't show up on the *exact day* that you are expected to, you forfeit all your rights and your site will be rented to someone else. Suppose none of this happens and you get to your campsite alright. But once there, you decide that there just is so much to see and do and you want to stay a bit longer. Too bad. Your departure date will be fixed in time. Go home, without seeing all that you wanted to see. You could try another approach by not even bothering to get a reservation. This is not an irrational choice, since many "by reservation" campgrounds have a few "first come first served" sites. (But some don't!) Or you could hope for an unexpected vacancy due to someone else's last minute cancellation. If unsuccessful, you might have to go home, without seeing *anything* that you wanted to see (at that location). Being shut out, due to full campgrounds, is just another price to be paid for grid dependency. For campers who choose primitive off-grid sites, being shut out, although possible, is rarely a problem. This is partly because so many more off-grid places to camp exist than on-grid places to camp. It also is because when both types of campsites are available, those with electricity almost always get occupied first, leaving those without electricity available much longer (sometimes indefinitely).

Strangely enough, often plenty-of-vacancy off-grid camping places exist in close proximity to high use no-vacancy camping, even in places that are heavily visited and times when visitation is heavy. Take the South Rim of the Grand Canyon, for example. Mather Campground, in the National Park, is by reservation only during peak season. Even during the intermediate season (between summer and winter), it sometimes gets filled to capacity. If I want to visit the Grand Canyon, I won't even bother trying to find a spot there. Instead, I will drive the main park road, hugging the rim of the canyon, then turn away from the rim. There is one gravel road, offshooting the main road in the National Park, on which one can drive less than five miles, and enter the nearby Kaibab National Forest, where there is plenty of free legal boondocking camping. How much vacancy was there when I visited one October? Even though Mather Campground was filled to capacity, I had the immediate vicinity in the Kaibab National Forest all to myself! No neighbors! No fees! Nobody bothering me! I could set up and vacate a campsite anytime, since any noise I might make would bother no one! The same is true of busy *times*, as busy *places*. Take Memorial Day Weekend, as an example. If the weather is good (making cancellations unlikely), good luck trying to find a site on public lands with electricity on that weekend without a reservation. I hope you are in the mood for a long round trip drive home, without camping anywhere. But finding nice off-grid sites, even on that weekend, without reservations, is possible. I have been successful at it. One hundred miles from my New York home, lies the Alleghany National Forest. There, a primitive campground exists, which doesn't even fill up on the holiday weekends. I have also gotten spots in Sequoia National Forest and other places, on such weekends, with no reservations. Is it *possible* for an off-grid camper to be shut out of an area due to "no-vacancy?" It is, especially if campers have highly specific geographic requirements. It just isn't very *likely*.

5) Being Able to Camp On *Public* Lands Rather than *Private* Lands: Suppose, in a certain town, that you normally don't live in, you owned a piece of land with a fully functional house on it, not rented to somebody else. Suppose that you decided to visit this town for awhile. Where would you stay? At the house that you own? Or at a motel, with comparable facilities, a couple blocks away? This seems like a stupid question. Of course you would stay at the house that belongs to you! The motel belongs to somebody else. Public land for camping is in some respects like private land belonging to you. Public land doesn't belong to *anybody*, it belongs to *everybody*. You have as much of a right to use it as anybody else.

All real estate in general, whether for temporary or permanent use, whether for camping or for any other purpose, can be divided into three types: A) private property belonging to you, B) private property belonging to somebody else and C) public property belonging to everyone. Let's discuss each.

A) *Private property belonging to you* is the type that allows you the most freedom and rights. Except for a few zoning laws, noise ordinances and some very reasonable anti-pollution laws, you can pretty much do what you want on your property. As long as you pay your taxes, you can live and/or work there, sleep as late as you want, decorate to your taste, and put up "no trespassing" signs if you like. You have the right to forbid people from entering your property and hassling you. Except under extreme circumstances, you usually aren't even required to give law enforcement permission to come onto your property without a warrant. But private property is expensive to buy, needs maintenance and taxes must be paid in order to keep it, whether it gets used or not. Private property doesn't move around, and unless you own several parcels in different parts of the country, neither can you. And, no matter how rich you might be, your private property is tiny, compared to the collective sum of private property owned by others.

B) *Private property belonging to somebody else* is the type of real estate most frequently encountered. Often, people may just not want you on their property, for any reason, and may put up "no trespassing" signs, which you are required to respect. But many times, people do allow you to be there, and may even encourage your presence. Shop or business owners are a good example of this. They want you to be there because your presence might cause you to patronize their establishments. In the camping world, private RV parks are the main example of private property belonging to somebody else. Private RV parks may want you there because they may be willing (or itching) to rent you a piece of what's theirs for awhile. But the key word here is "private." Most places to stay for RVs, that have full hookups, are on *private* property, not *public* property. RV parks are not public facilities nor charities, they are *businesses.* Their primary purposes are not providing beauty nor public service to their patrons, their primary purposes are *making money* for their owners. And although many RV parks (and the people running them) are nice, for some, the money making aspect manifests itself both in their general appearances and their management. Many private RV parks are high priced slums. Campsites are small. RVs are piled along side one another, parked so close that some of these places look more like parking lots than appealing places to stay at. Sometimes they have no ground, only asphalt. Those that do have vegetation often have no landscaping other than mowed lawns. Some places don't even have picnic tables. Even the places that are not really slums are often solely functional. All sites are exactly the same size, having as much individuality and character as the individual bricks in a brick wall. And there is one more aspect to consider when staying on someone else's property with an RV. When you stay on private property, there is almost always an owner/caretaker, who rarely misses an opportunity to let you know that, although you are welcome to stay there, as long as you keep paying the fees, you are on *someone else's turf.*

C) *Public property* is property that belongs to *everybody.* Anyone can use public property, almost no matter how poor that person might be. (And poor people with RVs often figure out how to make good use of this.) Sometimes fees are involved with use of public camping facilities, but those fees are usually not large enough to preclude anyone, especially in camping areas that are primitive.

Public lands, available for camping, are a part of a vast amount of "stuff" called the public domain. The public domain includes roads and bridges, library holdings, public recreational facilities, schools, radio and TV broadcasts, some literature and music not copyrighted, as well as public lands. Public Federal lands for camping are a bit like public roads are for driving and parking. Anybody can drive down any public road, as long as that person has a legal road vehicle and a driver's license. Any such person can park in any available parking space where and when parking is allowed. If you choose to drive on a road, nobody has the right to run you off the road or mess with you in any way. Although located on public property, your vehicle is your own private property. It temporarily claims a certain amount of space in and around it. No other citizens can enter or even touch your vehicle without your permission, and as long as your driving and parking are legal and not dangerous, even law enforcement can't. On public federal lands, anyone in this country has the legal right to temporarily occupation in certain places where and when camping is allowed. If you choose to legally camp, nobody has the right to make you leave or mess with you in any way. Although located on public property, your equipment is your own private property. It temporarily claims a certain amount of space in and around it. No other citizens can enter or even touch your RV or any of your stuff without your permission, and as long as your behavior is legal, even public servants managing these lands can't (at least they *shouldn't*). Public Federal lands are like public roads in another manner: *they must be shared.* Just as you cannot run someone else off the road nor harass a person out of a parking space, with the exception of reservable campsites in developed campgrounds, you cannot claim a piece of public lands for your own exclusive or future use. Claiming spaces to camp on

Federal lands is a bit like borrowing library books. Anybody has the right to borrow books for awhile, but they must be returned in conditions very near to their conditions when borrowed. Anyone has the right to "borrow" camping places but they must be "returned" (vacated) very near to their conditions when first "borrowed" (occupied).

Rules for camping on public Federal lands

National Parks almost universally allow camping only in established campgrounds. There are some exceptions to this, on lands that are administered by the National Park Service, but are not technically National Parks, like National Recreation Areas (usually reservoirs created by damming rivers) and National Preserves. (If unsure as to the rules, always assume dispersed camping is *not* allowed in any places run by the National Park Service.) With established campgrounds, you pay your fee and you camp. There's not much more to it than that. You may destroy or take nothing. You may dispose of waste only in containers or at facilities designed for such a purpose. You are expected to be quiet from 10:00 PM to 6:00 AM, and not engage in disruptive behavior that could unreasonably interfere with the enjoyment of camping neighbors.

National Forests and BLM lands typically have both established campgrounds and areas where dispersed camping is allowed. With established campgrounds, the rules are pretty much the same as they are in the National Parks. With dispersed camping, the rules are as follows: 1) *You may not camp anywhere, just at certain places.* This is usually not a problem, because public land holdings are so vast, that even if only a small portion of them are open to camping, this usually is enough to meet your needs. 2) *You are expected to practice "leave no trace" etiquette.* This means, you should not litter, dump toxic waste, nor alter, change or take anything. You should also not make and leave behind "improvements," such as picnic tables, treehouses, rock gardens, etcetera. Usually, only solid and liquid human waste may be left. However solid human waste must be buried. (Toilet paper should not be buried because animals may dig it up. It takes quite awhile to decompose and looks disgusting.) In some places, it is forbidden to leave solid human waste, even buried, so it must be carried out. Graywater (generally left from dishwashing) usually must be transported out and disposed of at dump stations or toilets. Tree limbs from live trees should not be cut off. In some areas, gathering of dead down fire wood is permitted, in other areas even taking dead wood is not permitted. Digging or building up ground to make roads or alter drainage patterns is not permitted. In some places, new fire pits should not be made, although ones that already exist may be used. Rocks should not be taken, carved or painted, although some invertebrate fossil or rock collecting is allowed on certain National Forest or BLM lands (never in National Parks). In some areas, the ground is "alive" with microscopic bacteria and plants, and very fragile. Off road driving and sometimes even off trail walking is prohibited in these areas. And: 3) *Public lands may be occupied only temporarily.* Almost universally, there is a time limitation of two weeks for continuous camping at the same spot (though some places allow longer occupations during off-season, for hunting or other reasons). Various regions have time periods that people and their equipment need to be gone for before returning previously occupies spots. And there are annual total maximum time limitations for occupancy (for individuals who return repeatedly to the same spots). The temporary occupancy rule is in place because without such a rule, it would be too easy for retirees to go live on public lands. Any person could just say, "I don't want to bother spending money buying a plot of land or paying taxes on it." That person could then just pull into a National Forest or BLM land holding, set up a double wide mobile home, dig a well, then "camp" there for the next fifteen years. Of course, during that time span, that person would accumulate several tons of very important personal "stuff" (actually rubbish), die, and leave behind a colossal mess for somebody else to clean up, which wouldn't be fair to the rest of us. Although you cannot stay a really long time at any *one* spot on public Federal lands, if you move around frequently, you *can* live for months, even years, on public Federal lands, paying very little or no actual rent. One can literally "get lost" from society, and live without an actual "address," if one so chooses.

Dispersed RV camping on public lands is like owning and using the ultimate vacation cottage. You provide the "cottage" (RV) and the government provides you with places to park it. When you set up your RV on public lands, usually nobody hassles you. Often you don't even see any Park/Forest Service/BLM rangers for weeks. You choose your own level and type of equipment, at a budget that's right for you. You can wake up as early as you want or sleep in as late as you want. You can enjoy the beauty of nature. In some ways, dispersed RV camping on public lands is even better than having the permanent vacation cottage. You don't need to buy the property. You don't need to pay taxes on property. The cost of camping (including your own equipment and commuting) can be a small fraction of the cost of maintaining a permanent vacation cabin on private property. Unlike private parcels of land, which may be tiny, public lands holdings available for camping are vast. You have almost an endless amount of "favorite spots," if you are willing to take the time to drive to them. (Like public roads for driving, public lands for camping are so extensive that it is impossible for any one person to use them all.) Finally, if you get bored with a spot, you can just drive away and never return--- you don't need to "sell" your spot. Of course, there are disadvantages. You must bring and remove all your equipment from spots. You need to relocate frequently. And many great camping places are likely to be far from your regular home, so commuting can take considerable time and money, thus be just be too difficult. Still, dispersed camping on public lands is a feat worthy of consideration.

<u>Conclusion</u>: Would like to spend *thousands* of extra dollars unnecessarily, camp at bland spots (or slums) rather than by beautiful beaches and rivers, eliminate 95% of your camping options right off the bat, be "shut out" of full campgrounds, and camp on *someone else's* property rather than *yours*? Then by all means, *choose to remain dependent upon municipal electricity*.

> I love camping off grid. I don't really understand why more people don't love it too. I don't know why so many people insist upon camping with electrical outlets. It never ceases to amaze me how many people will pass up a beautiful spot by the beach, a river or a creek, favoring a spot by a smelly garbage dumpster, or some noisy highway, just because that spot has that all-too-important electrical outlet.

The Case For Grid Dependency

Despite several wonderful advantages to camping off grid, many people choose grid dependency, because having electricity at campsites is just such a convenience. The term "convenience" may be an understatement, perhaps "dependence" would be better. All the individual reasons for grid dependency stem from this general principle of convenience/dependence.

> The "Age of Electricity" is older than the oldest living humans, thus many people have completely forgotten how to live without it. Some electricity users without electricity behave a bit like heroin users without heroin--- they become more than just uncomfortable, they become miserable, inept, completely dysfunctional. I've seen grown men and women almost reduced to crying babies, wanting to go "home," after rather short time periods with their RVs disconnected from the grid.

Reasons For Grid Dependency:

1) People who camp with electricity don't need to buy or install special equipment. They can use their electric lights, toasters, coffeepots, kitchen gadgets, fans, air conditioners, TVs, DVD players, stereos, computers, etcetera, just the way they do at home. People who camp off grid, on the other hand, must buy special equipment and install it, in order to function and be comfortable.

2) People who camp with electricity don't need to spend time buying special equipment. In addition to taking time to install the right equipment for off-grid use, it takes time just to buy it. Furthermore, inexperienced off-grid campers are likely to be inefficient in their purchasing and installing. They may mistakenly buy and struggle with some items that just don't "cut the mustard" in the field, while passing on a few items that would work well in the field. This wastes more time and money. Multiple shopping trips, many conversations with other off grid campers, considerable reading and lots of trial and error are probably needed, for those new to off-grid camping to get it right.

3) People who camp with electricity don't need to learn new habits. For off-grid campers, it's not enough to just to buy and install special equipment, they need to learn how to use that special equipment. Typically, off-grid campers need to learn how to accomplish tasks normally done with electricity using alternative methods, conserve electricity (this takes practice), estimate how quickly their batteries drain, make electricity, charge batteries, and troubleshoot their systems. Learning to live in an RV off-grid is a bit like learning to live in a foreign country where a different language is spoken. When learning a foreign language, one must relearn every word, grammar, and sounds. When learning to live in an RV off-grid, one must relearn usage of appliances, most household chores, and general habits. This can be done, but the task is formidable. It does not come easily nor quickly.

4) You can't just buy an RV that functions well off-grid without modifications. I don't know why. You can buy a car, step on the brakes, and be almost certain the car will slow down or stop. You don't need to know anything about how those brakes work nor what components are in them. RVs are unlike cars; RVs are designed to "die in the field" when taken off-grid. If you buy an RV, and try to use its electrical & appliance system, you can be almost certain that it *will not work* off-grid very well or very long. To get it working, you almost need to become an electrical engineer and redesign its system from A to Z. This task is way out of the leagues of most people.

5) Electricity is becoming more available in many public camping places. This is especially true of State Parks. This trend is likely to continue. It is thus becoming easier to remain grid dependent.

Who Should Choose Grid Dependency and Why?

1) Weekend or Short Term Campers: Often, young to middle-aged people with families and careers only have time to take weekend trips or occasional one or two week excursions. They don't have the time to buy special equipment and learn new habits. Limited time translates into limited geographic striking range. In many places, no Federal or State land with suitable primitive camping areas may even exist within that striking range, especially for people living in the East. Easterners may thus have to depend upon privately owned campgrounds or State Parks for their camping needs. Since almost all private campgrounds have electricity, and many State Parks now do too, equipping themselves to camp off-grid may be inadvisable, especially if they repeatedly return to a few "favorite spots" known to have electricity.

2) Seasonal Campers on Private Lands: Unlike Federal lands, which have rather short maximum time limits for occupancy, most private campgrounds are willing to rent out spaces long term. While daily or weekly campsite fees on private lands are usually outrageous, yearly rates can be very reasonable. (Owners realize that patrons don't live at sites full-time, and take this into account when determining prices.) Some of these private campgrounds are quite nice. Some are well landscaped, having big sites, trees, showers, swimming areas, playgrounds and other attractions. (One seasonal spot that my father rented had a golf course. Another had a large recreation hall, where country bands played live on Saturday nights.) Buying a trailer RV and parking it on a rented seasonal lot in a developed campground costs considerably less than buying a vacation cabin on a private piece of land. And if the trailer is never moved, it costs even less, because a tow vehicle isn't needed, as one can hire someone else to do the initial towing. Also, trailers alone usually need less maintenance than private vacation cabins and properties. Since private campgrounds that rent seasonal sites almost all have electricity, it makes no sense for people who only camp this way to gear up for off grid use.

painting of my parents' family on-the-grid seasonal campground
(the one with the country music), early 1990s

pictured left (with guitar): Joseph Childs, author's brother
pictured right: William Childs (1930-2001), author's father
artist: Frances Childs, author's mother

Conclusion: Although there are definite disadvantages to remaining dependent upon electrical outlets, for some, this is the best course of action. This is because camping habits vary greatly from one person to the next.

Bringing Municipal Electricity Into Campgrounds

There is one alternative camping strategy to redesigning RVs to make them functional in wilderness: this of course is to bring electricity into the campgrounds. Who is doing this and who isn't?

Trends: ***State Parks*** in many states have tended to electrify their campsites over the past few decades. ***National Parks*** tend to take the opposite approach, preferring to keep campsites primitive. ***National Forests*** tend to have both primitive and electrified campgrounds, but the vast majority are still primitive. (Their campgrounds with electricity tend to exist only in areas that are heavily visited and/or near substantial areas of human population.) ***BLM campsites*** are almost always primitive.

State Parks

When I was a young boy, very few (if any) State Parks had electricity in their campsites. People went to these parks with tents or small trailers, taking only their cloths, cookware, propane lanterns, flashlights and their ingenuity. Camping was simple and cheap and that's the way many of us liked it. Simplicity and creative problem solving were considered part of the fun of the whole camping experience. Now, many campsites in State Parks have electricity. Many people like this change but some do not.

There are four basic problems associated with adding electricity to campsites: 1) Universal electrification is a functional impossibility, as many areas are too remote or too lightly used to justify the cost. [This applies mostly to large Federal lands and mostly in the West.] 2) Electrification has caused runaway campsite price increases. 3) Runaway price increases has caused State Parks to "close" to those of modest incomes and become less accessible to the disabled. And: 4) Widespread electrification encourages poor off-grid RV designs.

Places and Thoughts:

Florida State Parks: In 1988, I took a three month camping trip to Florida, staying heavily in the State Parks. The price of a State Park campsite back then was $6.42/night. By 1990, somebody in charge decided that Florida Parks should be money making "resorts," so they more than doubled their fees virtually overnight. Outraged and being on limited income, I stopped camping in the Florida State Parks, until a brief visit in 2012 (with a friend to share the bill), when the price of a site was $27/night---more than quadruple the 1988 amount. I know inflation exists, but this is ridiculous. Nothing else I know of (not even crude oil) has quadrupled+ in that time span. $27 in 2012 had considerably more buying power than $6.42 in 1988, so much of this represents an *actual* price increase, not just an *adjusted-for-inflation* price increase. What has contributed to this price change? Back in the 1980s, very few of these park campsites had electricity. Today, very few (if any) of these park campsites don't have electricity and many have hookups to water as well. When facilities increase, prices inevitably increase too. These price increases have caused something to get lost. In the 1980s, many people, with modest incomes and simple equipment (often just vans), used to snowbird in Florida. They had no Florida addresses; they just bounced around between the State Parks. They were often nice friendly well educated people who loved nature. Poor people and rich people camped side by side and often accepted one another. Today, limited income people can no longer afford to snowbird in the Florida State Parks. This seems like a shame, even for the State of Florida, because many campsites go frequently unoccupied (especially during weekdays), doing no good to anyone. Also, these price increases have made the parks less accessible to the disabled, because many disabled people rely solely on modest government benefits.

Texas State Parks: In Texas State Parks, they divide their campsites into two categories: *RV sites* and *tent sites*. In one park in 2015, the tent sites were $17/night and the RV sites were $29/night (including a $5 entrance fee). While anticipating passing through, I telephoned them, trying to get a reservation for a tent site for my small trailer for a couple days. (Many people define "tent sites" as a non-electric sites, and allow small RVs to occupy them.) When I explained that I had a small trailer, I was told that I could not take a tent site; I must take an RV site, that all RV sites have electricity, and I must pay the higher $29/night rate ($12/night difference). I explained that I have a "non-electric RV" and wouldn't do that, because I didn't want to be forced to pay for an electrical outlet that I didn't need and couldn't even use. So I politely ended the phone call. I drove through the panhandle of Texas, without stopping at any of their State Parks. Thanks for the hospitality, Texas.

California State Parks: In California in 1986, I camped at a State Park within the coastal Redwoods for $6/night. Camping was cheaper back then--- $6/night was the most I paid at *any* State or Federal campground during my two month trip, and one park, Lassen Volcanic National Park, had a (non-discounted) price of $3.50/night. In 2015, I camped at a similar California State Park, also within the Redwoods, where the price was $36/night (non-electric site). What the hell? Sextuple the price? What is going on here? Well, the park I stayed at in 2015 had electrical outlets in most of their campsites; the park I stayed at in 1986 did not. Although the 2015 site I took did not have electricity, it nonetheless appears to me that fees from non-electric users had been used to "improve" the rest of the campsites, by putting electrical outlets in them.

The Poor Funding the Rich: Look, as a poor person, I have nothing against rich people--- really I don't. In fact, I camp next to people of greater affluence (than I) in big RVs all the time. They seem just as friendly, enthusiastic and intelligent as us poor folks. I make it a point to be nice to these people. But I draw the line on friendliness when it comes to the electrical outlet issue. If you have $100,000 or more to spend on a luxury RV, you can pay for your own electrical outlets. You don't need us poor people with vans and simple trailers (who don't have air conditioners, microwave ovens, electric coffeepots and toasters) to pay for your 50 Amp power outlets (so you can run your air conditioners, microwave ovens electric coffeepots and toasters).

Where does campground overdevelopment logically stop? If one can justify putting electrical outlets in every site, forcing everybody to take them, and quadrupling prices, why not add cable TV, landline internet, complimentary (not really) breakfast, and have tennis courts, swimming pools, golf courses and indoor ice skating rinks on every State Park grounds? Then, maybe the parks could (and would need to) septendecuple (multiply by 17) their prices within a human lifespan. Why not eliminate camping altogether and have expensive ($250/night) lodges as the only option for overnight stays? This would be a wonderful way to keep those damn poor people off public lands.

A Vicious Self-Feeding Cycle: Greater campsite electricity availability encourages poor off-grid RV electrical system design (people see less of a need for good off-grid RV performance), and poor off-grid RV design encourages greater electricity availability (more people want, request and eventually get campsite electrification). Where is this heading? Probably toward the demise of simple cheap camping in State Parks and those parks becoming exclusive country clubs only available to a rich aristocracy, rather than to everyone.

Final Thought on Accessibility: Is it ethical to limit accessibility of the poor and disabled to State Parks (or any public lands), if funding for those Parks comes from taxpayer dollars? Please be aware that many State Parks were developed during the Depression by the C.C.C. using Federal Taxpayer dollars.

National Parks:

(includes N.P.S. run National Monuments, Lakeshores and Recreation Areas)

National Parks seem to be taking the opposite approach (that many State Parks do) on the campsite electrification issue: they want to keep most campgrounds primitive. It looks like they are going to continue this trend in the future. Currently, only the most visited National Parks have some campsites with electricity, and those sites exist in limited numbers. The National Park Service seems opposed to supplying widespread electricity to their public campsites, both for practical and philosophical reasons. On the practical side, I think it is just the goal of the N.P.S. to keep facilities simple and prices cheap. The N.P.S. seems committed to keeping the Parks open to everyone. Their entrance fees are reasonable and they offer substantial camping (and entrance) discounts to senior citizens and disabled people. (This is in stark contrast to State Parks in many states, which seem almost proud of the fact that they offer no discounts to out-of-state senior or disabled persons.) On the philosophical side, N.P.S. has an almost "Amish" view of electricity in their campsites. Amish people often don't mind having electric lines run across their property, supplying electricity to other people, or taking their horse-drawn buggies to corner stores to use public telephones to communicate with non-Amish people. They just don't want electric lines or telephones *in their houses*. N.P.S. doesn't mind having miniature cities within their more well visited National Parks (such as Yosemite, Death Valley, Yellowstone), containing camp stores, gas stations, visitor centers, ranger/employee housing, medical clinics, all of which are electrified. They also don't mind having electricity supplied to their camping rest rooms, and even to one or two special campsites reserved for campground hosts or maintenance staff. But in the midst of these "cities," they often don't want electricity supplied to *their public campsites.*

National Forests:

National Forests are publicly owned Federal lands. Unlike National Parks, which are designated as protected wilderness, National Forests are designated as "multiple use." They allow logging, mining, hunting & fishing, some fruit/nut gathering, even some rock collecting; all of which activities are limited and regulated. They allow ATVing, snowmobiling, horseback riding and of course camping (both developed and dispersed). They do designate some areas as wilderness, to be protected, but outside these areas, they are much more lenient in terms of usage rules than National Parks, which are almost fanatical at times about preservation. Because National Forests are much more lenient in terms of usage rules, they don't have any philosophical dilemmas regarding campground electrification (or other man made improvements); their agenda is strictly practical. They do provide electricity to many well visited campgrounds. However, National Forests have vast acreages, compared to National Parks. Much of their land is remote, lightly visited and unremarkable. Developed campsites within these areas are primitive, because they need to be, in order to exist. National Forests, National Parks and the Bureau of Land Management are all subdivisions of the same general government office (the Department of Interior), therefore all these agencies offer the same generous camping discounts to senior and disabled persons. All D.O.I. agencies seem committed to keeping their public lands accessible to everyone.

One Place That Has "Gotten It Right"

Jefferson National Forest (Stony Fork Campground, Virginia): I have camped at this location occasionally for years, as they are conveniently located just a few miles off a main highway route between New York and Florida, which I frequently travel. When I first camped there, fees were small and few (if any) campsites had electricity. Then, visitors stuffed the comment box, with the (#1) suggestion that they wanted more hookups. So the campground obliged, digging shallow ditches, to bury electric and water lines. It was not too messy. When all was said and done, many sites had electricity, and a few also had water, *but not all of them*. They left a substantial number of sites primitive. They decided that electricity should cost an extra $6 per night. Those who want electricity can have it; those who don't need it and/or prefer low prices can be accommodated too. Everybody can camp side by side and everybody can get along fine with one other (hopefully). This campground has strived to keep their prices low. I commend them for that. This campground (management) should be a model for campgrounds everywhere.

Should RVs Be Designed to Function Without Municipal Electricity?

Some RV users dodge the issue of off-grid capability entirely by saying: "I think I'll just find campsites with electricity." Simple solution, right? Even if a few places don't have municipal electricity right now, very soon all of them will, right? This may seem like a valid viewpoint to people who live in 39th floor apartments in big cities---people who walk on concrete and asphalt outside only briefly, while moving from the insides of one building to another, usually stepping on actual ground/lawn no more than four or five times per year. This is the viewpoint of city people who are far removed from the camping experience. I cannot respect such a viewpoint, because I've spent too much time camping on public lands, and I know what's out there. I can tell you for certain that the vast network of non-electric campsites will not be disappearing anytime soon.

I must acknowledge however that some people do camp exclusively with electricity. Also, some people use RVs as *houses* rather than *campers*. (Some RVs are too big to take "camping;" they are intended to be houses parked long term with full hookups. Also, traveling carnival workers or musicians often use RVs as homes without really "camping" in them.) I therefore feel that it would be alright to sell some RVs without off-grid capability, if they had disclaimers like "NOT INTENDED FOR USE WITHOUT MUNICIPAL ELECTRICITY" and "NOT SUITABLE FOR CAMPING IN MOST NATIONAL PARKS." These disclaimers should be clearly posted both on the RVs and in bold print on their owners manuals' covers (not in fine print buried somewhere on page 54). But considering that the most beautiful lands in the U.S.A. are in National Parks, very few National Parks have campgrounds with electricity, the N.P.S. seems to have no plans for future widespread campground electrification, and National Parks are a frequent and reasonable destination for RVs, yes, general purpose RVs should be designed to function off-grid. Without such capability, RVs are somehow incomplete. Designing and selling them without off-grid capability is a bit like designing and selling automobiles without windshield wipers, because *sometimes it doesn't rain*.

What are some other reasons why RVs should be designed to function off-grid? First, *customers shouldn't be deceived*. Many first time RV buyers just *assume* that RVs can function in wilderness. Some salespeople feel that it's alright to remain silent while knowingly letting somebody assume something to be true that's false. I think this is not alright; it is dirty dealing, especially if the faulty assumption is somehow implied. (Some rented RVs actually have murals of National Park vistas painted on them. Isn't this a bit deceitful, if those RVs "melt down" in National Parks?) Second, *the decision whether to use RVs in RV parks or in wilderness should be made by RV* ***owners, not manufacturers or sellers***. If manufacturers design RVs that are dysfunctional in wilderness, and sellers don't disclose this, then they make that decision for the buyers. And finally, *RVs (especially motorhomes) cost too much to have their electrical systems "die" in the wilderness*. For the sky-high prices that RVs cost, designers should *engineer* electrical & appliance systems, not just *slap them together*.

So, excluding special purpose RVs with disclaimers, asking the question: "Should RVs be designed for use without municipal electricity?" is a bit like asking: "Should cars be able to drive down the road?", "Should airplanes be able to fly?", "Should cameras be able to take pictures?" or "Should submarines be able to dive beneath the surface without taking in water?". There is something inherent to the whole concept of RVing, that implies going places and surviving comfortably. And Federal lands, especially National Parks, are places that RVs often go. My answer: **Yes, they should**!

Suppose you were considering taking in a foreign exchange student for a year, who had a $750/month illegal drug habit, which you were expected to pay for, if you invited this student into your home. Wouldn't you feel you had a *right to know* about this drug habit? Isn't municipal electricity kind of the "drug of choice" for RVs? Isn't dependence upon it kind of like some unnecessary drug habit?

above: My campsite in Allegheny National Forest during autumn. This site costs $12/night ($6/night with GA discount).

Electricity Basics

What Is Electricity?

Often, children's books define electricity as *moving electrons*. This is a good start, but I believe that the definition should also include *stored electrons* (ready to move easily), and all the nifty acrobatics that electrons do. Electrons are parts of atoms. Atoms are the basic building blocks of all material we encounter, such as the ground we stand on, the food we eat, the water we drink and the air we breath. There are literally septillions of electrons all around us. So why doesn't electricity occur more often? Because most of the time, electrons just sort of stay with their host atoms. It is only when electrons move, without taking their host atoms along, that electricity occurs. Electrons usually move through circuits by migrating from one atom to another, but they can also move through complete vacuums.

Voltage:

Look at *figure 1* below, an illustration of two metal wires. The dots represent electrons. In metals, such as these wires, electrons very easily migrate between atoms, almost "floating" through the wires, like gas molecules in sealed tubes. Notice any difference between these two wires?

Voltage

figure 1

You should. In the wire on the right, the electrons exist in greater density (they are more crowded) than they do in the wire on the left. What do you suppose would happen if these two wires were connected, in a manner that would freely permit electrons to exchange, either directly or indirectly through an appliance? If you answered:

Electrons would flow from the right wire, where they are more dense, to the left wire, where they are less dense, until the densities of the two wires equalize, you answered correctly. *Figure 1* shows a voltage, which can be defined as: *a difference in electron density (pressure) between two points in a circuit.* Note that a voltage is not a measurement of one thing, it is a comparison between two things. Many people get the concept of "voltage" wrong (even on some scientific TV documentaries), so let's also describe what a voltage is *not*. A voltage is NOT a description (nor measurement) of flowing electrons. A voltage is NOT a description (nor measurement) of electrons stored in batteries nor used by appliances. A voltage is NOT a description of stored, used or flowing energy.

> Once, I watched a TV news report about the electric chair. The reporter said: "The switch is thrown, and 9000 Volts of electric current flows through the condemned inmate's body, thus killing the inmate instantly." That reporter completely didn't know what he was talking about. There is no such thing as a *volt* of electric current!

A voltage is measured by placing the two leads of a *voltmeter* onto two different points of a circuit (usually wires or the terminals of a battery). This device compares the electron densities, then assigns a number to that difference. Voltages (some people call these "voltage differences") are measured in *Volts*. Either circuit point can have the greater electron density, therefore the voltage can be a positive or a negative number. Electrons carry a negative (–) charge, therefore the wire which has the greater density of electrons is also negative. Another common term is "voltage drops." This term is often used to describe voltages as measured between two points of a live circuit. If the electron densities of two wires (of the same metal alloy and temperature) are equal, as in *figure 2*, the voltage is zero.

zero volts

figure 2

Current and Charge:

Consider an analogy: a swimming pool with water in it, being filled by a hose. *Quick problem:* If the water in the hose flows at a rate of 3 gallons per minute, for 5 minutes, how much water gets added to the pool? If you solved this, you just used a very common mathematical formula:

(i.) Capacity = rate × time.

In this case, the *capacity* describes *the amount of water* in or added to the pool. The *rate* describes the *flow of water* through the hose. Notice that capacity and rate are two separate non-interchangeable quantities, each having their own separate units of measurement. In this "pool" case, capacity is measured in *gallons*; rate is measured in *gallons per minute.* One does NOT measure the amount of water in the pool in *gallons per minute*, nor does one measure the flow of water through the hose in *gallons*. I hope you correctly answered the above problem with "15 gallons." If so, you substituted the specific values for this problem into the general formula (i.) yielding:

(ii.) (15 gallons) = (3 gallons/minute) × (5 minutes).

Now, let's get back to electricity. *Charge* is an electron capacity, like the water capacity in the pool. It describes a *quantity of electrons* stored in, added to or drained from batteries (or capacitors), or used by gadgets *accumulated over time.* (Electric) *Current* is like the water current flowing through the hose. It describes the *flow of electrons* through wires, batteries or gadgets *not taking into account how long a circuit is left on.* A suitable unit of measurement to describe Charge is the *Coulomb*, which one of my electricity professors liked to call a "bushel of electrons" (about 6¼ quintillion electrons--- which is 6,240,000,000,000,000,000). A suitable unit of measurement for Current might be *Coulombs per second.* Like the pool analogy, Charge and Current are two separate non-interchangeable quantities. And like the pool analogy, one does NOT measure the quantity of electrons in *Coulombs per second*, nor the flow of electrons in *Coulombs*. *Quick problem:* If during charging, electrons flow into a battery at a rate of 3 Coulombs per second, for 5 seconds, how many electrons get added to the battery? If you solved this, you turned the general capacity-rate-time formula (i.) into an electrical formula:

(iii.) Charge (in Coulombs) = Current (in Coulombs/second) × time (in seconds).

I hope you correctly answered the above problem with "15 Coulombs." If so, you substituted the specific values for this problem into the general formula (iii.) yielding:

(iv.) (15 Coulombs) = (3 Coulombs/second) × (5 seconds).

Unlike the current of water, the unit for the current of electricity has its own name: the *Amp*, which is just another name for a Coulomb per second. (Also stated: *1 Coulomb* is the amount of electrons that move when a Current of *1 Amp* flows for *1 second.*) Thus:

(v.) 1 Amp = 1 Coulomb/second and
(vi.) 1 Coulomb = 1 Amp second.

Thus, formula (iii.) is rewritten as:

(vii.) Charge (in Coulombs) = Current (in Amps) × time (in seconds).

(Remember this formula. I've put the important equations in bold print.)
The specific electrical problem (equation iv.), where 3 Amps of Current flow for 5 seconds resulting in 15 Coulombs getting moved, is thus rewritten as:

(viii.) (15 Coulombs) = (3 Amps) × (5 seconds).

There is another unit of measurement for Charge: the *Amp hour*. The Amp hour is the favored unit in RV and automotive literature, because of it's bigger size. The smaller Coulomb is the favored unit of chemists and electrical engineers, especially when dealing with small internal components of electronic devices (especially capacitors). Since a Coulomb is an Amp second, and there are 3600 seconds in an hour,

(ix.) 1 Amp hour = 3600 Coulombs.

(This makes the Amp hour a much larger "bushel of electrons," containing about 22½ sextillion electrons, which is 22,500,000,000,000,000,000,000.)

Another version of formula (vii.) can be obtained by multiplying each side of the equation by 3600 (in algebra, if both sides of an equation are multiplied or divided by the same number, the truthfulness of the equation gets retained),

(x.) Charge (in 3600 Coulombs) = Current (in Amps) × time (in 3600 seconds),

then substituting (ix.) (1 Amp hour) = (3600 Coulombs) and (1 hour) = (3600 seconds), yielding:

(xi.) Charge (in Amp hours) = Current (in Amps) × time (in hours).

(Definitely remember this bold print formula.)

Some people use the term *capacity* instead of *Charge* to refer to the quantity of electrons. I do not like this term, because the word "capacity" already has too many other meanings in the English language. Confusion gets created when a very general term is attempted to be recoined to have a very specific meaning. The term "capacity" can be used to refer to quantities of water or gasoline stored in tanks or even gumballs in gumball machines. Furthermore, the term "capacity" can refer to the amount of something *actually stored* in a tank at the moment, or the maximum amount that a tank *can hold* when full. (I like to use the term "Charge" to describe the quantity of electrons *actually stored* in a battery or used by gadgets accumulated over time, and "storage capacity" to describe the amount of Charge that a battery *can hold* when full.) Also, the words "electrons" or "electricity" are sometimes used to describe Charge, or sometimes the unit name is just omitted. Thus, the following sentences all have the same meaning:

> Note that Current *DOES NOT take into account how long a gadget is left on,* whereas Charge *DOES take into account how long a gadget is left on.*

That 100 Amp hour battery currently has *80 Amp hours of Charge* stored in it.
That 100 Amp hour battery currently has *80 Amp hours of capacity* stored in it.
That 100 Amp hour battery currently has *80 Amp hours of electrons* stored in it.
That 100 Amp hour battery currently has *80 Amp hours of electricity* stored in it.
That 100 Amp hour battery currently has *80 Amp hours* stored in it.

Power and Energy:

For those who are becoming a bit overwhelmed, there is both bad news and good news. First, the bad news: *there are more terms, concepts and formulas, associated with Energy and Power.* Now, the good news: *the mathematical relationships and concepts relating Energy to Power are exactly the same as those for Current and Charge.* Only the names of the units and what they represent are different.

Let's start with a simple definition of Energy (also sometimes called "work"). *Energy* is the ability to *accomplish something.* It is the ability to move something, lift something, heat something or "zap" something. Energy is the amount of "kick" that something has. Energy is a value that *accumulates with time*, like the water in a swimming pool. *Power* is the *flow of Energy*, and like the water flowing through a hose, this value *does not accumulate with time.* The unit of measurement for Energy is the *Joule*, (a "bushel of zap") whereas Power might be measured in *Joules per second.* Like the swimming pool analogy, one does NOT measure Energy in *Joules per second*, nor Power in *Joules. Quick problem:* If the Energy used by a gadget flows at a rate of 3 Joules per second, for 5 seconds, how much Energy gets used? If you solved this, you turned the general capacity-rate-time formula (i.) into a formula relating Energy to Power:

(xii.) Energy (in Joules) = Power (in Joules/second) × time (in seconds).

I hope you correctly answered the above problem with "15 Joules." If so, you substituted the specific values for this problem into the general formula (xii.) yielding:

(xiii.) (15 Joules) = (3 Joules/second) × (5 seconds).

Just as the flow of Charge has its own name (the Amp), the unit of measurement for the flow of energy flow has its own name: the *Watt*, which is just another name for a Joule per second. (Also stated: *1 Joule* is the amount of energy transferred when a Power of *1 Watt* flows *1 second.*) Thus:

(xiv.) 1 Watt = 1 Joule/second and
(xv.) 1 Joule = 1 Watt second.

Thus, formula (xii.) is rewritten as:

(xvi.) Energy (in Joules) = Power (in Watts) × time (in seconds).

(Remember this bold print formula.)
In the specific problem above: a circuit that consumes 3 Watts of Power, for 5 seconds, uses 15 Joules of Energy. Equation (xiii.) is thus rewritten as:

(xvii.) (15 Joules) = (3 Watts) × (5 seconds).

There is another unit of measurement for Energy: the *Watt hour*. The larger Watt hour is the favored unit when discussing RVs and regular houses (which also use the kilowatt hour). The smaller Joule (and sometimes kilojoule) is the favored unit of chemists (chemical reactions usually release or absorb energy). Since a Joule is a Watt second, and there are 3600 seconds in an hour,

(xviii.) 1 Watt hour = 3600 Joules.

Another version of formula (xvi.) can be obtained by multiplying each side of the equation by 3600,

(xix.) Energy (in 3600 Joules) = Power (in Watts) × time (in 3600 seconds),

then making the appropriate substitutions (xviii.) (1 Watt hour) = (3600 Joules) and (1 hour) = (3600 seconds), yielding:

(xx.) Energy (in Watt hours) = Power (in Watts) × time (in hours).

(Definitely remember this formula.)

Power and energy are electrical terms but they are not exclusively so. They apply to other areas of science, such as physics and chemistry as well. [Example: The sun radiates 388 septillion Joules of energy per second into outer space, which is 388 septillion Watts. (388,000,000,000,000,000,000,000,000. That's a big number.) It is technically correct to say this, even though the sun doesn't derive its energy from electricity.]

Note that Power *DOES NOT take into account how long a gadget is left on,* whereas Energy *DOES take into account how long a gadget is left on.*

Energy-Charge Relationship:

What is the difference between Energy and Charge? Consider an analogy: three groups of coins. Which two groups are the most similar?

Group A:	15 nickels,
Group B:	3 quarters,
Group C:	15 quarters.

There is more than one right answer. You may have concluded that groups A and C are the most similar, since each contains 15 coins (whereas B contains 3). You may have concluded that groups A and B are the most similar, since the total monetary value of each group is 75 cents (whereas C has a value of $3.75). You also may have concluded that groups B and C are the most similar, since each of these contain coins of equal value (quarters worth 25 cents--- group A contains nickels worth 5 cents). There are two ways of measuring and describing a total group of coins: you can consider the *number of coins* or *what those coins can buy*. These two quantities relate to the type of coin (value) by the following mathematical equation:

(xxi.) (value of coin group) = (value of 1 coin) × (number of coins).

When you calculated the monetary value of each coin group, you substituted the specific values for each group into the above equation, yielding:

(xxii.) Group A: (75 cents) = (5 cents/coin) × (15 coins)
(xxiii.) Group B: (75 cents) = (25 cents/coin) × (3 coins)
(xxiv.) Group C: (375 cents) = (25 cents/coin) × (15 coins).

What does this have to do with electricity? Plenty. Sometimes electrons act like "nickels" and sometimes they act like "quarters." And, like in the analogy, there are *two* ways of measuring groups of electrons. You can consider the *number of electrons* or *what those electrons can accomplish*. The first quantity is called *Charge*, which has already been covered. The second quantity is called the *Energy*, which has also been covered. The mathematical equation that relates these two electrical quantities to the energy value of one Coulomb, is very similar to equation (xxi.) used for calculating the monetary values of groups of coins:

(xxv.) (Energy of entire electron group) =
(energy value of 1 Coulomb) × (number of Coulombs).

Stated more mathematically:

(xxvi.) Energy (in Joules) =
Energy value (in Joules/Coulomb) × Charge (in Coulombs).

Now, let's use this equation to calculate the energy values of three groups of electrons: A) 15 Coulombs of electrons, each having an energy value of 5 Joules per Coulomb, B) 3 Coulombs of electrons, each having an energy value of 25 Joules per Coulomb and C) 15 Coulombs of electrons, each having an energy value 25 Joules per Coulomb. Substituting the individual values into the general formula (xxvi.) yields equations which should look very familiar:

(xxvii.) A: (75 Joules) = (5 Joules/Coulomb) × (15 Coulombs)
(xxviii.) B: (75 Joules) = (25 Joules/Coulomb) × (3 Coulombs)
(xxix.) C: (375 Joules) = (25 Joules/Coulomb) × (15 Coulombs)

Now the $1,000,000 question: *What determines the energy value of a single Coulomb of electrons?* Answer: *The* *__Voltage__* *that it drops.* Whenever *1 Coulomb* of electrons drops *1 Volt*, *1 Joule* of energy gets released. (So if 1 Coulomb drops 5 Volts, it behaves like a "nickel," having an energy value of 5 Joules. If 1 Coulomb drops 25 Volts, it behaves like a "quarter," having an energy value of 25 Joules.) Thus:

> 15 Coulombs of electrons, dropping 5 Volts, release the same energy as 3 Coulombs of electrons, dropping 25 Volts, for the same reason that 15 nickels can buy the same stuff as 3 quarters (which is not much at today's prices).

(xxx.) 1 Volt = 1 Joule/Coulomb
and
(xxxi.) 1 Joule = (1 Volt) × (1 Coulomb)

The final version of the Energy-Charge formula is obtained from (xxvi.), by substituting "Voltage" for "energy value of 1 Coulomb" and "Volts" for "Joules/Coulomb":

(xxxii.) Energy (in Joules) = Voltage (in Volts) × Charge (in Coulombs).

Equations (xxvii.), (xxviii.) and (xxix.) can be rewritten as:

(xxxiii.) A: (75 Joules) = (5 Volts) × (15 Coulombs),
(xxxiv.) B: (75 Joules) = (25 Volts) × (3 Coulombs),
(xxxv.) C: (375 Joules) = (25 Volts) × (15 Coulombs).

Another version of formula (xxxii.) can be obtained by multiplying each side of the equation by 3600,

(xxxvi.) Energy (in 3600 Joules) = Voltage (in Volts) × Charge (in 3600 Coulombs),

then making the appropriate substitutions (xviii.) (1 Watt hour) = (3600 Joules) and (ix.) (1 Amp hour) = (3600 Coulombs), yielding:

(xxxvii.) Energy (in Watt hours) = Voltage (in Volts) × Charge (in Amp hours).

Electrons "Dropping" Both High and Low Voltages

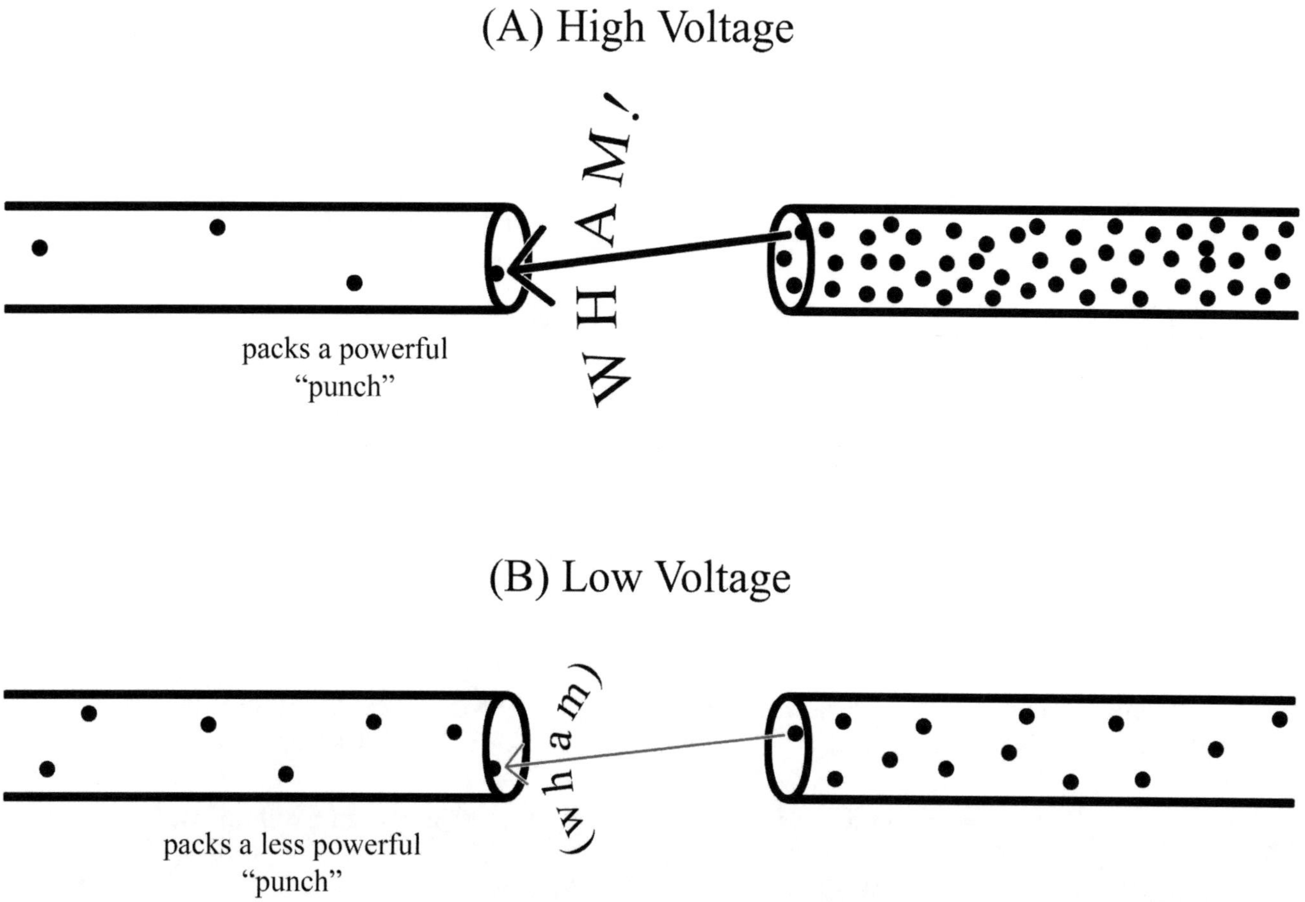

figure 3

Consider an analogy: *bowling balls "dropping" from various heights onto concrete.* A bowling ball dropped three feet would make a thud and kick up a bit of dust. A bowling ball dropped from a second story window would make a harder thud and kick up more dust. A bowling ball dropped from a window 65 stories high would come down hard enough to deform and might crack the concrete too. If the Earth had no atmosphere to slow the bowling ball down and cause it to burn up, a bowling ball dropped from outer space would impact so hard that it would disintegrate, evaporate, explode with more kick than dynamite, and leave behind a crater perhaps the size of a swimming pool after the dust settled. Electrons dropping voltages act a lot like bowling balls dropping heights: *the higher they "drop," the harder they "fall."*

Power-Current Relationship:

Recall formula (xxxii.):

Energy (in Joules) = Voltage (in Volts) × Charge (in Coulombs).

Energy and Charge are both *accumulated* quantities. This formula can be changed into a *flow rate* formula by simply dividing both sides of the equation by "1 second" (retaining the truthfulness of the equation):

(xxxviii.) Energy *flow rate* (in Joules/second) =
Voltage (in Volts) × Charge *flow rate* (in Coulombs/second).

Recall that *Energy flow rate* is called "Power," *Charge (electron) flow rate* is called "Current," (xiv.) (1 Watt) = (1 Joule/second) and (v.) (1 Amp) = (1 Coulomb/ second). When all four substitutions are made into the above formula, it simplifies into:

(xxxix.) Power (in Watts) = Voltage (in Volts) × Current (in Amps).

This is perhaps the most important electrical formula of them all. Remember it. In plain English (DC circuits), Power is obtained simply by multiplying Voltage with Current.

Note that a Current of *1 Amp* dropping *1 Volt*, uses a Power of *1 Watt*. Therefore:

(xl.) 1 Watt = (1 Amp) × (1 Volt).

The power-voltage-current formula (xxxix.) has real world consequences in electricity. Whatever the application, whatever the appliances are, and for however long things are left on, less electrons are needed to "do the deeds" when voltages are high than when voltages are low. Thus: *low voltage systems need to use greater amounts of current in order to achieve the same power as high voltage systems*. Thus, *low voltage systems need to use bigger thicker wires*, in order to carry all that extra current, than high voltage systems do.

> Example of formula (xxxix.):
>
> If each second, 5 Coulombs of Charge (electrons) leaves a 6 Volt battery, then each second 30 Joules of energy leave that battery. Stated another way: 5 Amps of Current leaving a 6 Volt battery results in 30 Watts of Power leaving that battery. Stated mathematically:
>
> 30 Watts = (6 Volts) × (5 Amps).

Shortcut Formula: Because RVs use 12 Volt systems, Power and Energy values can be obtained from Current and Charge values (respectively) by multiplying by 12.

Decades ago, some automobiles used 6 Volt systems for their lighting and other electrical needs. Big thick wires were needed. Eventually, these 6 Volt systems were abandoned. Today, almost all automobiles use 12 Volt systems, so thinner wires can be used.

Voltage-Current-Resistance Relationship:
(Ohm's Law)

Consider an analogy: blowing up a balloon then opening the end to let the air out. After a balloon is inflated, there is a *pressure difference* between the air inside the balloon and the air outside the balloon. (The rubber membrane pushes the air molecules inside the balloon closer together, making them more crowded than the air molecules outside the balloon.) When the end is opened up, a *resistance* is formed. A *resistance* is a path which permits a flow, but with some degree of difficulty. When a *pressure difference* is applied to a *resistance*, it permits a *flow*. (See *figure 4*.)

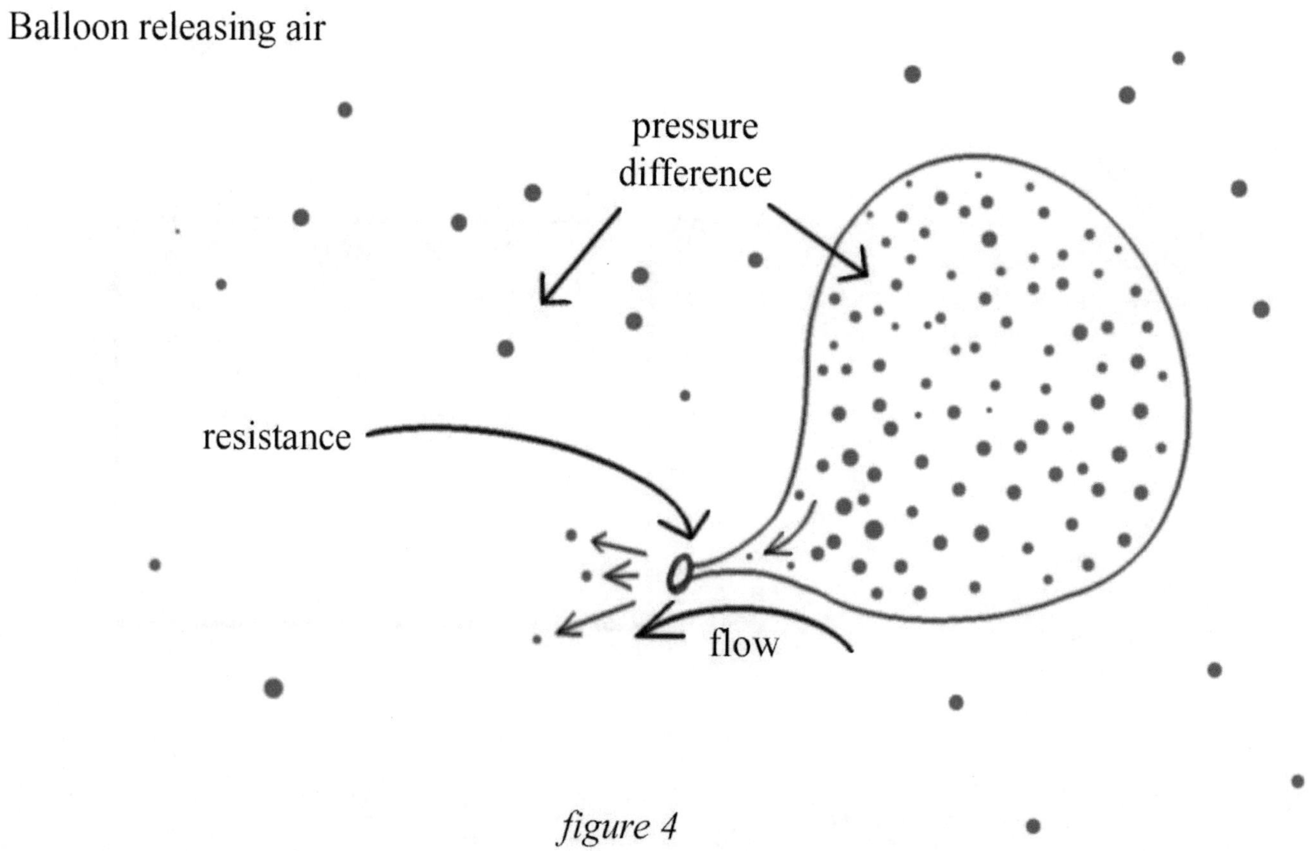

figure 4

A mathematical formula describing this is:

(xli.) pressure difference = (flow) × (resistance).

The same situation happens with electricity, except that the pressure difference and the flow involve electrons, rather than gas molecules. The above formula can be converted into an electrical formula, by substituting "Voltage" for "pressure difference" and "Current (of electrons)" for "flow (of gas)," yielding a formula known as *ohm's law*:

(xlii.) Voltage (in Volts) = Current (in Amps) × Resistance (in ohms)

Electrical resistance is measured in *ohms*--- the symbol being Ω. When *1 Volt* is applied across a resistance of *1 ohm*, *1 Amp* of current flows through that resistance. Thus:

(xliii.) 1 Volt = (1 Amp) × (1 Ω).

Ohm's law is very important for all electrical applications, including RVs. If wires leading up to a gadget are too thin or if connections are loose or corroded, then there is a high value of *resistance* leading up to the gadget. If a gadget draws low current, then this high resistance usually poses no problem. If a gadget draws high current and wires and connections have low resistance, then there also is usually no problem. But if *both* the current needed for the device and the resistance in the wires/connections are high, then there is a substantial voltage drop between the source (usually the battery) and the gadget (meaning the voltage at the gadget is substantially lower than at the battery or power converter). When gadgets are attempted to be operated from voltages too low, the gadgets struggle. They may work poorly or not at all or may even be damaged/destroyed by such operation.

Slang Terms: Some people use the terms "Wattage" to mean Power and "Amperage" to mean Current. This is rather logical, since Power is measured in Watts and Current is measured in Amps. (After all, "Voltage" is measured in "Volts," so why not do the same for these other two terms?) Though logical, these terms are slang, so I do not use them in this book. However I always understand these terms when used by others. You should be aware of these terms when reading material by other authors.

Summary of Terms:

Voltage (measured in Volts) describes an *electron pressure difference* between two different points in a circuit.

Current (measured in Amps) describes a *flow of electrons* through a wire or a gadget.

Charge (also called "capacity" or "electrons") (measured in Coulombs or Amp hours) describes a *quantity of electrons* stored in a battery or used by a gadget *accumulated over time.*

Power (measured in Watts) describes a *flow of energy.*

Energy (measured in Joules or Watt hours) describes the amount of "kick" that something has. In electricity, energy describes *what electrons can accomplish.* (Like Charge, this value *accumulates over time.*)

Resistance [measured in ohms (Ω)] describes the *difficulty* that electrons encounter when moving though some part of a circuit.

Summary of Formulas:

Charge (in Coulombs) = Current (in Amps) × time (in seconds) or
Charge (in Amp hours) = Current (in Amps) × time (in hours)

Energy (in Joules) = Power (in Watts) × time (in seconds) or
Energy (in Watt hours) = Power (in Watts) × time (in hours)

Energy (in Joules) = Voltage (in Volts) × Charge (in Coulombs) or
Energy (in Watt hours) = Voltage (in Volts) × Charge (in Amp hours)

Power (in Watts) = Voltage (in Volts) × Current (in Amps)

Ohm's Law:
Voltage (in Volts) = Current (in Amps) × Resistance (in ohms)

Components of RV Electrical Systems

The typical RV electrical system has the following parts: 1) DC appliances, 2) DC wiring connecting the appliances to their electricity source, 3) AC appliances, 4) AC wiring connecting the appliances to their electricity source, 5) fuses/ circuit breakers, 6) a battery (or several) to store DC electricity, 7) two battery chargers (one used while driving and a second used while parked), 8) a generator (to make AC while parked), 9) a power converter (to transform AC into DC) and 10) a power inverter (to transform DC into AC). Simpler RVs may lack some components (like generators or power inverters), whereas very sophisticated RVs might have still more components, such as solar panels for battery charging.

Two RV "House" Electrical Systems (AC and DC)

To speak of "an RV electrical system" in singular rather than plural form is somewhat misleading, because there are two electrical systems in each RV: *the 120 Volts AC system* and *the 12 Volts DC system*. Each system uses its own kind of electricity, has its own power source(s), its own separate wiring, its own distinct types of electrical outlets, and its own appliances. (Appliances designed to be run by one type of electricity cannot be run by the other.)

The 120 Volts AC System

What is AC? "AC" stands for "Alternating Current." It is the type of electricity delivered from standard household wall outlets and used by household appliances. Both the current and the voltage rapidly alternate in an AC circuit; the current changes direction, the voltage changes orientation.

Where is AC usually made? AC is usually made in generators. In a generator, a coil moves toward the north magnetic pole, which *pushes* electrons through the exit wire and pulls them out of the return wire. A tiny fraction of a second later, that same coil heads for the south magnetic pole, in which it *pulls* electrons out of the exit wire and pushes them into the return wire. Then, the coil heads back for the north magnetic pole, and the cycle repeats. Half the time, the electrons are *more* crowded in the exit wire (than in the return wire), the other half of the time the electrons are *less* crowded in the exit wire. The alternating voltages in AC circuits cause electrons in the wires to move back and forth like a pendulum, always changing directions and speeds. [A child on a swing is a good example of pendulum motion. The child starts at maximum backward point. At this point, the child begins moving forward, picking up speed until reaching a maximum forward speed (at vertical position), at which point, (s)he begins slowing down until completely stopping at maximum forward point. The child then begins to move backward, picking up speed until reaching a maximum backward speed (also at vertical position), at which point, (s)he begins slowing down until completely stopping at maximum backward point, where the cycle began. The cycle then repeats.] This type of motion is called *sinusoidal* motion. (Also called *sine wave* motion. "Sinusoidal" is an adjective, whereas "sine wave" can be an adjective or a noun.)

> Many things in nature exhibit sinusoidal motion: pendulums, current in AC circuits, vibrating strings of guitars and pianos, people who jump from bridges tied to bungee cords, and handsaws cutting pieces of wood.

What is RMS Voltage? How can a number be placed on a voltage that is always rapidly changing? One could consider the *average* voltage, but this value is meaningless, because the average voltage of AC is actually zero (half the time the voltage is a positive number and the other half of the time the voltage is a negative number), yet the electrons can deliver energy to devices, both when moving forward or backward. So electricians use a quantity called the *RMS voltage*, which is a special "average" voltage, based on the ability of an AC source to deliver power to a device. The RMS voltage is the equivalent DC (steady) voltage needed to apply the same average power to a resistive load as the AC. ("RMS" stands for "root mean square," which is the square root of the average square.)

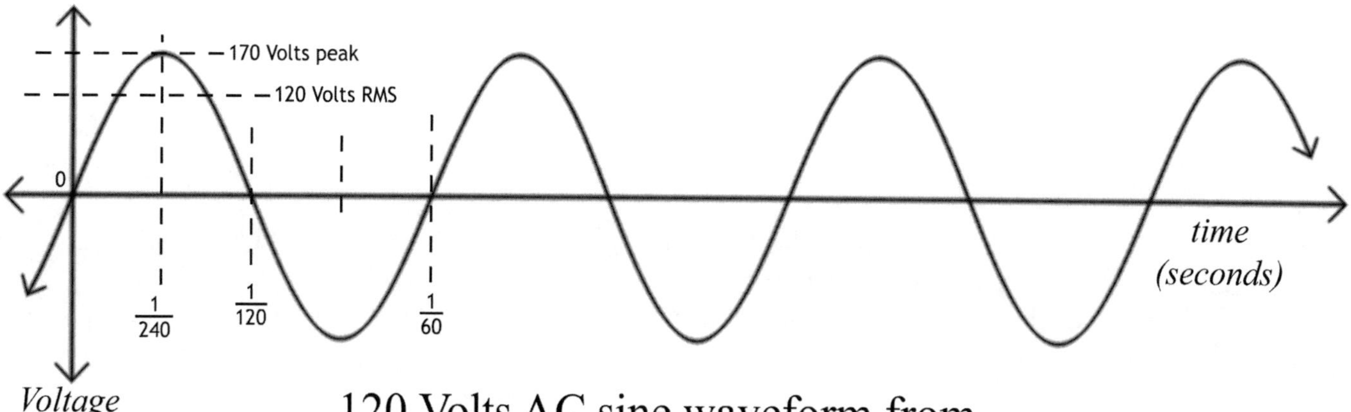

120 Volts AC sine waveform from standard household wall outlet

Consider an analogy: comparing wood being cut by a handsaw (with teeth capable of cutting in both directions) with wood being cut by a bandsaw. The handsaw's speed is always changing: sometimes the teeth move very fast, sometimes they don't move at all, most of the time, their speed is between these two extremes. The bandsaw moves at a steady speed. The "RMS speed" of the handsaw is the equivalent speed that the steady bandsaw blade must move, in order to cut the wood at the same rate as the alternating handsaw. Both AC voltages and AC currents can be expressed in RMS values.

Where can AC be found and what is it used for? 120 Volts (RMS) AC @ a frequency of 60 Hertz is the type of electricity that electric companies deliver to standard household wall outlets, to be used to run almost all electrical appliances found in regular houses and places of business. Other types of AC exist, such as 240 Volts (RMS) AC (used to run household stoves), three phase AC (used to run some industrial motors), very high frequency AC (to transmit and receive radio signals) and high voltage AC (used to send electricity long distances through wires). These others are beyond the scope of this book.

Where do RVs get AC? RVs can get 120 Volts AC from three sources:
1) municipal outlets [into which entire RVs are plugged in], 2) generators and
3) power inverters [which make AC from DC stored in batteries].

The 12 Volts DC System

What is DC? "DC" stands for "Direct Current." In a DC circuit, the voltage is kept constant (very little fluctuation), regardless of how much current the appliances use. The current in a live DC circuit moves steadily through the wires *in one direction only*, like a car in cruise control driving down the interstate or the blade of a bandsaw. In simple circuits with resistive loads (like incandescent lights and heating elements) almost no current fluctuations take place. More complicated circuits (like TVs or motors) fluctuate the current they use, because they draw current on an "as needed" basis, but that current always flows in one direction.

Where can DC be found and what is it used for? 12 Volts DC is the type of electricity that almost all appliances specifically made for RVs run from, because DC is the type of electricity that batteries store (thus appliances are ready for use anytime). 12 Volts DC is also the type of electricity that run things in cars. But DC is not just limited to just RVs and cars. All gadgets that run from any kind of batteries use DC, such as flashlights, portable stereos and cell phones, though the voltages used differ from one appliance to the next. Computers also use DC (which is made from AC, when plugged in). Specific examples of RV DC appliances include: most RV lights, fans, motors and circuit boards in appliances.

Where do RVs get DC? RVs can get 12 Volts DC from two sources: 1) big heavy deep cycle or automotive batteries and 2) power converters (which make DC from AC). RVs also could get DC from another type of electron-storing device called a supercapacitor, but supercapacitors are kind of exotic, and not used in today's RVs.

AC verses DC

Advantages/Disadvantages of AC and DC: AC advantage: AC can be transmitted very long distances through wires (DC cannot). This is the primary reason why AC was chosen as the municipal electricity type, because the generators that make it are often hundreds of miles away from the gadgets that use it. AC voltages can be easily transformed (either up or down) with almost 100% efficiency. Transforming AC voltages up then back down again is a necessary part of sending it long distances. DC voltage transformation is less efficient and more complicated. DC advantage: DC is the type of electricity that is stored in batteries. AC cannot be stored by batteries or directly by any other means; it must be made upon demand. Although DC cannot be sent long distances through wires, this is not a problem for automobiles and RVs, because the wiring distances in such vehicles are short.

Differentiating Between AC and DC Appliances:

1) Appliances that plug into electrical outlets: Differentiating between AC and DC appliances is easy, because the two plug/outlet types look completely different. Any gadget that plugs into a regular household wall electrical outlet is a 120 Volts AC appliance, since 120 Volts AC is the type of electricity delivered from those outlets. Any gadget that plugs into a "cigarette lighter" outlet of a car is a 12 Volts DC appliance, since 12 Volts DC is the type of electricity delivered from those outlets.

left: 120 Volts AC outlet and plug
middle: 12 Volts DC "cigarette lighter" outlet and plug in car
right: 12 Volts DC outlet on wall of RV

left: 12 Volts DC splitter (allows two devices to be plugged into one outlet)

right: 12 Volts DC appliance (GPS for car)
(recognize it by its distinctive plug,
which looks completely different from household plug)

2) Appliances wired directly to houses or cars: All appliances (like lights and fans) connected directly to the wiring of regular houses are AC appliances (since that is the only electricity type that houses have/use), whereas all appliances connected directly to the wiring of cars are DC appliances (since that is the only type of electricity that cars have/use).

3) Appliances connected directly to the wiring systems of RVs: These could be either AC or DC, since RVs are typically wired for both electricity types. (Helpful hint: Remember that when the electricity is on, the gadget works, when the electricity is off, the gadget doesn't.) Perform a simple test: Try the appliances both with the RV plugged in (to 120 Volts AC) and with the RV unplugged. AC appliances will work when the RV is plugged in but will be off when the RV is unplugged. The DC appliances connect to the battery, so they will work regardless of whether the RV is plugged in or not (if the battery is good and charged), or they will be off regardless of whether the RV is plugged in or not (if the battery is dead or there is no battery).

Note: Turn all power converters and power inverters off for the above tests, since these devices convert one form of electricity to the other. Leaving them on will confuse test results.

Power converters (AC to DC)

These turn 120 Volts AC into 12 Volts DC. AC is frequently converted into DC not only in RVs, but all over the place, in the world of electronics. Internal components of stereos, TVs, computers all run from DC, which is usually manufactured internally by these devices from AC. In RVs, 120 Volts AC is converted to 12 Volts DC for two purposes: 1) running DC appliances and 2) charging RV batteries. Usually, both these functions are combined in a single RV appliance called a *power converter/charger*. Here's how it works. When the RV is plugged in, or the generator is running, the entire part of the DC system that *uses* DC (all the DC appliances and the wires leading up to them) gets disconnected from the battery, and gets connected to the converter's own 12 Volts DC, which it makes from AC. (Sometimes an audible "click" occurs when this happens.) The 12 Volts DC made by power converters: 1) is smooth and almost completely free of voltage fluctuations, 2) is virtually indistinguishable from the DC provided by batteries, 3) can be (and is) used to power any of the RV's 12 Volts DC appliances usually run from its battery, including those not built into the RV [which plug into cigarette lighter type receptacles] and 4) causes no drain to the RV battery(s) when used to power appliances. The converter/charger also converts some of the 120 Volts AC it receives into DC and sends it through the battery, for the purposes of battery charging.

Power inverters (DC to AC)

These make 120 Volts AC from 12 Volts DC. Power inverters are really great gadgets to have while RVing. But they should be used sparingly, because they pose special problems.

photo: This 100 Watt portable power inverter is great for charging cell phones and other light uses. Note the 12 Volts DC male plug and the 120 Volts AC receptacle.

Power Inverter Problems:

#1: Power inverters have difficulty providing large amounts of power to appliances. Recall the formula: (Power) = (Voltage) × (Current). The power needed for a gadget to do its specific job is the same, regardless of the voltage that it runs from. Thus if the voltage is high, the current drawn is low, whereas if the voltage is low, the current drawn is high. Note that 120 Volts is ten times the voltage that 12 Volts is. Therefore, if two gadgets do the same job, and one runs from 12 Volts and the other runs from 120 Volts, the 12 Volts gadget draws ten times the current that the 120 Volts gadget draws. [Example: A 600 Watt appliance, running from 120 Volts AC, draws 5 Amps, because (600 Watts) = (120 Volts) × (5 Amps). A 600 Watt appliance, running from 12 Volts DC, draws 50 Amps, because (600 Watts) = (12 Volts) × (50 Amps).] If a power inverter is used to change 12 Volts DC into 120 Volts AC, nothing about this equation changes: 10 Amps @ 12 Volts must be used to make each 1 Amp @ 120 Volts. (And this would only be true if power inverters were 100% efficient. Power inverters never are 100% efficient.) To understand why huge currents through wires can cause problems, recall ohm's law: (Voltage) = (Current) × (Resistance), and consider the 12 Volt wiring leading up to and passing through the power inverter. If current is large and resistance is also too high, then there is a significant voltage drop leading up the power inverter (meaning the voltage between the leads of the power inverter is significantly less than between the battery terminals). In order for the voltage drop to not be too high, resistance must be very low. Thus big thick wires are needed, both externally (from the battery to the power inverter) and within the power inverter itself. If wires too thin are used, or if there is the slightest amount of corrosion, dirt or looseness at non-soldered electrical connections, then resistance is too high (not enough electrons can get through). This causes problems both for the appliance and the 12 Volt wiring connecting the battery to the power inverter. The problem for the appliance is: insufficient power may be made to run it. Insufficient power may not only cause the appliance to malfunction, it may also damage it. The appliance may even try to compensate for the power shortage by attempting to draw still more current from already current-stressed wires. The problem for the 12 Volt wiring is: heat will be formed when high currents pass through thin wires or dirty/corroded/loose connections. Enough heat may be formed to melt metal wires or start fires. The bottom line is: If you intend to use an expensive power inverter to run high energy consuming AC appliances, 1) use an inverter designed for heavy loads [its power rating must exceed the power consumption of the appliance], 2) make sure you use big thick wires on the DC side of the circuit and 3) make sure all electrical connections are secure [solder permanent connections, whenever possible, and sand corrosion off connecting rings/hardware]. Check for heat often. If too much heat forms, do not try to use high energy AC appliances from the power inverter. Unless you are very knowledgeable with electricity, it is best to consult a professional on electricity systems to ensure that wiring and connections are proper for high current/power systems.

#2: Many AC appliances use so much electricity, that they would drain RV batteries way too fast, if powered by them. A good example of this is a compression type air conditioner. Such a device could *theoretically* be run from an RV battery, but it would drain such a fully charged battery completely in only maybe 15 to 20 minutes, so nobody in their right mind attempts to run such air conditioners from RV batteries. If you have a house somewhere in the boonies, a shack for storing about 15 RV batteries, and a really large solar panel to charge them, (you need a lot of money to buy such equipment), it may be feasible to make significant amounts of 120 Volts AC from 12 Volts DC. But for RVs, where space and weight are important considerations, it is not. When using high power consuming appliances, such as an air conditioners, microwave ovens, vacuum cleaners or electric heaters, it is best to have a generator running at the time of operation to power them.

#3: Power inverters waste energy. Unlike generators, which manufacture AC quite naturally, power inverters manufacture AC quite artificially, so this process ends up being somewhat wasteful. (I've got an efficiency factor of 5/6, from: RV electrical systems, by Bill and Jan Moeller, meaning that it takes 1.2 Watts of DC to make every 1 Watt of AC.) Power inverters are wasteful for another reason: they use energy *even when no AC appliances are plugged into and running from them.* Power inverters capable of high outputs waste more electricity than those with low outputs, when nothing is plugged into them, or when using light-duty appliances, so some serious RVers have more than one power inverter, each with a different maximum power rating, to match the various needs of appliances. Often, it is best to turn a power inverter on just before using an AC appliance, then turn it off shortly after turning the AC appliance off, to save electricity. However, leaving a light-duty power inverter on all through the preparation of a meal, to occasionally run motorized kitchen gadgets, such as blenders and mixers, usually doesn't waste enough electricity to cause problems.

#4: Not all power inverters make true sine wave AC. Some older ones make square wave AC, which can cause problems with certain motors. *Quasi-sine-wave* (also called *modified sine wave*) AC is an improvement though still not as good as true sine wave AC. Quasi-sine-wave AC can cause problems with certain sensitive electronic equipment. (I've read that computer printers should have true sine wave AC. Check with manufacturers for specifics before attempting to run an expensive piece of equipment from a power inverter.) From what I have read, power inverters have dramatically improved with the passage of time. It may be best to get a power inverter that makes the truest sine wave possible (though my 450 Watt quasi-sine-wave power inverter seem to run every AC appliance I've got in my RV without problems). Don't bother trying to use old obsolete square wave power inverters, since new ones don't cost that much.

Two Operation Modes: "Generator On" and "Generator Off"

When a typical RV is parked away from a municipal electrical outlet and lived in, there are two modes of electrical system operation: "generator on" and "generator off." The difference between the ways RVs function in each of these modes is about as great as the difference between night and day. When an RV's generator is on, it makes electricity, and lots of it, which (with few exceptions) can run just about any electric appliance that would be found in a typical RV or a regular house. Because the generator naturally makes true-sine-wave AC, almost all AC appliances can be run from it. Because 12 Volts DC is also made from the generator's AC by the power converter (which usually turns on automatically as soon as AC becomes available), any of the 12 Volts DC RV appliances can be run as well, with absolutely no battery drain. When the generator is off, all electric appliances need to be run from the RV coach battery(s). The 12 Volts DC appliances run by the 12 Volts RV batteries directly. Small amounts of AC can be made from a power inverter to run some 120 Volts AC appliances as well. But certain AC appliances just consume so much electricity, that it is not practical (or even possible) to try to run them from the RV battery(s).

Batteries

Battery Basics

What is a battery? Simply defined: *A battery is a chemical demand pump which moves electrons.*

What are the important components of a battery? Every battery has three important parts, used for moving electrons: two electrodes and an electrolyte. The two electrodes are highly electrically conductive (often metal) rods or flattened plates, not in physical contact with one another, protruding deep into the battery. The two electrodes must be made of different materials. This allows one of them to become an electron donor and the other to become an electron acceptor. The electrolyte, which can be a liquid, a gel or a paste, is a chemical within which the electrodes are immersed, that reacts with the electrodes, to move electrons from one electrode to the other. The terminals are the end points of the battery, which connect to the outside circuit. All of this is housed in a physical casing, which holds everything together and may also seal the inside of the battery from the outside world.

How does a battery work?

Consider an analogy: blowing air into an empty pop bottle. If you put the end of an empty pop bottle in your mouth (with your lips sealed around it so no air can escape) and blow into it, initially, air flows from your lungs into that bottle. But you can't blow for very long, because as you add air to the bottle, pressure builds, and this pressure opposes further air from being added. The strength of your lungs (which tries to blow air into the bottle) stays the same, whereas the strength of the air pressure inside the bottle (which tries to blow air back out the bottle) grows as you add air. When the two pressures equal and oppose each other, a stalemate gets reached and no more air flows into or out of the bottle. Now, consider what might happen if you added a valve to the end of the bottle (away from your mouth) and released a small amount of the compressed air. Releasing air would cause a slight drop in the bottle's air pressure. This would make the pressure exerted by your lungs relatively *stronger*. This would cause your lungs to blow the *exact same amount of air* into the bottle as got released, to restore the pressure equilibrium. Now, consider what might happen if you attached a pump to the valve that were stronger than your lungs and forced some air into the bottle. Adding air would cause a slight rise in the bottle's air pressure. This would make the pressure exerted by your lungs relatively *weaker*. This would cause the bottle to force the *exact same amount of air* out of that bottle and into your lungs as got added, to restore the pressure equilibrium. These three conditions (zero air flow, air escaping the bottle, air being added to the bottle) closely replicate what happens to a battery in its three states: when not in use, when discharging through gadgets and when being recharged. In a battery, electrons are like the gas, a chemical reaction is like your lungs, and an external circuit (through which electrons leave and run though appliances or are forced backward during charging) is like the valve on the end of the bottle.

So inside a battery, a chemical reaction tries to move electrons from (+) to (–). A pressure difference (voltage) tries to move the electrons through the battery in the other direction from (–) to (+). When the voltage builds to a certain strength, these two forces equal and oppose each other, and a stalemate gets reached. Although electrons can't flow *inside* a resting battery from (–) to (+), they can and do flow in that direction *outside* of the battery, through an external circuit, when the battery discharges. This flow occurs naturally, kind of like the way a skier on a snowy mountainside slides downhill--- he doesn't need to be pushed. Discharging the battery causes the voltage to lower, which in turn causes the chemical reaction to move the *exact same number of electrons* from (+) to (–) as those that get lost through the circuit. This electron replacement process keeps the battery's voltage relatively constant through the process. When a battery gets recharged, current is sent through it in the opposite direction as takes place when the battery gets discharged through a circuit. An external electron moving device (called a charger), stronger than the battery's internal chemical pump, removes electrons from the positive (+) terminal and adds them to the negative (–) terminal. This raises the voltage between the terminals, making the chemical pump *weaker* by comparison than the voltage. This causes the chemical reaction to run in reverse and remove the *exact same number of electrons* from (–) to (+) as those that get added by the charger. This electron removal process keeps the battery's voltage relatively constant through the process.

<u>What happens to the chemicals within a battery</u>? When not being used, the chemicals inside the battery don't change. This explains why batteries retain their electricity for very long time periods when not being used. When being discharged, the pumping of electrons causes individual molecules of the chemicals to change from a high energy (ready for action) state to a low energy (tired and used up) state. When being recharged, reverse flowing electrons reverse the chemical reaction, so individual molecules change from a low energy state back to a high energy state.

How is a battery's voltage affected by charge state? When a battery's charge is high, there are plenty of high energy chemicals, and few low energy chemicals, therefore the chemical pump is strong. When a battery's charge is low, there are few high energy chemicals and plenty of low energy chemicals, therefore the chemical pump is weak. This affects a battery's voltage both while in use (actively draining) and while resting (nothing turned on). While in use, there is always a slight temporary voltage drop that occurs, because the chemical pump never completely keeps pace with the electrons lost through the circuit. The resting voltage of a battery also lowers slightly as the battery's charge state lowers, also due to lowering amounts of high energy chemicals. When a battery's charge is high and the chemical pump strong, the "in use" voltage drops tend to be small. When a battery's charge state is low and the chemical pump weak, substantial "in use" voltage drops may occur, especially if the current is high (although the battery may still be able to provide low currents with only small voltage drops).

What happens if gadgets are attempted to be run from a low battery? Batteries with really low charges might not supply enough voltage or current to the gadgets. This can cause gadgets to not work right, remain off or even be damaged.

Four things happen when a battery discharges:

1) electrons move from (–) to (+) outside the battery (through the circuit),
2) a chemical reaction takes place that moves electrons from (+) to (–) inside the battery (to replace those lost electrons),
3) chemicals within the battery become exhausted, and
4) the voltage gradually drops.

Do batteries *make* electricity or do they *store* electricity? For rechargeables, (like those used in RVs) the answer is simple: batteries *store* electricity. A rechargeable battery is like a "tank" that stores electrons. It is like a freshwater tank that stores water. Just as any water that comes out of a freshwater tank must have been added to that tank at some earlier time, any electrons that come out of a rechargeable battery must have been added to that battery at some earlier time. Water tanks do not *make* water and batteries do not *make* electricity. What about disposables? I think the answer is the same: disposable batteries *store* electricity. Yeah, maybe technically speaking, all batteries *make* electricity from chemical reactions that take place within them, but those chemicals need to be put there during the battery's manufacture. Those chemicals need to be very pure. Electricity is often (if not always) used to manufacture those chemicals. If more electricity is used to make a battery, than the battery liberates during its lifetime, then it is incorrect to say that the battery *makes* electricity. (I do not know if any battery ever made has ever *made* any electricity.) So why do people make batteries and love them so much? Because they allow us to carry electricity around to places where we can't plug stuff in.

Is an RV battery really a single battery? No. Many batteries (such as automotive and deep cycle RV) are actually several "batteries" within a "battery." The individual internal "batteries" are stacked end to end in a "chain" with opposite terminals attached, in which the two free ends attach to the outside circuit, a configuration known as a series circuit. In these cases, the internal individual "batteries" are called "cells," whereas the sum total of these cells is called the "battery."

What do batteries store? There are two correct answers: 1) Batteries store *charge* (also called electrons or capacity), usually measured in *Amp hours*. And: 2) Batteries store *energy*, usually measured in *Watt hours*. [Analogy: If a piggybank has 100 nickels in it, it is correct to say either that it has "100 coins" or "5 dollars.")

Problems With Batteries

Problem #1 (very important): Batteries don't store all that much energy. There are two modes of comparison that can be used to qualify this statement. One is to compare the amounts of energy that batteries *store* with the amounts of energy that electrical gadgets *use*. The other is to compare the energy-storing capability of batteries with other energy-storage mediums.

Let's start by comparing energy stored verses energy used. Notice that batteries work fine for running small things like cell phones, portable radio/CD players, smoke detectors, calculators, non-luminescent digital clocks and small flashlights. That's because these devices don't use that much electricity. However, you don't see too many people running big electrical devices from batteries like stoves, electric coffeepots, toasters, microwave ovens, household vacuum cleaners, washing machines, or refrigerators. Why not? It is technically possible to run big gadgets from batteries, but big gadgets use so much electricity, that hundreds of pounds of batteries would be needed to run them for reasonable time periods. Those batteries would take up too much space to truly be portable, be extremely heavy and would cost a fortune. Batteries store so little energy that they don't work terribly well even for some of the gadgets that they successfully run. (Cell phones need to be charged often. Flashlights are not terribly bright, especially when compared to household electric lighting plugged into the grid. And incandescent flashlights go dead kind of quickly unless used very intermittently.)

Now, let's consider energy stored in batteries verses energy stored with other mediums. There are a number of ways that energy can be stored: in springs (as in "wind up" toys or alarm clocks), in tanks of compressed air, in different water levels on the upstream and downstream sides of dams (often converted to electricity), or in chemicals that burn. Let's compare battery energy storage to a common energy storage chemical: *gasoline*. (Diesel, propane, butane and natural gas are similar to gasoline.) A gallon of gasoline, when completely burned with oxygen in the air, releases about 130,000,000 Joules of energy. That gallon weighs 6.3 pounds (2.86 kg) and occupies 3.79 Liters of space (volume). Storing that same amount of energy in lead-acid batteries instead would require 3010 pounds and occupy 695 Liters of space. That's over 475 times the weight and over 175 times the space.

Equivalent Battery Weight Calculation

A 75 Amp hour battery weighs about 38 pounds. If discharged 50% (recommended), then 37.5 Amp hours are usable. That's about 1 Amp hour per pound. Multiplying by 12 Volts gives an energy/weight ratio of (12 Watt hours)/pound. Formula:

(Energy)/weight = (Voltage) × (Charge)/weight,
thus,
(12 Watt hours)/pound = (12 Volts) × (1 Amp hour)/pound.

The 12 Watt hours can be converted to Joules by multiplying by 3600, since:

3600 Joules = 1 Watt hour.

(12 (Watt hours)/pound) × (3600 Joules/ (Watt hour)) = 43,000 Joules/pound.

To find out how many pounds of batteries are required to store the same energy as a gallon of gasoline, the energy value of a gallon of gasoline (130,000,000 Joules) is divided by the energy/weight ratio for batteries:

Equivalent battery weight for energy stored in 1 gallon of gasoline =
(Energy of 1 gallon of gasoline) / (Energy/weight ratio for batteries) =
(130,000,000 Joules) / (43,000 Joules/pound) = 3010 pounds.

That's a little bit heavier than the 6.3 pounds that a gallon of gasoline weighs!

Weight and space are usually not problems for stationary applications (like houses in the "boonies" located on large acreages of land), but for things that move, like RVs and electric cars, they are important considerations. The poor energy/mass ratio (called *specific energy*) and poor energy/space ratio (called *energy density*) of battery storage is a primary reason why electric cars have never been put into widespread production. There are intelligent people trying to solve these problems and design better batteries. However, improvements may be only marginal, and improvements may not be that significant unless the cost of improved batteries can compete with the lead-acid "cheapies." I doubt that a "superbattery" is on the horizon of technological progress anytime soon. So for now, this problem must be accepted and lived with.

Equivalent Battery Space (Volume) Calculation

A 75Amp hour battery, which weighs 38 pounds, storing 43,000 Joules/pound, stores about 1,640,000 Joules, because:

(38 pounds/battery) × (43,200 Joules/pound) = 1,640,000 Joules/battery

The battery's dimensions are:

Height = 8 inches
Length = 10 inches
Width = 6.5 inches,

each of which when multiplied by 2.54, to convert to centimeters, is:

H = 20.3 cm
L = 25.4 cm
W = 16.5 cm.

Multiplying all these together, yields the battery's volume in cubic centimeters:

8500 cubic centimeters = (20.3 cm.) x (25.4 cm.) x (16.5 cm.).

Since: 1000 cubic centimeters = 1 Liter, that's 8.5 Liters/battery.

To get the energy to volume ratio, the energy of one battery is divided by volume of one battery:

(1,640,000 Joules/battery) / (8.5 Liters/battery) = (193,000 Joules/Liter).

To find out how many Liters of batteries are required to store the same energy as a gallon of gasoline, the energy value of gasoline (130,000,000 Joules) is divided by the energy/weight ratio for batteries:

Equivalent battery volume for energy stored in 1 gallon of gasoline =
(Energy of 1 gallon of gasoline) / (Energy/Volume ratio for batteries) =
(130,000,000 Joules) / (193,000 Joules/Liter) = 673 Liters.

That's a bit more than the 3.79 Liters required for the gallon of gasoline!

Problem #2 (very important): Batteries are slow to recharge. This is because batteries just cannot accept large amounts of current flowing through them. Slow charging is not be much of a problem to people who charge at locations having continuous supplies of municipal electricity (like their homes), but for those "on the go" without much time (like RV or electric car users far from "home") this is a real issue. Also, lead-acid batteries can *release* their electricity considerably faster (about 5 times) than they can *absorb* it while charging. [The 5x figure is taken from "RV Electrical Systems A Basic Guide to Troubleshooting, Repair and Improvement" by Bill and Jan Moeller.] This means that usually battery chargers must operate for considerably longer time periods than the running times of high current using gadgets, just to "break even."

Problem #3: <u>Batteries are sensitive to many usage and environmental conditions</u>. Batteries can degrade from being overcharged, undercharged, charged with too much current, left standing for long time periods in only a partially charged condition or cycled too deeply (discharged too substantially before being recharged). They can be damaged from adding too much or too little water to the electrolyte, contaminated by adding tap water instead of distilled water to the electrolyte, or by contact with other chemicals. They can be destroyed by shock (being dropped), which can cause cells to become shorted or open. They are sensitive to temperature. They are prone to certain amounts of self discharge. (Frequently, RVs put into storage are not started, therefore their batteries aren't brought back to full charge. The batteries self discharge after awhile, then sit in partially charged conditions, which trashes them.) And even normal use of batteries degrades them. Even if used under optimal conditions, after a certain number of discharge/recharge cycles, batteries need to be replaced. Batteries just don't last forever.

Problem #4: <u>It is difficult to measure how "full" or "empty" a battery is</u>. The usual method of approximating a battery's charge state is to measure its voltage after it has rested awhile. (The voltage between the two terminals gradually drops as the battery discharges.) But several things can throw this test off: A) Most notably, the chemicals within the battery are often not evenly distributed. Immediately after partial discharge, low energy chemicals exist in disproportionately high concentrations near the electrodes. After sitting for awhile, the chemicals within the battery homogenize. [Excess low energy chemicals near the electrode surfaces move deep within the battery (away from those surfaces) while excess high energy chemicals deep within the battery move in the other direction (toward those electrode surfaces).] The uneven distribution of chemicals, during and for awhile after discharge, causes a battery's voltage to be lower, so it appears to have less charge than it really does. After sitting awhile, the battery's voltage, and its ability to run things, seems to magically "recover." The same phenomena occurs when a battery recharges, except with reverse ramifications. Immediately after charging, high energy chemicals exist in disproportionately high concentrations near the electrodes, causing the battery to have a *higher* voltage measurement than it should, and appear to be more charged than it really is. After it rests for awhile, the excess voltage seems to mysteriously "disappear." B) The resting voltage of a battery doesn't measure the amount of charge stored, it only measures its percentage between full charge and dead. The size of a battery (storage capacity in Amp hours) must be considered, in order to calculate its charge stored (in Amp hours). [Multiply its storage capacity by its % charge state to get its charge stored. Example: A 100 Amp hour battery that's 75% full has 75 Amp hours of charge stored.] C) As a battery ages (or sits around only partially charged), its electrode plates may become sulfated, thus reducing the effective areas of those plates. This causes the battery to become "smaller" (have less storage capacity) with time or use. D) The most accurate way of measuring a battery's charge, is to use an Amp hour meter and a computer. An Amp hour meter measures and keeps track of both *currents* and *time lengths* those currents flow, both for currents going into and out of the battery. If voltages are also measured and kept track of, then it can actually *measure* the storage capacity of the battery, and even note how that figure declines as the battery ages. But even Amp hour meters aren't completely accurate. For some inefficiency reasons, batteries drained at high currents give up less Amp hours than batteries drained at low currents. Amp hour meters, coupled with computers, using complicated mathematical formulas, are needed to accurately calculate the charge states of batteries. I have some doubts though that this can really be done accurately.

Problem #5: Batteries come in many varieties. This is not necessary a problem for electrical engineers, but it can be frustrating to the average person, who just doesn't want to take the time to become familiar with all the idiosyncrasies of each battery type. There are lead-acid batteries (typically used in automobiles and RVs), nickel-cadmium batteries, nickel metal-hydride batteries, lithium batteries, etcetera. Each major battery type may have multiple sub-types. [Within the lead-acid category (used in RVs), there are several varieties: automotive batteries, deep cycle batteries, maintenance free batteries, gel batteries, absorbed glass mat batteries.] There may be as many "recipes" for making batteries as there are for cooking pots of chili! Some battery types haven't even been developed yet. Each battery type has its own characteristics. For instance: A) Lead-acid batteries degrade when left in discharged condition for awhile, but some other battery types are not phased by this practice. B) Deep cycling is bad for lead-acid batteries, but Ni-Cads actually benefit from the practice. C) Lead-acid batteries can last a long time if not used but maintained in fully charged conditions (like with solar panels), whereas lithium batteries have definite shelf lives (they degrade after a couple years even if never used). I don't know what type of battery would be ideal for RVs, but there are many possibilities to choose from. Probably, the best battery for RV coaches is *not* what RV manufacturers put into them.

Battery Charging Systems

RVs typically have two house battery charging systems: the "charge as you drive" system and the 120 Volt AC charging system. These are very basic charging systems, which almost all RVs have and use. Advanced charging strategies, using equipment not normally sold as standard with RVs (such as solar panels, wind turbines, fuel cells, etcetera), will be covered in Chapter 3, "Making Electricity & Charging Batteries."

The "Charge As You Drive" System

This is a system for charging batteries, that most (if not all) motorhomes have. A variation of this system, fundamentally not very different from the motorhome version, can be used by campers with trailer-type RVs that separate from their tow vehicles.

Since every motorhome is both an automobile and a house, let's start by describing the alternator based battery charging systems of automobiles (including motorhomes). When an automobile is regularly used, the battery in that car just keeps starting it and starting it and almost never seems to run out of energy. That battery can last several years, giving hundreds or even thousands of starts to that car. Other electrical items may be used while driving, such as running lights, headlights, windshield wipers, electric fans for heating or cooling systems, radios, etcetera. How is this possible? Batteries of the size, weight and cost that are put in today's cars don't store nearly enough energy to give that many starts and run all that equipment while driving, in fact, they don't even come close. Such a battery capable of that level of energy storage is absolutely impossible with today's technology. So where does all that electricity come from? And why does the battery never seem to go dead? The answer is that the electricity, that runs the car and provides so many starts, does not come from the battery, it comes from a device under the hood, called an alternator, which is driven (rotated) by the car's engine by means of a fan belt. When the car is started, the battery supplies a considerable amount of energy to the starter motor to get the engine moving. Naturally, the transfer of energy, from battery to engine, puts a considerable drain on the battery. The battery does not stay drained for very long. As soon as the engine gets going, the alternator starts making electricity and putting it back into the battery. It continues to do so until the battery is brought back up to full charge. During each ride, the alternator both replaces the electricity used to start the engine, and makes considerably more to run all the cars' electrical devices. Strictly speaking, a car doesn't really "spend" electricity from its battery to get itself started, it merely "borrows" it. And since the electricity from the alternator is made from the energy of the rotating engine (which gets its energy from burning gasoline, diesel or other combustible fuel), as long as fuel keeps getting added to the tank, the car just keeps starting, driving and making electricity for a good long time. What a good workable system! (Genuinely)

Now let's consider the motorhome. When functioning as an "automobile," the electricity is made, stored and used by the exact same method just described for regular cars and trucks. When functioning as a "house" (while parked) another approach to electricity is needed. Let's consider several possibilities.

Possibility #1: Have all the "house" items hooked up to the "automobile" battery. This could theoretically run all the household items. But there is a big problem. When a motorhome is parked, the engine is off, therefore, the alternator does not make any charge to put back into the battery. Draining a battery while parked without charging it might not pose a problem, if only small amounts of electricity were to be used for the "house," from low energy consuming appliances not left on for very long. If the motorhome's battery has enough charge to start it for its next ride, it's alternator would recharge that battery completely by the time it parks again (provided the ride taken were not too short). But if too much charge were depleted from that battery while parked (a situation that occurs very easily and frequently), then the engine would be unable to start. At that point, the only method available, both to start the engine and charge the battery, would be to hook up jumper cables to the battery of somebody else's automobile (preferably while the engine were running). This, of course, would require finding a cooperative stranger and imposing on that person's time and equipment. Overusing the battery could be prevented with some sort of a guard circuit, which would sense the battery's lowering voltage and disconnect and shut down all the house items, if the battery got too low. (Such guard circuits do exist. My conversion van has one.) Still, this plan would present two problems: First, it would leave the house in the dark sometimes. And second, starting the motorhome's engine on a moderately low battery is risky, especially in very cold weather. The engine might still not start. This is not a good plan.

Possibility #2: Have two separate unconnected batteries: one to run the "automobile" and the other to run the "house." Having a second battery to run the household stuff while parked would certainly solve one problem, because it would always leave the unused "automobile" battery completely charged, therefore always ready to start the engine. But this strategy would create another problem: there would be no way to charge the "house" battery. It would go dead quickly. This method would only allow a single discharge of that house battery, each time the RV went into the wilderness. (Automotive and RV batteries are not meant to be discharged only once, they are meant to be charged and discharged hundreds of times.) A single discharge doesn't last very long. After the battery got too low, the user would have to take it home, to a service station, to an RV park or some other place where an electrical outlet is available, and charge it there with an electric battery charger. This would be a royal pain and an imposition to others. Even if used in an ultra-conservation mode, a single discharge of an automotive or RV battery would only run things in the "house" for a few days or a week at the most. This plan is also not very good.

Possibility #3: Have two separate batteries, one for the "automobile" and the other for the "house." Connect both to the alternator, while the engine runs, but disconnect the two of them from each other while the vehicle is parked. Here is a plan that gives users the best of both worlds. While parked, the "house" would be powered by its own separate battery. If the appliances were overused, only the "house" battery would get drained, so the unused "automobile" battery would always be able to start the engine. After starting the RV, both batteries would be connected while driving, so the vehicle's alternator would charge them both. As soon as the engine were stopped, the batteries would disconnect again. By that time, hopefully the "house" battery would be completely charged, but even if it weren't, it would still have quite a bit more charge stored than prior to that ride. This is a really good plan.

In motorhomes, fortunately possibility #3 (two separate batteries which connect only while driving) is used. In most RV literature. the "automobile" battery is referred to as the "chassis battery" and the "house" battery is referred to as the "coach battery."

In detachable trailer-type RVs, possibility #3 is also used, but only while tow vehicle and trailer are hooked together. The chassis battery is located in the tow vehicle, the coach battery is located in the trailer. They connect by means of long wires, which plug into one another when the vehicles get hooked together and unplug when the vehicles detach (the same way the trailer lights/brakes plug in when hitched and unplug when unhitched). Unfortunately for trailer-type RVs, this method provides for no charging when the trailer is detached and the tow vehicle is driving around. This creates a problem when people unhitch and stay put in areas for extended periods of time.

There are four methods by which the two batteries can connect while driving and disconnect while parked: A) Solenoid switches: A solenoid switch automatically connects the positive terminals of the two batteries together when the ignition key is turned to the "drive" position and disconnects them when the key is turned to the "park" (off) position. Solenoid switches are frequently found in older motorhomes. They work and don't cost too much. But if a switch gets old/corroded or a fuse blows in the line that sends current to the switch (to turn it on), that switch may become locked in the "off" position, causing the coach battery to never charge. B) Battery isolators: Technically, a battery isolator doesn't connect the two batteries to each other. It does use diodes to distribute current from the alternator into both batteries, so both batteries charge while the engine runs. Battery isolators are probably the best way to go, though the most expensive to install in vehicles not already having them. Installing an isolator can cost several hundred dollars. C) Manually operated high current switches: These switches are cheap. You can buy one for about $6, or you can even make one. Users must be disciplined enough to turn them on and off at the right times though. Leaving the switch on when parked can result in discharge of the chassis battery and the motorhome being unable to start. Leaving the switch off while driving results in absolutely no charging of the coach battery during that ride. And: D) Manually connecting/disconnecting wires: When connected, wires can be held together with a clothspin or with a small nut and bolt. It's kind of a pain to connect and disconnect wires this way, and not really much of an advantage, since switches are cheap. This method works great in emergencies if switches break or malfunction though. As with manual switches, users must be disciplined enough to connect and disconnect wires at the right times.

Having separate chassis and coach batteries, to handle the "automobile" and the "house" needs of motorhomes, while having both charge while driving, was a brilliant idea. Unfortunately, in my opinion, it was the last of any truly good RV electrical system ideas for many years.

The 120 Volt AC Charging System

This is a second system that both motorhomes and camping trailers have for charging coach batteries. The charger is powered by 120 Volts AC, which is obtained either by plugging the whole RV into a municipal electrical outlet, or by running the RV's generator. Such battery chargers can either be standard parts sold with the RVs themselves (as converter/chargers) or they can be separate units which plug into wall outlets (as chargers only). Built-in chargers have wires built into the RVs and permanently attached to their batteries. Separate chargers have cables which must clamp and attach to the battery terminals.

A 120 Volts charger does several things: 1) It transforms the 120 Volts into a lower voltage [about 14 Volts, which is correct for the battery]. 2) It changes AC into DC. [A battery needs to have DC pushed through it in order to charge. Pushing AC through a battery would do no charging at all, since the average current is zero.]
3) It filters out the voltage fluctuations, to provide smooth even charging. And:
4) Computer control, in more expensive models, decides proper voltages, currents and times for best charging, (hopefully) taking into account the characteristics of the individual battery type, including number of batteries to be charged simultaneously.

120 Volts AC battery chargers tend to work alright when RVs are parked in RV parks for awhile, plugged into municipal electrical outlets. They also work well when RVs are "home" between trips, parked in driveways and plugged in. But they tend to function horribly in the wilderness when generators are used to power them, for reasons that will be discussed shortly in Chapter 2.

Generators

I had an Aunt, who owned some recreational forested property with a simple cabin on it, connected to municipal electricity. She was complaining once, because the electric company was charging her a minimum of about $12/month, which she didn't like paying, because many months, the cabin went completely unused. So one day she said: "Maybe I should just have the electricity shut off and get a generator." BAD IDEA. (Thankfully, she didn't go through it.) This situation typifies a false belief that if one has a generator, any and all of one's electrical needs will somehow magically be taken care of. In reality, there are a number of problems and special considerations associated with making electricity using portable generators, both for remote real estate pieces and for RVs.

What Can They Do? Are They the Answer?

What are generators? Small portable personal generators (like the kind used in RVs) are internal combustion engine driven electricity producing machines. They make 120 Volts AC, in true sine waves, the same type that comes out of standard household wall outlets. (They are also sometimes called *dynamos.*)

What are the main components of a generator? 1) An internal combustion engine and 2) a rotating shaft of magnets [and stationary coils of wire to make electricity]. The engine keeps the shaft rotating, which enables it to produce electricity.

What can generators do? While running, a generator can power up just about any electrical gadget that would be found and used in a regular household, such as: lights, fans, air conditioners, microwave ovens, toasters, electric coffeepots, refrigerators, vacuum cleaners, TVs, and computers.

What can't generators do? A generator can't allow an RVer to function the same way as in a regular traditional house, because houses have almost limitless quantities of electricity available 24 hours per day. In order to replicate this, the generator would have to be very powerful and be left running 24 hours per day. Often generators are very powerful, but they just can't be run 24 hours per day, for reasons soon to be discussed.

When might a generator fail to provide enough power? 1) *When the power rating of the generator is too small.* 3000 or 4000 Watt ratings are good overall sizes for general RV use. 4000 Watt minimum rating is needed for a compression air conditioner. 2) *When attempting to operate a high power consuming appliance.* Electric space heaters and electric stoves are good examples of massive energy gobblers. Choose propane gas for these heating needs. Microwave ovens and vacuum cleaners consume quite a bit of power but can usually still be run from a generator. 3) *When attempting to operate more than one high power consuming device simultaneously.* Don't try to run a vacuum cleaner and air conditioner at the same time. Blown circuit breakers or generator shutdowns may occur. This limitation isn't too problematic for most RVers, because choosing to use only one high power consuming item at a time isn't that inconvenient, and it is a much cheaper option than buying a super powerful generator. 4) *When it is off.* Remember that generators don't run all the time.

The compression air conditioner is a good example of a common RV appliance that requires a high power rating generator. I tried to run such an air conditioner from a 3000 Watt (built in) generator. It whizzed quite a bit, but didn't cool very well. Later, a reputable RV parts dealer told me that at least a 4000 Watt generator is needed to properly run of one of these units. He also told me that many RV customizing shops won't even install a 3000 Watt generator in an RV having an air conditioner.

What should one do if a generator (or appliance) seems to be struggling? One can increase generator output by decreasing its load of other gadgets. Unplug the refrigerator or switch it to gas (it switches to gas automatically when AC electricity goes off) and turn the power converter/charger off. Also, make sure nothing else is being run while attempting to run a high energy gobbler. If still struggling, the only other options are to either cease using the problem gadget altogether or get a more powerful generator.

Why shouldn't generators be run 24 hours per day? 1) *Generators make noise.* This can annoy both the users and their neighbors. 2) *Generator use is prohibited by law* at certain times and places [especially at night in campgrounds]. This is because of their noise. 3) *Generators consume fuel while operating.* Fuel costs money. 4) *Generators have moving parts.* These require maintenance and wear out approximately in proportion to their use. And: 5) *Generators make deadly carbon monoxide exhaust gas.* This gas could under some circumstances find its way into the RV's interior, and *kill* the occupants.

Generators have a big problem: *they need to be running in order to make electricity.* The moment an RV generator goes off, all the RV's gadgets must run from 12 Volt batteries. If the batteries are too small or the charger too weak, even a generator big enough to power the entire city of Houston Texas won't help the RV, once that generator turns off.

Portable generators have a legitimate place in providing electricity to RVers and others not attached to the grid. But they cannot come close to making electricity as cheaply, as reliably, as continuously or in as much of a problem free manner as municipal electrical companies. If they could, people everywhere would be disconnecting from the grid and switching to generators.

Generators Make Noise!

During the considerable time that I've spent in campgrounds on public lands, I've had numerous conversations with other campers. If you think that generator noise is some small issue, think again. Few subjects evoke more passion in campers (on both sides of the fence) than generators and their noise. Generator noise, and the rules as to where, when and for how long generators should be allowed to operate, sharply divides campers up politically and socially. The noise that a generator makes can bother both the user/owner of the unit, as well as camping neighbors. It is likely to bother the neighbors more, because those neighbors don't benefit from the electricity that is made, and because those neighbors don't have control over when or for how long the unit operates. Let's examine the physical reasons why this noise can be so bothersome, and how people have responded socially and politically to this issue.

Physical Factors

There are six different "irritation" variables that work together, causing generator noise to bother people so much. There are only two "mitigation" variables that tend to ease this situation.

Irritation Variable #1: Generators are driven by internal combustion engines. They fall into the same mechanical category as automobiles, lawnmowers, weedwackers and chainsaws. The amount of noise that a generator makes can vary, depending upon its make, model, size, age and condition of the unit, where the unit is placed and whether or not the unit has a muffler. Well designed properly muffled units may produce noise about equal to idling automobile engines. Poorly designed improperly muffled units can be as noisy as lawnmowers or even chainsaws. But all generators make at least some noise while operating.

Irritation Variable #2: Generator noise is amplified by using them in RVs or other vehicles, due to the *sounding board effect*. If a shaking vibrating object is placed on the solid ground or suspended from a string or cable in the air, much of that vibration stays within the object itself, because that vibration transfers poorly into the solid ground or air. It will make some noise or sound, but the noise may be relatively faint. But if that same vibrating object is placed in contact with another object that is also free to shake and vibrate, some of that vibration transfers into the secondary object, then gets sent out into the air, amplified. This *sounding board effect* is put to good use in many musical instruments, such as pianos and acoustic guitars, where strings are intentionally put into contact with large wooden boards called "sounding boards." If pianos and acoustic guitars didn't have sounding boards, they would be kind of quiet. Generators built into RVs unintentionally and inevitably use those RVs as sounding boards, amplifying their noise. RVs make good sounding boards because they are not attached to the ground, so they are free to shake and vibrate, especially if they sit on tires. There is a remedy to the RV generator sounding board effect. If a portable generator is small enough for the owner to safely lift and move, its noise can be reduced by putting the unit on the ground, instead of in a vehicle, at times of operation. However, if a unit is left unattended, there is always a risk of it getting stolen.

The "sounding board effect" can be demonstrated with an electric shaver. In the air, the shaver makes a hum. But put the shaver down on a vanity and notice how much louder that hum gets.

Irritation Variable #3: Campsites are often close together. The closer a listener is a sound source, the louder that sound is. (The volume is proportional to the inverse square of the distance. Thus, if you are 1/2 the distance from a sound source that I am, you will find it 4 times as loud as I do, if you are 1/3 the distance from a sound source that I am, you will find it 9 times as loud as I do.) Because neighboring campsites are often considerably closer to each other than neighboring regular houses, noise transferring between campsites is much louder that noise transferring between regular houses.

Irritation Variable #4: RVs and (especially) tents offer poor noise protection. A person camping in an RV is protected from incoming sound only by thin aluminum, fiberglass or plastic sheeting on the outside, some thermal insulation, and a thin (usually 1/4 inch) piece of panel board on the inside. This offers some noise protection, but not much. It certainly doesn't block sound like a brick or stone house does (which can render even construction noise from next door almost inaudible). Campers is tents have it even worse, as thin nylon tents have virtually zero ability to block incoming sounds.

Irritation Variable #5: Generators are often operated for long periods of time. The longer they operate, the more annoying they are. Some RVers operate generators several hours per day.

Irritation Variable #6: Multiple RV generators often operate in the same camping areas. They don't all turn on and off at the same time, so in crowded campgrounds, their noise may be present from morning till night.

Mitigation Variable #1: Most generators have mufflers. Anyone who's had a muffler detach from a car knows that mufflers are very effective at quieting loud engines.

Mitigation Variable #2: Many times, campgrounds are not filled to capacity. If only a few of a campground's sites are occupied, those sites may be separated by considerable distances. Noise received by distant neighbors is quite a bit quieter than noise received by nearby neighbors. Also, sparsely occupied campgrounds don't have that many generators to operate. Fewer generators mean fewer generator operating sessions, thus increasing the probability of reasonable time periods of quiet.

Social Aspects

Some people, especially campers in simple RVs or tents, passionately hate generator noise. They consider extensive generator use to be incompatible with the "wilderness experience" purpose of National Park, National Forest and other remote Federal campgrounds; part of this experience includes the *sounds* of nature. These people want to hear birds singing, brooks babbling, coyotes yowling or wind blowing through leaves of trees, not engines idling from morning till night. They feel that big heavy noisy industrial machinery belongs in cities, not in areas set aside for wilderness and beauty. These people are further infuriated by what generators are often used for, which is powering unnecessary or energy wasteful items or for tasks that could be accomplished using no electricity at all. [Examples: 1) Great big energy gobbling TVs are often used, instead of smaller ones that use less power. 2) Super inefficient battery chargers are frequently found in RVs, even though faster better ones exist. 3) Some people insist upon cooking coffee in electric coffeepots, even though coffee tasting just as good can be cooked just as fast, on gas stoves, consuming zero electricity, less fuel, and bothering no one.] I call the people who hate generators "generaphobes." I call the people who love (to turn on) generators "generaphiles." (I'm pretty those two words are not in the English language, but they *should* be, so I'm adding them now.)

How do radical "generaphobes" behave? They scout out entire campgrounds, attempting to select their campsites based on one criteria alone: *How can they be furthest away from those big RVs and their noisy generators?* Such people often spend considerable amounts of time changing campsites, if they think that other sites might be less noisy. Generaphobes start off with strong feelings of dislike for generators, but as time passes, they find themselves disliking the *people* who use generators as well. They end up being enemies with people, whom they wouldn't otherwise be enemies with, and might even be friendly toward, under different circumstances. Naturally, because generaphobes often dislike generaphiles, the generaphiles usually respond by disliking generaphobes as well, feeling that generaphobes are unreasonably intruding upon their rights to use their hi-tech high cost gadgetry. (Will non-smokers and smokers ever be able to get along in the same place? I rest my case!) Strong psychological and personal divisions form between these two groups of people.

> Quotes From Generaphobes:
>
> A) "I don't want to be camping next to miniature factories. I feel like I am in an industrial zone."
> B) "I am a small simple camper. These people are bringing their entire houses. I am being charged the same camping fee that they are. I feel like I am being forced to subsidize their luxury housing."
> C) "I have absolutely no use for people (generator users) like that."
>
> Quotes From Generaphiles:
>
> A) "I paid for this generator. I think I should be able to use it!"
> B) "How am I supposed to cook my coffee early in the morning if I'm not allowed to use my generator at that hour?" *(My Answer: Use fire. It was discovered--- what--- maybe 110,000 years ago? Are you an idiot?)*

Political Aspects

The National Park Service, National Forest Service, BLM and public campground managers are highly conscious of the incompatibilities presented by generators in public campgrounds. And they have responded with some laws. First, generator use is prohibited in public campgrounds during the nighttime hours. This "quiet time" is generally between 10:00 PM and 6:00 AM. Although specific rules and exact times may vary (some places are more restrictive), this common sense rule seems to be almost universal, for all campgrounds in National Parks, Monuments, Forests, Recreation Areas, BLM lands, and even State and County campgrounds. Second, generators need to have mufflers, in order for campground operation to be allowed.

With the passage of time, there seems to have been a trend toward stricter regulation of generator use in public campgrounds. In 1988, when I visited Everglades National Park, there were three rules posted on the bulletin board: 1) Keep generators off at night. 2) Generators must have mufflers. And: 3) Generators "must not be too loud." Now, the climate is different. 1) There are currently actual scientific regulations [measured in decibels] on the books that clearly define how loud they can be, which have replaced the highly subjective "must not be too loud" rule. 2) Some campgrounds have instituted a "zoning" policy in which generator use is permitted in some campsites, but prohibited in others, even during the day. 3) Some entire campgrounds are designated as "generator free" [no generator operation anytime]. 4) Some campgrounds restrict generator use to only two or three narrow time bands during the day [usually two hour time periods]. All other times they must be off. Typically "generator allowed" time bands occur when most people have breakfast or supper. 5) Campgrounds that have electricity in some sites, but not others, usually expect that electric sites be taken by those who need electricity--- big RVs should not overflow into tent/ simple trailer areas and annoy simple campers by "chugging and a lugging away." 6) Some primitive campgrounds state that if you really must use a generator, its operation should be confined to no more than two hours per day. But they don't specify which two hour period is acceptable or how they intend to enforce this. 7) Fines can now be imposed for violators, and there are clear written warnings that such fines *will* be imposed, if rules are broken. This trend toward greater restriction goes beyond just Federal lands in the United States. I've read that in Europe (never been there myself), *any* generator use in a public campground is considered unacceptable, because it interferes and is incompatible with the "nature experience," which includes enjoying nature's sounds. And one State Park campground I visited (Adirondack State Park in New York) went further than our National Parks, posting a very ominous written warning: Anyone caught operating a generator during the nighttime hours, in addition to being issued a summons, *would be evicted from the premises immediately*. If you have an RV dependent upon heavy generator use, you need to be very aware of your surroundings and the local rules. You should also exercise some restraint and good judgment in consideration of neighbors.

Of course, generaphiles sometimes cheat on the rules a little, under certain circumstances, such as in campgrounds that are barely occupied (especially if the generator user is the only one there), or on unpatrolled public lands with undesignated campsites, where camping neighbors are far away. Also, sometimes big RVs cluster together, forming unofficial "suburbs." If all the neighbors have and use generators, they aren't very likely to be annoyed by a little more "chugging and a lugging away" from an additional RV moving in. They certainly don't have the same dislike for generator noise that tenters and small generator-free RVers have. But even generator users, who cheat a bit in unpatrolled areas, should try to use good judgment and show some consideration for others. It is a good idea to consider at least the "keep them off at night" rule to be universal anyplace where neighbors exist, even if not posted and even if no law enforcement is around.

Easing Tensions

The sad part of all of this is the hostility that gets created. I've met some very nice people, on both sides of the fence, who fundamentally don't seem all that different from each other. Both generaphobes and generaphiles seem friendly. Both groups like to go places and see things.

My advice for generaphiles: If you find yourself in close proximity to simple van or tent campers, consider using some restraint. 1) Talk to your neighbors, find out how they feel about generator noise. Find out what time of day they are less likely to be around, and do your battery charging or heavy appliance usage then. 2) If you really must generate for long time periods, consider driving your motorhome out of the campground, to a nearby unoccupied picnic or parking area, and doing so there, instead of making all that racket in a crowded campground. 3) Ease tensions by getting to know your neighbors a bit. Offer them an occasional cup of coffee, tea, hot chocolate or a cookie. Let them know that you are living breathing human beings with feelings, just like they are. There is no great advantage to making enemies out of people if you don't need to. 4) Offer to share some electricity with them when you do generate. They probably don't need it and can't use it, but they may appreciate the offer anyway. 5) For heaven's sake, *obey the posted rules* as to "keep them off" places and times.

My advice for generaphobes: 1) If someone operates a generator in a place/time of prohibition, knock on their door and politely let them know that such use is illegal. They almost always shut the unit off without confrontation. 2) Try getting to know these people a bit. If you have become good at "boondock" living without a generator, share some of your knowledge that enables you to do this successfully. They often like such conversation. 3) Try to persuade them into using some restraint, in exchange for your having tolerance for at least some of that noise. 4) Let them know that you are living breathing human beings with feelings, just like they are. Once again, there is no great advantage to making enemies out of people if you don't need to.

My Personal Feelings

On noise in general: I don't mind *some* noise, especially if I'm making it and its absolutely necessary. I don't mind the engine noise of my car while I'm driving it, because it's absolutely necessary. I don't like the noise from my lawnmower or weedwacker, but I am willing to put up with it in order to do the job, though I do plug my ears with paper towel pieces while using these. I also am quite tolerant of noises of others, when they really need to do or use something. But I wouldn't want to run a lawnmower all day long, when it's unattended and not cutting lawn, just in case I *might* want to mow three rows every now and then. So I also don't like the notion of idling a generator for hours on end, when nothing is apparently being run from it.

On generator noise in Federal campgrounds: I camp in a small simple trailer. I get all my electricity from batteries, which my tow vehicle charges while driving. I don't even own a generator. I survive in the wilderness for months at a time just fine. Naturally, I am a bit of a generaphobe. However, I honestly don't mind *some* generator noise. Other people need to have their fun too. But I do get angry when generator use is excessive. It infuriates me when people come into parks and start "chugging and a lugging away," for hours on end, for reasons as frivolous as running a couple lights (that could be run from batteries), charging cell phones (a cell phone can be charged by plugging into a car's battery or with even a small solar panel), making toast in electric toasters (toast can be made on frying pans on gas stoves) or coffee in electric coffeepots. And some people leave generators on, even when not running electric appliances, *just in case they might decide to turn something on.* I also become quite angry when I find a generator is running in a campsite near me and *nobody is even there occupying the RV.* Some people feel that if they buy a generator, they have some sort of constitutional right to unlimited use anywhere anytime--- to hell with the feelings of others. The laws do not permit such a right. Generaphiles inconsiderately pushing the limits too far is perhaps a bigger cause of hostility toward them than the noise itself.

Some Disregard the Feelings of Others

Some generaphiles not only disregard laws and the feelings of others, they do so quite aggressively. Here are three actual cases of such behavior observed by me.

Case #1: Once, a couple in an RV drove right by a sign prominently displayed at a campground entrance stating: "Generators are not permitted at this campground anytime." They occupied a site there with their RV. The next morning, they were "chugging and a lugging away."

Case #2: Once, a campground allowed up to six hours of generator use during a single day, which must occur within three specific two hour time windows. The camper next door to me operated his generator for the entire six hours, two days in a row. His timing clearly indicated to me that he had read and was familiar with the rules posted on the bulletin board at the entrance. Still, he wasn't satisfied. The third day, he started pushing the boundaries further, by "chugging and a lugging away" during posted "no generator times," *in addition to the six hours.* Excuse me neighbor, but if you need to do more than six hours of generation per day, maybe you don't belong in a National Park!

Case #3 (Granddaddy of them all. This case was so outrageous, I remember it completely and feel compelled to write it down, even though it happened more than 29 years ago.): In Everglades National Park in 1988, there was a sign at the campground stating: 1) "Generators must be off at night." And: 2) "Generators must have mufflers." Some guy, probably in his late twenties, with his wife and kids, entered the campground, set up only a screen tent, then turned on a generator that had no muffler, which was about as noisy as a chainsaw and had a high pitched whine besides, at about 2:00 in the morning, amongst a campground well occupied by people in tents trying to sleep. What was he running? From my observations, apparently only one single light bulb. He left it on for quite awhile, seemingly having no plans to shut it off. He seemed oblivious to everyone around him and didn't quite look right. I wondered if he was on drugs or crazy or something. I was somewhat concerned for my safety but I decided to approach him anyway. I walked up and said, "Excuse me, but generator hours end at 10:00 PM." He shut it off, without confrontation. Immediately thereafter, someone in a tent screamed, "WILL YOU TURN THAT THING OFF!"

<u>How I Deal With Generator Users</u>

If operating legally but not excessively, I often just tolerate the noise. I try to see the world from others' points of view as well as my own. (And, based on my experience with my mother's RV, which had a generator, I must admit, sometimes it in nice to watch TV, run an air conditioner or charge a battery.) If operating legally but excessively, I often respond by going over and getting to know the people a bit. I start offering my advice on how to better manage electricity, and share my knowledge about the subject with these people. Despite being a generaphobe, I often find myself sympathetic to the struggles of people with generators and frequently find myself liking them. They often like me too and go easy on the generation without my even asking. Their behavior is not completely their fault, I believe that the main reasons for excessive generation are poor electricity and appliance knowledge and poorly designed RVs. If operating illegally but not excessively, I often just knock on doors and politely let the people know what the rules are, in a very calm voice. If operating both illegally and excessively, I sometimes do the same (still calm but with no smile), but sometimes I skip the polite reminders and report offenders directly to park personnel. Generaphiles blatantly disregarding laws and other people's feelings is becoming more rampant, and they are becoming more aggressive at doing so. I am not always in the mood to be friendly and understanding toward those who show extreme inconsideration for both others and the law.

Chapter Conclusion: No Easy Answers

In RVs, managing the system of electricity and the appliances it normally runs, is tricky business. This is probably because most people usually live in houses and aren't accustomed to disconnecting from the grid. There are no solutions that are easy and problem free. Bringing electricity into every campground imaginable in not the answer. Generators are not the answer. Designing and customizing RVs to function off-grid can be an answer, but RV designers and customizers may be unwilling to do this or they just might not know how. You can customize yourself and learn to function off-grid, but this takes monumental effort. Is it possible to just "grin and bear it" and use RVs *as they are* off-grid? What happens when RVs are attempted to be used *as they are*? The next chapter discusses this in detail.

Chapter 2
RV Electrical Systems Are Defective!
(Problems & Solutions)

Finding problems with the *general* design of RV electrical systems is easy. Finding the biggest *specific* problem of RV electrical systems is not so easy, as problems are everywhere. These systems are messy complicated mixtures of unreliable, ill-suited and sometimes incompatible devices, so much so, that if one list were made of what is right with RV electrical systems and a second list were made of what is wrong with RV electrical systems, little would fall onto the *right* list, while nearly everything would fall onto the *wrong* list. So poorly thought out these are, that I seriously wonder if RV designers ever bother studying electricity at all. The five specific design problems articulated in this chapter are all part of one great big general design problem: *RVs are just loaded with electricity-hogging appliances, but there is no reasonable plan as to where all that electricity is supposed to come from.*

An Analogy: The "Stupercar"

I'd like to tell you about a car that I have. It is absolutely not used for towing or trailer camping in any capacity. I bought this car as a second vehicle. I wanted something small, with a really fuel efficient engine, that I could easily maneuver around in city traffic, so I wouldn't have to use (and wear out) my big gas guzzling trailer-towing van all the time. I really like this car. I drive it almost exclusively, when I am living in my regular traditional house, and not on one of my camping trips. It gets over 30 miles per gallon on the highway. Of course, I have to put gasoline in the tank once in a while. I may occasionally complain about the price of fuel, especially when that price is high, but I understand and accept that "filling up" is just something that I must do, once in a while. The laws of physics absolutely demand it. I do not consider "filling up" the car to be a major task or inconvenience for two reasons: 1) I don't need to fill it too often, and 2) filling the tank doesn't take very long. On the first point: The gasoline tank holds about 13 gallons. That is big enough for the car to drive well over 300 miles on the highway between fill-ups. It is also enough to last me about three weeks, when the car is only used for small drives in the city, to places that are close to home. On the second point: When I do fill up, it takes but a few minutes. Then I am on my way. This is true because gasoline gets dispensed from the pump at a hefty flow rate. Even after the time it takes to drive into and out of the service station and to pay for the gas is added to the equation, the procedure still takes less than fifteen minutes.

Now let's change something. Suppose that my car were to be taken away from me, and replaced with a different model, called the "Stupercar" (short for "super stupid car"). This newer model of car would be identical to the old one it replaced except for two things: 1) the gasoline tank would be much smaller and 2) it would take a really long time to fill that tank. On the first point: The gasoline tank on my "Stupercar" would hold only 3 quarts of fuel, instead of the 13 gallons that my current car holds. With the "Stupercar," I could barely drive 20 miles down the highway after filling up, before being dangerously close to running out of fuel. Even a trip to a destination 10 miles away, through city traffic, would require a fill up upon arriving and another fill up upon returning home. On the second point: Suppose the gas tank on my "Stupercar" had a different type of intake valve that only allowed a small trickle of fuel to get through at once--- like maybe at the rate of a child's toy water squirt gun. My current car can be filled in minutes, but my "Stupercar" would take several hours to fill. This would be a major inconvenience. I suppose, in frustration, I could try redesigning the car myself. I could take out the 3 quart gas tank and replace it with a 13 gallon tank. But if I didn't also replace the intake valve, I would still have problems. My improved "Stupercar" would be able to drive reasonable distances between fill-ups, but without an intake valve able to accept fuel at a reasonable rate, it would take so long to fill the gas tank, that I would need to drop the car off at the service station one day, hook it up to a fuel delivery pump, and then pick the car up the next day. What a bother! In order to turn the "Stupercar" into a decent workable car, I would need to replace *both* the fuel tank and the intake valve.

By now, you may be thinking something like: "Why is this author wasting my time with such stupidity? Nobody would ever design, build nor buy anything like that. Nobody could ever be that stupid." With cars, this is true. Working cars, like the one I own, are real. "Stupercars" are only imaginary. RVs however are a different story. RVs have the same two design flaws with their electrical systems that the hypothetical "Stupercars" have with their gasoline systems. Yes, somebody would design, build and buy something like that. Yes, somebody could be (and is) that stupid. "StuperRV"s are real. RVs, with factory installed intelligent electrical systems that actually work, are probably only imaginary at this point in human technological history.

The Defects:

Defect #1: Insufficient Batteries

When any electrical gadget/system is powered from batteries, it works for awhile, until the batteries run down, then it shuts off. If use beyond a single discharge cycle is desired, sometime before that run down point, the batteries need to be either replaced with fresh ones (if disposables are used) or recharged (if rechargeables are used). Thus, a very fundamental engineering question for anything battery powered is: *How long can the device/system run before the batteries go dead?* RVs, when not plugged in, are complex battery operated systems. The central battery(s) typically runs lights, furnaces, water pumps, small electric fans, TVs, stereos, DVD & CD players, gas refrigerator circuit boards, and hot water heater & gas refrigerator ignition systems. Now you would think, with all these gadgets that use electricity, that somewhere within the chain of the many people who design RVs, somebody would've had the presence of mind to make sure that RVs have enough batteries to actually run all this stuff, for reasonable time periods before "going dead," but if you have made this assumption, you are wrong. Correlating the amounts of electricity *used* with the amounts of electricity *in storage* is not something RV designers see much of a need for doing--- at least not in any reasonable manner.

Oh sure, RVs do typically have *some* provisions for running electrical gadgets away from municipal electrical outlets, they just do not have *adequate* provisions. The typical RV, despite being just loaded with energy-hogging electrical appliances, has only one single central coach battery to run them all. How long does it take to run this one battery down? Excluding neurotic self deprivation (using candles instead of electric lights and leaving furnaces off, no matter how cold it gets), usually only *one or two days*, with normal to moderately heavy use. Astonishingly, many RV designers seem to feel that it's alright to have only enough batteries to last *one or two days*. I would like to know what kind of psychedelic mushrooms these RV designers ate or what batches of LSD they did, before coming up with this "one or two day" standard, because they sure aren't living in the real world! Who camps for only a day or two at a time? (Certainly I don't. In my own personal experience, after driving hundreds of miles, I need one day just to rest, before doing any National Park touring.) Would it be reasonable to take an RV into some remote place for a week, with only enough food, water, propane or clean cloths, to last just a day or two? I certainly don't think so. So why do these designers think the standard for RV electricity should be different?

I can't speak for nor comprehend what takes place within the minds of most RV designers. I can however offer speculation as to how and why the "one battery" practice may have arisen and perpetuated. A long time ago, RVs were smaller and simpler. They had fewer devices that used electricity. Maybe back then, one battery was actually enough to run all the household electrical stuff, which may have consisted only of a couple of lights. As time passed, RVs got more complex. More and more electrical devices were added, and more devices were redesigned to use electricity, that formerly did not. The logical thing to do, of course, would have been to add more batteries, as the needs of these RVs grew. But somebody just didn't think this situation out clearly and logically. RV systems for *providing* electricity just haven't caught up to RV systems for *using* electricity. That's my best guess as to why, even today, many RVs have only one coach battery.

So how long should an RV battery (bank) be able to run appliances before "going dead?" A lot depends upon how often, how easily and how reliably the battery is to be charged. (There must be a working charging system in place. If battery charging is impossible, then the system doesn't work, no matter how long its run-down time is.) For now, the quick answer, based on my personal camping experience: I recommend a minimum run-down time of four days (with heavy usage) to seven days (with moderate usage). If run-down times too much less are in effect, then RV users need to spend too much time trying to charge batteries. RVers are busy people. They have more important things to do.

> In the early days when I camped in my ultra-energy-conserving trailer, I typically got about a week out of a battery (with normal use), four days (with very heavy use) or ten days (with light use), before needing to switch it with a charged one.

Defect #2:
Inadequate Chargers

Before getting into the specifics of RV battery charging, one very general forever true principal for all rechargeable systems needs to be pointed out: When gadgets run by batteries, electricity flows *out of* the batteries. When batteries recharge, electricity flows *into* the batteries. On average, a charger must be able to put electricity into the batteries at least as fast as the appliances draw electricity out of the batteries, otherwise the system goes dead, and usually quite quickly. This is such a simple concept, I think even the average fourth grader could understand it. Yet I am convinced that few RV designers actually get this. Most of the specific design problems described in this section ultimately boil down to violations of this one simple general "in/out" principle.

> Electricity is like money. If you usually spend $1000 per week but your only source of income is a part time job flipping burgers that pays $100 per week, what happens?

There is another problem that charging systems sometime have: *inadequate wiring.* If wires are too thin or distances too long (or a combination of both), then the wires have too much *resistance*, meaning electrons have a hard time getting through. (This is true for both standing and moving chargers. This situation can occur if a house battery is located too far away from the vehicle's alternator or AC charger). When a high current passes through a resistance, a voltage difference forms between the two ends (Ohm's Law). This means that *the voltage difference between the battery terminals will be less than the voltage difference between the output terminals at the charger.* Since the charger usually measures voltages at its own output, it can be fooled into "thinking" that the battery has a greater charge than it actually has. Having wires that are too thin or too long is not a fatal mistake. It does not stop the battery from charging, but it does cause charging to occur more slowly, sometimes unacceptably so.

I.) Standing Systems (120 Volts AC chargers):

Standing systems tend to be more important to long-term RVers (than moving systems) and are also more problematic. These systems can work alright when RVs are plugged into municipal electrical outlets, but in wilderness, they have two major problems:

Problem #1: Slow Chargers

Like the hypothetical Stupercar, with an awful gas intake valve that took almost forever to fill, RV standing chargers only send small amounts of current through the batteries, so the batteries take seemingly forever to properly recharge. Is charging batteries slowly with AC electric chargers a problem? It might not be, for weekend or short-term campers or campers who stay frequently in RV parks with municipal electricity. Weekend campers usually don't need to do any field charging. They can usually survive on the electricity that they take from home. Once home again, they can simply "plug in" for awhile, to slowly recharge their batteries for their next trip. Campers who stay in RV parks frequently can do the same at those RV parks. While there, such campers can sit around, drink wine or coffee, eat, go sightseeing, shop, or just get that good night's sleep, while their batteries charge to completion. For campers who spend more than just a day or two at a time in wilderness areas however, slow charging is a problem, because 120 Volts AC isn't available long enough (if at all) to effectively run the battery chargers.

Problem #2: Where is the electricity to run the chargers supposed to come from? When some electrical gadget is run absent from a municipal outlet, a battery is provided to power it. Unless that gadget uses only a tiny amount of electricity (example: a smoke detector), the battery discharges and becomes unusable after a short while, at which point, that battery needs recharging (or replacement). So a charger is provided. But if that charger depends upon municipal electricity to work, the basic "lack of available electricity" problem is not solved, only "the buck is passed." It is kind of like having a problem with an item recently bought, taking that item back to the store, explaining the problem to some low level clerk, and having that clerk pass the problem off to a high level store manager, who also doesn't solve the problem. Just as a chain is only as strong as its weakest link, any electricity-dependent charger is only as reliable as the electricity source that runs it. If that electricity source is only occasionally available, then the charger only occasionally works, and is inadequate because *it can't be on very often or for very long.* All an inadequate unreliable charger can do is prolong the time span before everything goes dead--- and not by much, if the batteries don't store enough charge to run things very long (usually the case). There are some answers to the question: "Where is the electricity to run the charger supposed to come from?". They are just not very good answers. The electricity can come from: A) municipal outlets or B) generators. Both present serious difficulties. Let's cover each.

> Isn't lack of available municipal electricity the reason for the battery's very existence? If so, how could a battery charging plan that requires municipal electricity possibly work?

A) Charging from municipal outlets. Doesn't charging by "plugging in" work? People charge cell phones, laptops or cordless power tools all the time, using this method. They rarely complain that these devices don't get charged. Couldn't such a strategy also work for RVs? In theory it could but in practice it doesn't--- at least not very well, because there are substantial differences between charging small things like cell phones, laptops or cordless power tools, while living in houses, and charging big RVs, while living out in the boonies.

Difference #1: RV batteries have a different purpose (than cell phone batteries). With cell phones and cordless power tools, the main purpose of the batteries is really *portability*. People can walk around or go places with cell phones, without having the inconvenience of being attached by a cord to some wall. People can also take cordless power tools quickly to locations, without dealing with dozens of feet of electrical cord, which they must untangle, place, then wind up and put away. With RVs in the wilderness, the purpose of house batteries is to *substitute for unavailable municipal electricity.*

Difference #2: RVs may be miles away from electrical outlets (whereas cell phones or cordless power tools are usually close). Charging small items like cell phones or cordless power tools by "plugging in" is rarely a problem because almost all people in the technological world have live municipal electrical outlets close by. We have such outlets in our houses, work places, libraries and schools. We can plug in and charge cell phones and cordless power tools while we sleep or even when we're not using these items. With RVs, parked in wilderness areas, municipal electrical outlets may be entirely unavailable. This can mean no electricity during the day, the night, for days or weeks at a time. The nearest electrical outlets may all be on private property, therefore not accessible to the RV, unless those property owners happen to be friends (rather unlikely).

Difference #3: RVs are heavy-duty (whereas cell phones are light-duty). Cell phones and laptops use only very small amounts of energy to operate and to recharge. (They are light-duty enough to be charged from other bigger batteries by plugging into 12 Volt cigarette lighter receptacles in cars and RVs.) A traveler may not have a hard time finding friendly store owners or librarians willing to allow plugging in of a small cell phone or laptop to recharge its battery. That same traveler might have a hard time finding friendly people willing to allow plugging in of an entire RV to recharge its batteries. When people see a big RV, they wonder what heavy-duty industrial strength apparatus is in there. People worry that plugging in an entire RV will cause their electric wires to melt and burn, trip circuit breakers, or add $20 to their next month's electric bill.

Difference #4: Charging RV batteries takes longer (than charging cell phones). An RV with substantially drained batteries would have to be parked for quite awhile to recharge--- probably at least overnight. Cell phones, laptops and cordless power tools usually require only a couple of hours to recharge.

Difference #5: RVs are large and therefore hard to park. Even small RVs are quite a bit bigger than cars and have blind spots. An RV might have a difficult time maneuvering into and out of driveways and parking lots in well populated areas. Also, many cities and towns have local ordinances prohibiting overnight RV parking (or any RV parking at all) in certain lots. Owners of such lots probably are unlikely to be willing to give anyone permission to break the law.

Realistically, the only places with electrical outlets suitable for RVs to charge their batteries, are RV parks. An RV park costs lots of money and almost certainly requires an overnight stay. Charging a cell phone by plugging in at home is easy. Charging an RV by plugging in far away from home is difficult.

B) Charging by generator. It is possible for a generator to supply electricity for battery charging. But to succeed, chargers must put enough current through the batteries. If currents are too small, proper charger would require extraordinarily long generator run times, or be downright impossible. Many converter/chargers in RVs send only 3 Amps through the batteries, which is a ridiculously small amount of current. (I've even seen chargers that send less than 2 Amps.) Let's put some numbers on this charging rate.

The 3 Amps Charging Rate (put into perspective)

How does this charging rate compare with appliance discharging rates? The 3 Amps of current often used to charge batteries is equal only to the current drawn by two 18 Watt incandescent 12 Volt light bulbs (these are common standard RV bulbs). That's not very much. Some fans in energy-hogging forced air furnaces draw 8 Amps (almost three times the charging rate). Even fans in furnaces claiming to be "energy efficient" often draw about 3 Amps (equal to the charging rate). Water pumps can draw about 8 Amps, while pumping (almost three times the charging rate). A small TV, run by a power inverter, might draw 4 Amps (slightly greater than the charging rate). In other words, charging takes place at a rate often less than that drawn by many RV appliances, when those appliances are actually on and running.

How do charging times compare with appliance run times? The ratios are the same as the above ratios for the rates, plus 20% more charging time due to inefficiency. For every hour that the two light bulbs are on, the charger and generator must run for *one hour* (plus 20%) at some other time to replace the electricity used. For every hour the "energy efficient" furnace fan runs, the charger and generator must run for *one hour* (plus 20%) at some other time to replace the electricity used. For every hour the TV runs, the charger and generator must run for over *an hour and a half* (including the 20% inefficiency) to replace the electricity used. For every hour the non-energy-efficient forced air furnace fan or the water pump runs, the charger and generator must run for *over three hours* (including the 20% inefficiency) at some other time to replace the electricity used. With the water pump, electricity consumption is not much of a problem, since water pumps aren't likely to be on for very long. (RV users develop natural tendencies to conserve water, because the tanks only store limited quantities and require effort to fill and drain. The pump may thus be used for just seconds or only a couple minutes at a time.) But even a mere three hours of the 8 Amps energy-hogging furnace fan use, on a moderately cold night (a very realistic conservative estimate), would draw 24 Amp hours of electrons. Charging inefficiency included, such a drain would require a whopping **9½ hours** of charger/generator run time to replace the electricity used.

How long would the charger need to run to match overall daily electrical usage? Daily electrical usage is highly subjective and dependent upon many variables. I have seen daily RV electrical usage estimates in some books as high as 50 Amp hours. A less "high off the hog" estimate might be 30 Amp hours. A 3 Amp charger, which effectively charges at a 2.5 Amps rate (due to inefficiency) would take **twelve hours per day** to replace the 30 Amp hours or over **twenty hours per day** to replace the 50 Amp hours.

Is it even possible for a 3 Amps charger to keep pace with RV electrical usage? Probably not. Running a generator for even two hours per day annoys many people. If you try running a generator for ten to sixteen hours per day, to power the charger, in any campground where there are other people, somebody will hate you. You or your equipment might be subject to attack. Since generator use is banned between the hours of 10 PM. and 6 AM., there are at most sixteen hours per day when use is permitted, making twenty hours of generator use per day even theoretically impossible, for anyone obeying the law and considering others.

How long would the charger need to run to recharge a battery? A single 100 Amp hour battery, discharged 50% (meaning 50 Amp hours are drained), would require *twenty hours* of generator run time to charge back up to 100%. [(50 Amp hours) = (2.5 Amps) × (20 hours).] But that figure is just for one battery! If the RV has several batteries, that figure needs to be multiplied by the number of batteries in the system!

How does this charging rate compare with typical battery chargers? You can buy an inexpensive charger at Wal-Mart for about $40, that charges at a rate of 5 Amps. That's nearly twice the rate of an RV 3 Amps charger.

Efficiency of 3 Amps Charging (by generator):

Efficiency is a simple equation. It is the amount of something that actually gets put to good use, divided by the total amount consumed in the process. The total amount consumed equals the amount put to good use plus the amount wasted. Efficiency can be measured as a percentage between 0% and 100%, as a decimal number between 0 and 1 or as a fraction (expressed as "parts per").

Consider the efficiency of a charity, as an analogy. If you give money to a charity, some of it will be used for the cause (hopefully) and the rest for administrative costs, waste and corruption. What percentage for the cause is considered acceptable? 90%? 80%? 50%? 20%? Less that 10%?

Here is a hypothetical situation. Suppose you are reading this very book, and you hear a knock on your front door. (Interruptions! Interruptions!) You answer that door. Someone is collecting money for the "Send Food To All Those Poor Orphaned Starving Children in Africa, Because We Are So Kind" charity. (No intended similarities to any real situations or charities that might be legitimate.) This person gives you a real sob story about the horrid conditions in the African country of ___________. Civil war, drought and grasshoppers have ravaged their food supplies. Furthermore, almost everybody has casual unprotected sex in that country, so there aren't many live healthy adults left to care for the children, as many have succumbed to AIDS. In the end, you just can't say "no" to a cause like that. So you write out a check for $300. Now, suppose that six months later, you watch the TV news. Somebody has done an audit and investigation of the "Send Food To All Those Poor Orphaned Starving Children in Africa, Because We Are So Kind" charity. It seems that for every $300 received by that charity, only $1 actually gets used to buy food for those kids. To make matters worse, ¾ of that food spoils, gets eaten by insects/rodents or gets hijacked and stolen by corrupt soldiers before ever getting to those kids. (So for your $300, those kids only got to eat 25 cents worth of food .) Would you be happy with the performance of that charity? Would you give to that charity again? I certainly wouldn't! If someone came around collecting for that charity again, I might be tempted to find out who and where the C.E.O. is, find that C.E.O. and greet that person with "pepper spray" or blow one of those really loud portable boat horns in his/her ear.

Now, let's consider the efficiency RV battery charging. If you burn gasoline in a generator for the sole purpose of battery charging, using a standard RV 3 Amps charger, how much of that energy transfers into the battery? (Hopefully it will be better than that stupid hypothetical charity!) [If you have an extreme aversion to numbers or don't have the time for details, skip right to the result (step #4).]

Efficiency Calculation (of 3 Amps charging)

Technical Data:

Energy Density of Gasoline = 130,000,000 Joules/Gallon (Wikipedia)
Generator gasoline consumption rate = 0.3 gallons/hour for a 4000 Watt Onan unit under no load conditions (According to its manufacturer: www.onanindiana.com)
Efficiency Factor for battery discharge/charge = 5/6 ("RV Electrical Systems A Basic Guide to Troubleshooting, Repair and Improvement" by Bill and Jan Moeller)

Note: Since the energy density of the gasoline is stated in Joules, the usable energy put into the battery must also be stated in Joules in order to make a valid mathematical comparison. Since: (1 Joule) = (1 Watt second), generator run time must be given in seconds and charge rate must be given in Watts.

Step #1: The length of time that a generator will run from one gallon of gasoline needs to be calculated. A 3 Amps battery charger uses very little power, therefore the no-load consumption rate of 0.3 gallons per hour is appropriate.

Capacity = Rate × Time,
therefore:
Time = Capacity/Rate,
specifically:
$3\frac{1}{3}$ hours = (1 Gallon) / (0.3 Gallons/hour).
Converting to seconds, this value is:
(Generator run time from burning 1 gallon of gasoline) = ($3\frac{1}{3}$ hours) × (3600 seconds/hour) = 12,000 seconds.

Step #2: The effective charging rate needs to be calculated. This rate, given in *current*, describes the flow of *charge* or *electrons* (3 Amps). However the charge rate must be given in the flow of *energy* (defined as *power).*

Since:
(Power) = (Voltage) × (Current),
(charge rate in power) = (12 Volts) × (3 Amps) = (36 Watts).

Since not all the electricity that flows into the battery flows out of it (some gets wasted as heat), the 5/6 efficiency factor must be multiplied in. (For every 5 Amp hours that flows *out of* the battery during use, approximately 6 Amp hours must be put *into* the battery during the recharging process.) Thus:

(effective charging rate in power) = (charging rate in power) × (efficiency factor) = (36 Watts) × (5/6) = 30 Watts.

Step #3: The amount of usable energy put into the battery needs to be calculated. This value is simply the product of the two values from steps #1 and #2. Thus:

(usable energy put into the battery from burning one gallon of gasoline) =
(effective charging rate in power) × (run time of the generator) =
(30 Watts) × (12,000 seconds) = (360,000 Watt seconds) = 360,000 Joules.

Step #4: Finally, efficiency needs to be calculated by dividing the energy value for a gallon of gasoline given above by the energy value transferred into the battery calculated in step #3.

(Efficiency) = (usable energy) / (maximum possible theoretical energy) =
(360,000 Joules) / (130,000,000 Joules) = .00277.

Look at that final number: .00277. (All values are approximate.) Notice what a truly horrible percentage this is: about ¼ of 1%. This efficiency is only about one part in 360, which is worse than the efficiency of the fictitious "Send Food To All Those Poor Orphaned Starving Children in Africa, Because We Are So Kind" charity. This means that for every 360 joules of energy in the gasoline burned, only about one joule gets put into the battery. The rest gets thrown away. Imagine life in a world where everything operated at such a low efficiency--- a world where you had to throw away 360 of everything, just to use one. In such a world, you would pay about $1440 to eat just one hamburger, normally costing $4 in our world. You would need to put $90 into a gumball machine, for a gumball that normally costs 25 cents in our world. And a major appliance, such as a television set, in that other world, would cost about as much as a house normally does in our world. I personally find such low efficiencies revolting. I wouldn't want to destroy money like that, even if I had it to burn. I'd rather just give it away to some good cause, like cancer research. Think the figures can't get any worse? Think again. The 360:1 ratio occurs as the energy transfers from the gasoline in the gas tank to the battery. More energy gets lost later as the electricity transfers from the battery into the appliances, if inefficient appliances are used (which they are). If the electricity stored in the battery from the charging process later gets used for lighting, and incandescent lights get used (standard in RVs--- they unnecessarily waste over 75% of the electricity), then only about ¼ of every joule leaving the battery makes the light. So if you decide to turn on your RV generator, for the sole purpose of charging a battery with a 3 Amps charger, to later run your incandescent lights, remember this: for every $360.00 you spend on gasoline, only about 25 cents worth actually gets used to make the light. Are you beginning to understand why I don't like, don't recommend and don't use such systems? Anyone for candles or disposable batteries?

One thing is for certain. Somebody needs to do some better engineering of RV wilderness battery charging systems.

II.) Moving Systems ("Charge As You Drive"):

These alternator-based systems can work well, for small motorhomes, but for larger motorhomes and trailer type RVs, they can fall short as primary charging systems.

Small Motorhomes (without extra vehicles): Doing all battery charging while driving can work well in these because: 1) *Small motorhomes are likely to be driven frequently*. They are quite maneuverable, unlikely to have their insides cluttered with lots of objects (which need to be carefully packed away when traveling) and must be used for all road travel, due to the absence of extra vehicles. 2) *Alternators can do substantial battery charging*. They put current through batteries considerably faster than many 120 Volt AC chargers. And: 3) *Small motorhomes usually have lower electricity requirements than big ones*. Thus, any charge stored in batteries is likely to run their systems longer than systems in big motorhomes.

Large Motorhomes and Trailer RVs: These are usually driven only intermittently. Typically they put on huge mileages, between which are long periods (days, maybe more than a week) of driving nowhere. Motorhomes are likely to tow extra vehicles, enabling them to stay put. Trailers are likely to be detached, enabling them to stay put. Also, large RVs may have slide-outs, leveling jacks and lots of room for loose objects, making moving them major deals. House batteries, located in stationary RVs, do not charge while detached vehicles move about. Sometimes RVs are driven long distances. While they are, alternators might do substantial charging, but once batteries completely charge, there simply is nowhere to store any more electricity, even though their alternators are capable of making it. So from the point of full battery charge on, alternators only make enough electricity to satisfy immediate driving needs. Since a typical RV only has enough batteries to store electricity to last a day or two (parked) before going dead, that's exactly what it does--- it goes dead. Large RVs may put on *sufficient mileage* to do substantial charging using the "charge as you drive" method, they just aren't usually driven *often enough* for this system to work really well.

Conclusion

In life, there are many chores. When it comes to chores, the following truths are universal: 1) chores that must be done often must not be too difficult and 2) chores that are difficult must not need to be done too often. If a chore is both too difficult and must be done too often, it becomes unfeasible and simply does not get done. (The level of "difficulty" of a chore depends upon how much *effort, time* and *money* is required for completion, how much *irritation/aggravation* the process involves and how *reliably* the task can be performed.)

> *Washing dishes* is an example of a chore that must be done often but is not too difficult.
>
> *Re-roofing a house* is an example of a chore that is difficult but doesn't need to be done too often.

RVers have a few special chores that are unique to their lifestyle. Charging RV batteries is one such chore. Charging by finding municipal electrical outlets, while parked primarily in wilderness, using typical equipment, is a chore that is difficult, because it requires an enormous amount of time and effort to pack/relocate/unpack the RV, sit in an RV park for awhile, then reverse the process. And it must be done very often. This task requires 24-36 hours of time, and needs to be done every two or three days, so performing this task on a regular basis is almost impossible. Charging by generator, using wimpy chargers, is a chore that is "difficult" because extraordinarily long amounts of irritating "chugging and a lugging away" are needed, and must be done so often to keep up with quickly discharging batteries, that it too borders upon the impossible. And charging by driving with typical equipment doesn't occur often enough to be reliable. Thus: *Most RVs have no working charging systems when those RVs are parked in wilderness.*

Defect #3: Generators that Start From "Dead" Batteries

Suppose you were an RV designer, engineering a system for charging a battery. You decided to use a battery charger powered by a generator. Generators must begin spinning to get started, and usually electricity is used to initiate their spin. Suppose also, that your level of intelligence were approximately that of a hamster. *Question:* How would the generator start spinning? *Answer:* Using electricity from the same "dead" battery that the system is supposed to charge! If you think that this is a bizarre design plan, you are not alone, I do too. Please note however, RV designers have no problems with this plan. They love this plan and use it all the time. (One RV service technician who told me that the vast majority of RVs are wired *exactly this way.*) Starting the generator from *the same battery that it charges* (the coach battery) has the following practical consequence: When an RV's coach battery charging system is needed the least, it works just fine. (The fully charged battery starts the generator which operates the charger.) When an RV's coach battery charging system is needed the most, it doesn't work at all. (The dead battery can't start the generator, so the charger has no electricity to operate, therefore the battery stays dead.) I have three things to say to the person who first designed such a plan as well as all those who have chosen to copy it: 1) <u>You're a moron</u>. 2) <u>You're a moron</u>. And: 3) <u>You're a moron</u>. There just are not two ways of looking at this, nor are there any polite or kind ways of saying this. You have violated common sense. This plan makes about as much sense as grabbing your own ankles and pulling upward to try to lift yourself off of the ground. It just doesn't work.

In general, when you design a plan for fixing something, you don't rely on the broken thing, to somehow "rescue" or fix itself. If you designed an emergency automobile road service plan, in which patrons would be required to *drive their disabled cars* to service stations for repairs, I would probably want to know if you were on drugs. If those cars are *disabled*, how are patrons supposed to drive them? The correct way to do this, would be to have a second fully functional vehicle (a tow truck) tow the disabled vehicle in. If you were a surgeon, and you planned to use *the same patient you were going to operate on* to hand you surgical tools during the operation, I'd probably suggest that you have your head examined. How is a patient who is cut wide open and unconscious supposed to be able to hand you surgical tools? The correct way to do this, would be to assign the tool handing task to an awake surgical assistant standing beside you. So, if you were going to design a system of components to charge a dead battery, and you decided to use *that same dead battery* to start the system, I would probably want to know what is wrong with your brain. The correct design of any charging system must use something *other than its own discharged battery* to get that system started.

Defect #4: Energy Hogging Appliances

Whenever selecting appliances, it is important to note that not all identical purpose appliances are alike. Some are energy "hogs" whereas others are energy "misers." What happens when "hogs" are selected over "misers" capable of performing the very same tasks? In regular houses, not much. Since municipal electricity is delivered to houses in virtually limitless quantities, the only effect of this decision is slightly higher electric bills. In RVs, energy gluttony causes more problems than just depleted money. Batteries don't store all that much electricity, and charging can be a challenge. In an RV, selecting "hogs" over "misers," leads to a choice between rapidly depleted batteries and a whole lot of shut down appliances, or the necessity for hundreds of pounds of extra batteries and super heavy duty chargers, costing thousands of extra dollars. Selecting "misers" over "hogs," on the other hand, can lead to a success in the field--- in fact, this can be the difference between devices that work well and devices which run down so quickly that their only real "use" is pointing to them and saying: "Look at these nifty expensive (currently off) gadgets that I've got--- aren't they cool!".

The past few decades and to this day, there has been an irresponsible trend with RV designs: RV gadgets are becoming more and more wasteful, in terms of their electricity consumption. Of the appliances that absolutely need to use electricity: usually the least energy efficient are chosen. Of the appliances that don't even need electricity: more reliable appliances of older design, that used no electricity at all, have been replaced with newer less reliable appliances, redesigned to use electricity, use lots of it, and shut down when the electricity goes off. Perhaps this design trend stems from an attitude that: "RVs are supposed to be 'luxury' items.", "RVs are for rich people, who've got money to burn, so they might as well spend it!" and "RVs are in the same category as private airplanes, where 'conservation' simply isn't an issue.". Whatever the case, the result of this design trend is the same: *RV gadgets and systems that simply don't work.*

Examples of Energy Gluttony:

Lights: There are actually *four types* of lights that can be connected to an RV's system. Almost universally, and almost exclusively, the type that wastes the *most electricity* to make light is chosen. The two types that use the *least electricity* are almost *never* used.

Furnaces: Decades ago, RV furnaces used no electricity at all. Somewhere along the line, somebody decided that all RV furnaces needed to have electric fans. These electric fans run batteries down so fast, that these furnaces are almost useless in wilderness locations where municipal electricity is unavailable.

RV users can also be a problem. While it is true that failure can result from smart people using equipment of dumb designs, it is also true that failure can result from dumb people using equipment of smart designs. We all know what usually happens when a four year old is given a flashlight. The kid plays with it so much, that almost certainly its batteries die within half a day. An adult, mindful of conservation, who only uses the flashlight when really needed, can get those same batteries to last for months. Clueless RV users are a lot like four year olds playing with flashlights. They can manage to run even multiple battery systems down quickly, almost no matter how many batteries such systems have. RV users need to become familiar with their appliances and develop good conservation habits, in order to be successful.

When it comes to RVs, some mindfulness of "conservation," both on the design and use levels, is more than just a good idea. It is an absolute survival necessity.

Defect #5: The "Octopus Effect"

Long ago, for whatever reason, be it for curiosity, adventure, commerce or advantage in war, people started building ships and taking them out over oceans and lakes. Regardless of a ship's purpose, century in which it was built, the material of which it's made or its size, all ships have something in common: *damaging or poking holes in them is really bad.* Poking holes in ships, from hitting things, tends to make them take on water and sink. As is the case with almost all human inventions, the longer people make things, the better they get at it. As centuries passed, people developed the capability of building bigger and bigger ships. A problem is encountered with building a large ship, having only one compartment. With a one compartment design, *one hole anywhere sinks the whole ship.* So somebody came up with a genuinely brilliant engineering idea. If a larger ship were to be divided into several compartments, with each sealed water tight from the others, then one hole or crack somewhere would not sink the whole ship, only one compartment would flood. Thus, a damaged ship would still be able to float, and could potentially be brought into port and repaired, instead of ending up as useless trash at the bottom of some body of water. And even if it did sink, sinking would be delayed, which might give people on board time to escape. Ship builders are smart. RV electrical designers are not so smart. RV electrical designers have not yet caught on to what ship builders realized a long time ago.

In biology, an "octopus" is a creature with one brain and mouth, and many arms attached to its central solitary brain/mouth/body. In electricity, the term "octopus" refers to a whole bunch of wires (too many), all converging or plugged into a single outlet or source. RV electrical systems are typically exactly this--- big messy electrical octopi. Each consists of one battery connected to a complicated entanglement of wires and a bunch of appliances. In regular houses hooked up to the grid, an "octopus" causes problems when too many gadgets are attempted to be operated simultaneously from a single outlet. (Too much current flows through the central wire, causing that wire to get hot, which can start a fire.) In RVs, the "octopus" problem is different: *everything is ultimately connected to everything else.* (The problem has nothing to do with heat or fires.) Therefore, one substantial problem anywhere in such an electrical system can shut that entire electrical system down, and effectively "sink the whole ship." To understand the severity of the problem, consider the following text box, which lists 20 separate problems that can shut an entire RV electrical system down.

<u>20 Things That Can Shut an RV Electrical System Down</u>:

A) A discharged battery.

This can occur from:

A1) accidental overusage due to inexperienced campers (#1),

A2) accidentally leaving an appliance on (like a light) (#2),
(This can happen to the best of us. Nobody's perfect.)

A3) too few batteries and too many appliances, which can cause significant drain even under "normal" usage conditions (#3), or

A4) a short, somewhere else in the system, not severe enough to blow a fuse (#4).

B) A defective battery, which is unable to hold a charge.

This can occur if:

B1) it is too old (#5),

B2) it has been left standing for long periods of time in discharged condition (very common for RVs put into storage) (#6),

B3) it is shorted, due to being dropped or handled too roughly (#7),

B4) it is open, for same reason as above (#8),

B5) the electrolyte in the battery gets contaminated by adding tap water, the wrong acid/water ratio, or some other contamination (#9), or

B6) it has been improperly charged on somewhat of a regular basis [overcharged (#10) or undercharged (#11) or too much current used (#12)].

C) A defective charger.

This can occur from:

C1) internal problems (#13),

C2) loose or corroded wire connections (#14) or

C3) it just doesn't send enough current through the battery to charge it in in a reasonable time frame (very common with older RVs) (#15).

D) A defective generator, which can't make electricity for the charger.

This can occur from:

D1) internal problems (generators have lots of parts) (#16),

D2) not starting due to a "dead" coach battery (very common) (#17),

D3) not running long enough to adequately charge (#18),

D4) users forgetting to turn it on altogether (#19) or

D5) running out of fuel (#20).

Actual Horror Story: The electricity was "off" more often than it was "on"

My mother bought a used motorhome in 2005. Within less than a month of camping without hookups, eight of the problems on this above list *actually occurred* [(#1), (#3), (#4), (#8), (#14), (#15), (#17) and (#18)]. Since arriving home, on subsequent short trips (that's all she takes now), four more problems on this list *occurred* [(#2), (#5), (#6) and (#17)], bringing the total to twelve. Repeated chronic failure of the electrical system was the primary reason why, after spending over $12,000 on a used motorhome, and driving as far west as Colorado (from Florida), my mother decided to abort the trip, almost before starting it, and give up long-term RVing, after spending only *one day* in *one National Park*, while camping only three weeks in the woods. (Most of this "camping" time was spent on futile attempts to get the electrical system working, which was "off" more than it was "on.")

Long ago, when RV electrical systems were quite simple, it might not have been much of a problem to have everything connected to everything else. But when systems became more and more complex, they just became big messy electrical octopi. When an RV's electrical system shuts down, the "whole ship sinks." The occupants have no lights, no heat, no refrigeration, no water comes out of any faucets (the hot water tanks shut off too), no motor action (to move slide outs or levelers), no air conditioning and no 120 Volts AC electricity. The propane stove and oven are the only appliances that still work during such shutdowns. Somebody really should put some thought into either simplifying such systems or putting some safeguards to prevent frequent total system shutdowns.

The Defects Working Together

Defects present problems not only in isolation, but they can interact to produce disastrous results. Consider how batteries and chargers work together. A system with a battery that runs down quickly might still work alright if that battery can be charged often and reliably. On the other hand, a system with a battery (bank) that can only be charged occasionally might still work alright if that battery (bank) has a very long run-down time. The worst combination is a small battery with a short run-down time, that can only be charged occasionally and not reliably--- and this is the combination that RVs usually have. Consider how appliances choices and batteries work together. Having a battery with only a small storage capacity is asking for trouble, but might still produce acceptably long run-down times if appliances that conserve electricity are used. On the other hand, connecting energy-hogging appliances to a battery (bank) with a very large storage capacity might also produce acceptably long run-down times. But connecting appliances, that waste electricity like there's no tomorrow, to a battery with a storage capacity that's already too small, produces an unacceptably short battery run-down time--- again, this is the combination that RVs usually have. Add to these three interacting defects a generator that can't start when the battery is low, and the result is a tossed salad from hell.

What Happens When Existing RV Electrical Systems are Used?
(from inexperienced RV owners to seasoned "pros")

(*fictional, but based on personal experiences and stories related by others*)

The First Few Days Out: Ah, the excitement of the maiden voyage. The freedom. The adventure. The new owners are finally getting the chance to try out their recently purchased RV. The inexperienced owners realize that learning how to live in an RV will be challenging, but they have faith that those nice RV designers took the time to design their system right.

The Bubble Bursts: About three days after they park somewhere without electricity, their lights go dim, their furnace stops heating, water doesn't come out of their faucets (even though there is plenty of freshwater in the tank), their toilet won't flush and their refrigerator shuts down. They soon realize the problem: *their coach battery has run down.* Being in the cold and in the dark at night is not their idea of luxury living. Having a toilet that won't flush is annoying (though they soon realize they can flush manually with a bucket of water). But having to throw away the contents of that expensive new refrigerator, only days after its first field use, really bums them out.

They Try to Turn On Their Generator: Of course, it doesn't start! So their 120 Volts AC system is down, as well as their 12 Volts DC system, which means they can't use any electrical appliances at all. Too bad.

It's a Hot Sunny Summer Day: With no electricity to run their air conditioner, their plastic and fiberglass sun-absorbing RV turns into a wheeled oven, leaving them with two choices: stay outdoors or bake like raisins indoors. They take the third choice: **They find a motel to cool off for awhile.**

Their First Attempts at Battery Charging: They realize that the RV starts just fine, regardless of the charge state of the coach battery, so they continue with their journey the next day. They notice that driving the RV does charge their coach battery, because after driving awhile, the RV seems to have magically come back to life. Once again, their lights glow, their furnace heats, water comes out of their faucets, their (by then empty) refrigerator cools, their generator starts just fine and their air conditioner cools. But any bliss is short lived.

It Happens Again: Soon they get tired and park for awhile. Just as before, their coach battery runs down and everything shuts off. Thankfully, they still have cold food, but not in their non-functioning refrigerator--- they keep cold food in recently bought coolers, stocked with ice. Obviously, "going dead" was not a one-time occurrence, this seems to happen every time they park for awhile.

They Try to Charge By Driving: But they soon realize that charging a coach battery by driving doesn't work particularly well when parked for awhile. They have to drive long periods nearly everyday to keep up with electrical demands. Charging while driving seems to work only while "in transit" (when they drive long distances each day, stopping for only one night at a time).

They Try to Charge While Parked: They decide to "jump start" their generator. One of them starts the RV's engine. The other one attaches jumper cables to the positive terminals of each battery, thus connecting them. (I hope this person wears safety goggles, because a spark at the battery terminals can ignite hydrogen gas made by a battery, causing an explosion. This can splash sulfuric acid in this person's face, causing severe burns. Of course, if this happens, no water will come out of the faucet to wash the sulfuric acid off. Hopefully this person has a jug of water nearby! Isn't it amazing how everything works so well together!) What do you know! The jump has been successful. As the generator runs, the RVers can use any of their electrical appliances, and hopefully do some battery charging too.

They Turn Their Generator Off: After doing so, they discover that their coach battery still seems almost dead. What did that couple hours of generator running do for the battery? Not much. The generator made lots of noise and smelly exhaust, and got some camping neighbors upset, but it didn't do a whole lot of battery charging. They soon realize that the AC charger works too slowly to be effective. They are puzzled as to why designers chose such a slow AC charger, because the DC (alternator) charger works much quicker.

In Desperation, They Head For an RV Park: There, they plug in overnight. Once plugged in, their RV comes back to life immediately. Their "state of the art" refrigerator starts working too, although they can't seem to figure out what to do with it, as by then, they have chosen to continue keeping their perishable foods in ice chests. The next day when they take off, they notice that after a night of being plugged in, their coach battery has become fully charged. But how long will this happiness last?

Decisions, Decisions: There are two ways that people can respond when unpleasant situations present themselves. They can accept defeat or they can become even more determined to move forward. These rapidly-learning RVers have realized that their RV's electrical system functions very poorly in wilderness, so they need to decide what to do next. They may accept defeat by choosing to stay exclusively in RV parks that have electricity. If so, they give up most really nice camping spots and sign a never-ending contract with the devil, to pay thousands of extra dollars to RV park slumlords. Many people however find the prospect of dependence upon private RV parks so unappealing, that they accept defeat another way: they drive home, give up RVing altogether, and put their RV up for sale (like my mother did). Then, the dark reality of RVing sets in, as they realize how quickly their RV has depreciated. (Rapid depreciation may be partly due to poor field performance.) The RVers may however choose to get tough and move forward by starting to modify their RV to try to correct its defects. However, this task can be monumental. They need to do most of the modifications themselves, as they are likely to get little help or good advice from RV repair shops. The RV owners may become overwhelmed by their system, and choose to make just a *few* modifications, rather than completely overhauling their system. Making just a few modifications will make their system only minimally functional. Or the RV owners may choose to move forward by continuing to RV, while learning to use their RV as it is. In a way, this is moving forward, but really, it is just another way of accepting defeat.

Resolution: Assuming the RVers don't take on the task of redesigning their RV from A to Z, nor do they give up and go home, they eventually adjust by becoming some combination of "generaphiles" (generator lovers) and "electrophobes" (people afraid to use anything electrical).

They may become "generaphiles:" Even those who are "challenged" at understanding electricity, quickly learn that while a generator runs, almost any electrical appliance can be used, with no drain to the battery. So the RVers turn their generator on very frequently and for prolonged time periods. They turn it on in the morning, all through the preparation eating and cleanup of breakfast. They turn it on again in the evening, all through the preparation eating and cleanup of supper. They turn it on at night while watching TV. They turn it on for hours sometimes, in the hopes that their batteries charge. They turn it on sometimes for no apparent reason at all. Can anyone really blame them? They are sick and tired of having everything go dead. Then somebody knocks on their door. It might be the polite neighbor requesting that they go easy on the generator use (and explaining local rules). It might be the irate neighbor cursing at them to do the same. Or, it might be a park ranger, issuing them a $50 summons, for generator use after hours or in a prohibited zone. That friendly enthusiastic dealer who sold them the RV never told them about any of this! Can anyone blame the neighbors? It's one matter to turn on a noisy generator once in awhile, it's another matter to run that device nearly all the time.

They may become "electrophobes:" Perhaps the RVers dislike their own generator noise. Perhaps they are considerate enough to not want to annoy neighbors. Perhaps they don't want to run their generator too much because they don't want to wear it out or waste money on gasoline. (Some RVs don't even have a generator.) Perhaps their generator is broken and they don't want to shell out big bucks to fix it. Whatever the case, they become terrified to turn on even the smallest electrical thing for more than ten minutes, fearing their RV will go dead again. They take drastic action. At night, they light their RV with candles and oil lamps. They refuse to use their furnace, almost no matter how cold it gets (they buy warm sleeping bags or heat their RV with portable catalytic heaters). They leave their TV off. They may even choose to draw water from plastic jugs instead of using the faucets. Here is a question to ponder: *What good are all those "high-tech" RV electrical appliances, if people are afraid to use them?*

They may become both "generaphiles" and "electrophobes:" These are people who love to "generate," and use lots of electrical gadgets while doing so, but develop an alternate personality while the generator is off, in which they are terrified to turn anything electrical on for more than ten minutes.

<u>The RV Owners are Now Seasoned</u>: Whatever decision they have made, the journey has been quite an ordeal. There are no options which are both easy and good.

The "Official Plan" For RV Electricity

What exactly is the "official plan" for RV electricity? Spend ten hours a day running a noisy generator (which can't even start when the battery is low)? Stay only in expensive private RV parks? Find and plug into (non-existent) electrical outlets in the wilderness? Spend seven hours per day driving around, searching for places to plug in on private lands? Is this the "plan?" I don't call it a "plan." I call it: "somebody showed up for work drunk and took home a large pile of money for designing an electrical system, but never bothered reading books on electricity beforehand." Maybe RV designers plan on having people plug into electrical outlets growing out of trees or emanating from rocks or soil. Maybe RV designers believe that electricity comes from leprechauns. (Forget about finding a leprechaun and demanding his pot of gold, the new trend for the 21st century RVer is: *find a leprechaun and demand his pot of Amp hours!*) Tell you what--- If you figure out how to successfully use RV electrical systems of current designs in wilderness, without running generators for ridiculous time lengths, write to me and explain it. Despite RVing (off and on) for over three decades now, reading books on electricity, having hundreds of conversations with other RVers, and even writing a long book about RV electricity, **I still don't get what the official "plan" is!**

There is a tree species, called *coniferous electricus*, that grows live electrical outlets out of its trunk and branches. Plugging into such trees might seem like a viable option, however these trees aren't very common--- in fact, despite several years of RV camping, I have never encountered one. I wanted to include a photograph, but since I never came across one, I instead made and included a drawing (next page) of what I think a coniferous electricus tree might look like.

I'll be happy to believe that leprechauns that produce electricity exist, if compelling evidence can be found supporting this. However, I have yet to meet a single RVer who gets his electricity from leprechauns. So for now, I hold the opinion that they do not exist.

coniferous electricus tree

(RV designers "official plan" for RV electricity)

<u>Proposed Solutions to Defects</u>:

<u>Solution #1</u> (to: Insufficient Batteries):
<u>Have More Batteries</u>

Yes, the answer, that is the most obvious and simple enough for anyone to understand, also happens to be correct. Having more batteries does for electrical systems what having more water tanks does for water systems; it allows more to be stored, thus it permits longer use periods between refills. The amounts of time that multiple batteries can operate appliances is directly proportional to the number of identical batteries in the system. (Two batteries run things twice as long, three--- three times as long, four--- four times as long, etcetera.)

How can multiple batteries be configured? How should they be configured? Batteries can be used either *sequentially* or *simultaneously*. If used sequentially, only one battery is connected to the appliance system at a time. That battery is used until it runs low, then it is disconnected, then the next fully charged battery in line is connected and used until it runs low, etcetera. (Switches are often used, rather than connecting and disconnecting wires.) The procedure continues until the last of the batteries runs low. If used simultaneously, multiple batteries all share the electrical work load in a configuration known as *parallel* (*figure 1*). In a parallel configuration, all "like" terminals of the batteries are connected by wires to all other "like" terminals.

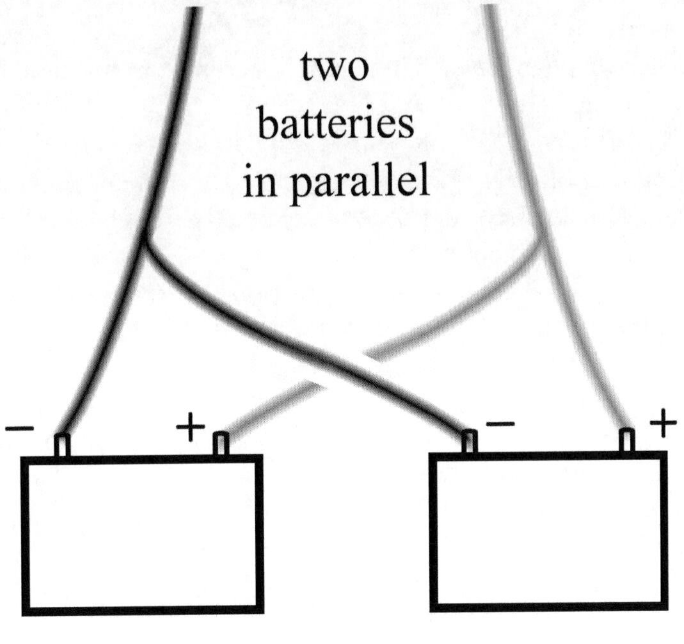

figure 1

[In parallel configurations, all the negative terminals (–) are connected to each other, and all the positive terminals (+) are connected to each other,.]

Sequentially battery use: 1) Each battery discharges more completely [than parallel]. *Advantage:* Batteries can accept charging at quicker rates when deeply discharged. This means that if a generator powers the charger, the generator doesn't need to run as long. *Disadvantage:* Deeply discharging batteries shortens their lifespans. 2) Each battery is isolated from the others. *Advantage:* When batteries are separated, if one goes bad, it doesn't harm the others. *Disadvantage:* Having multiple separate batteries, with multiple levels of discharge, means that multiple battery chargers are needed for simultaneous charging (recommended in wilderness). Since there are several methods of charging batteries, that number needs to be multiplied by the number of batteries, which could get messy and complicated. [Example: A system having four batteries and three charging methods (by generator, while driving or by solar) would require twelve chargers--- or at least a pretty sophisticated computer to decide where and how to direct the available current for charging.] For this reason, sequential battery use is not a good practice for long term off-grid RV use, except for those who take low batteries and chargers to service stations to charge them remotely overnight. However, this may be a good option, as well as the quickest effective fix, for RVers who spend only short time periods in wilderness, and charge all their batteries "at home."

Simultaneous (parallel) battery use: This greatly simplifies the system, both for charging and use. However, one bad battery can ruin the rest of the bunch. For this reason, almost everybody, who knows anything about batteries (including me), recommends that all batteries to be used together be purchased new, and be of the same size and type from the same manufacturer.

Let's consider what happens when a gadget is run from multiple batteries in parallel, instead of a single battery setup. For this example, we'll consider a 72 Watt gadget that draws 6 Amps from 12 Volt sources. [(72 Watts) = (12 Volts) × (6 Amps)]. In a one battery configuration (*figure 2A*), the entire 6 Amps of current is supplied from the single battery. In a two battery configuration (*figure 2B*), each battery supplies half of the total 6 Amps of current, so each battery supplies 3 Amps of current. In a three battery configuration (*figure 2C*), each battery supplies one third of the 6 Amps of current, so each battery supplies 2 Amps of current. The currents from each battery add at the wire junctions heading toward the appliance, then divide at the other junctions heading back toward the batteries, kind of like the way water currents flowing through pipes would. (If the batteries and the wire lengths are alike, then the currents divide equally.) It should be noted that connecting identical batteries in parallel has no effect on the *voltage* applied to any gadget--- the voltages of *figures 2A – C* are identical (all 12 Volts, in this case). Since voltages are what gadgets ultimately respond to, gadgets can't differentiate between single and multiple battery setups, so they respond exactly the same, with one important exception: *gadgets run twice as long from two parallel batteries (as the do with one), three times as long from three parallel batteries (as they do with one), etcetera.*

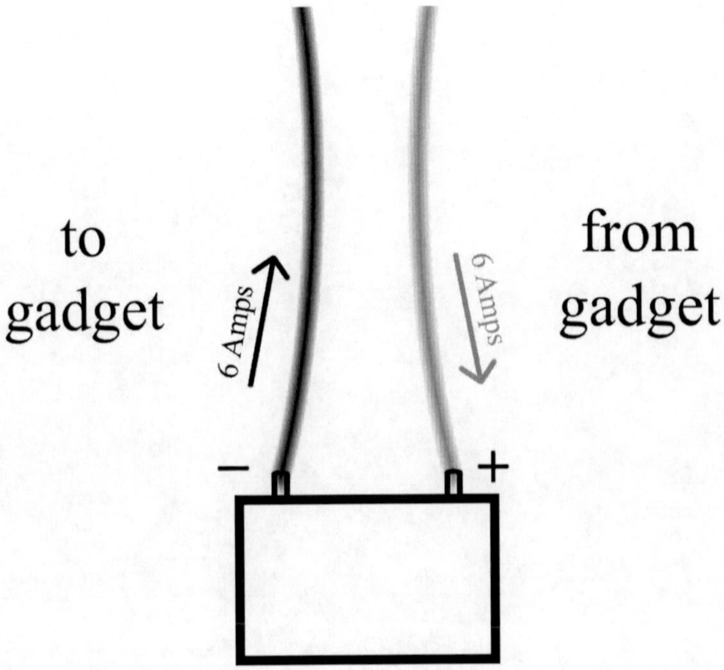

figure 2A

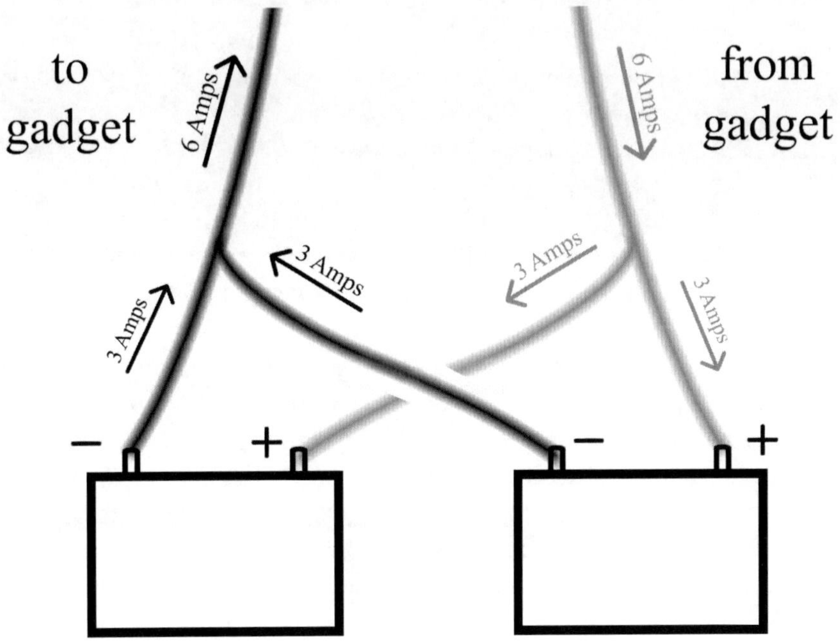

figure 2B

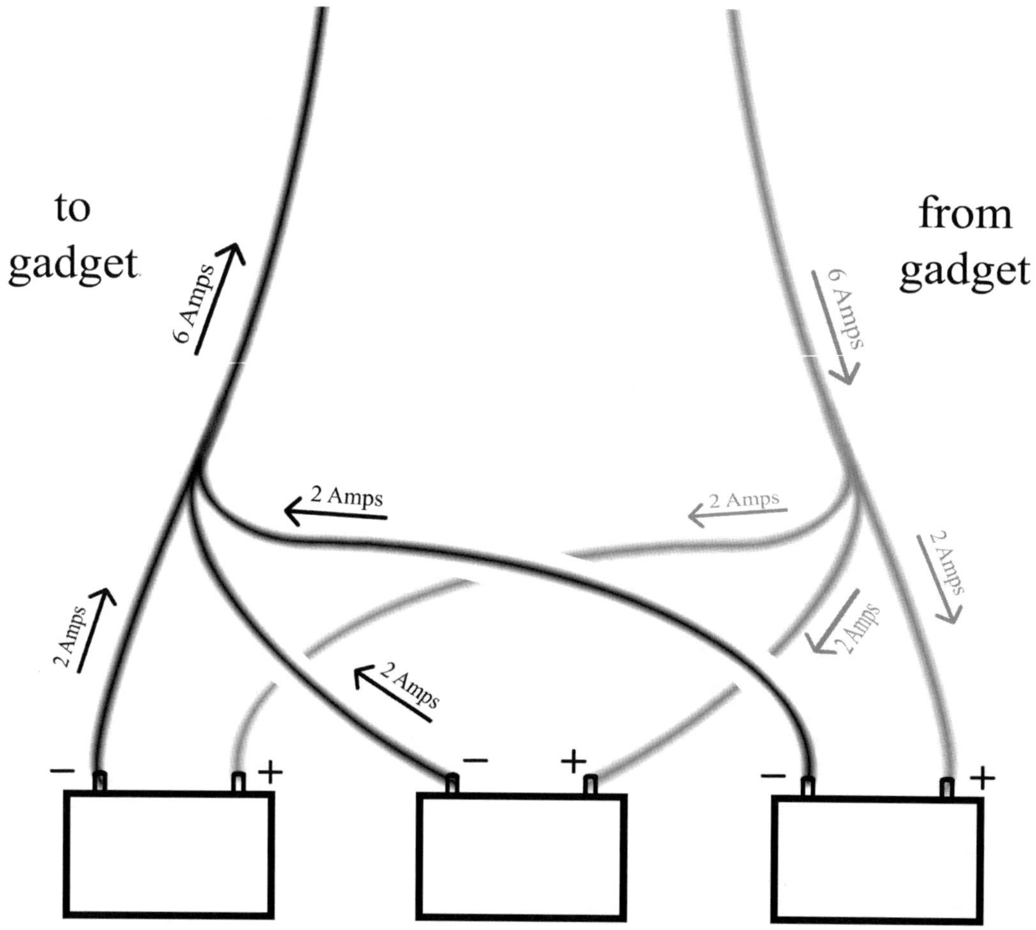

figure 2C

Any gadget, designed to be run by a single battery, can also be run by multiple batteries in parallel. Connecting multiple batteries in parallel is an effective, accepted and well established method of increasing the amounts of electrical energy put into storage, both for RVs and for other uses.

There are however some drawbacks to having lots of batteries in RVs. Batteries add weight and take up space. They cost money and do not last forever. Many RVers do choose to stay almost exclusively in trailer parks with electricity, thus there is a group of consumers who might be reluctant to sacrifice storage space and pay for batteries that they just don't use. RV manufacturers may be justifiably reluctant to add costs by installing these unwanted extra batteries. (One RV salesperson told me: "Many RVs have only a single coach battery. It is up to the owner install more.") Still, the average RV could have and fit a few more batteries than they've got. And I think they should. I think a medium-sized RV could fit 4 or 5 batteries, without adding too much weight or taking up too much space.

<u>Solution #2</u> (to: Inadequate Chargers):

> <u>Note</u>: This section considers common battery charging electricity sources, by equipment usually sold as standard on RVs (generators and alternators). Chapter 3, "Making Electricity & Charging Batteries," covers less common electricity sources (sun, wind, heat flow and fuel cells), harnessed by equipment not usually standard on RVs.

I.) (solution for standing systems): <u>Charge With More Current</u>

This sounds really easy. Charging a battery with more current should fill that battery with electrons faster, just as filling an empty swimming pool with water, gushing from a fat hose connected to a fire hydrant, fills that pool a lot faster than a slower flowing garden hose connected to house faucet. In reality, battery charging is considerably more complicated.

<u>Battery Charging Current Limiting Factors</u>:

<u>A.) Batteries Themselves</u>: Batteries can only accept limited currents during charging. The maximum allowable charging current that a battery (bank) can handle is dependent upon several variables:

1) <u>Battery Type</u>: Although wet lead-acid batteries are currently and almost exclusively used in RVs and automobiles, there are several other types of batteries that could be used in RVs, each with different properties. Examples are Nickel-Cadmium and Lithium-ion. Lithium-ion batteries, the type usually used in computers and electric cars, have some advantages. For identical energy storage capacity as lead-acid, lithium batteries can accept charging up to four times as fast. Lithium batteries also weigh less--- 1/3 to 1/4 as much. However, lithium-ion batteries have some disadvantages. They cost several times what lead-acid batteries do and are plagued by short shelf lives (they degrade after a couple of years, even if never used). Within the lead-acid category, there are several sub-types. Included in these are wet cell (most common and least expensive), gel (dry and sealed) and absorbed glass mat (dry and sealed). Gel and absorbed glass mat batteries can accept greater charging currents (than wet cell batteries), especially when discharged deeply, therefore can charge more quickly. But they must not be subjected to charging voltages too high when nearing full charge--- if they are, they can be destroyed. [Unlike wet cell batteries, which simply boil away their electrolyte if charged at too high a voltage (the remedy is adding distilled water), dry sealed batteries can't handle internal gas built up from charging with too much current.] Gel and absorbed glass mat batteries cost several times what wet lead-acid batteries do. Deep cycle batteries are designed to function better (than regular automotive batteries) when deeply discharged repeatedly. Regular automotive batteries are designed to supply large currents for starting engines, but not necessarily to withstand being deeply discharged repeatedly. There are entire books written on battery types, optimal use and charging. That information can be quite overwhelming. I do not wish to replicate this information here nor devote too much text to this subject, because many alternative battery types are so costly, that their use may be

inadvisable, even if they do have better properties. (It may be preferable to buy several wet lead-acid batteries, instead of just one of another kind.) Deep cycle wet lead-acid batteries work well for RV use, due to their low cost, long shelf lives, familiarity and availability. In addition, wet lead-acid batteries charge well from electricity sources other than generators and alternators (sun, wind, thermoelectric, fuel cells). The rest of this discussion on charging assumes use of lead-acid battery types.

2) The Charge State: When a battery discharges, high energy molecules convert to low energy molecules as the battery pumps electrons. When a battery's charge state is near 100%, high energy molecules are abundant and low energy molecules are rare, therefore the battery pushes current well. As a battery's charge state approaches 0%, the important high energy molecules become rarer and the useless low energy molecules become more abundant, therefore the battery struggles to push current and eventually goes dead altogether. When a battery charges, the opposite takes place: low energy molecules convert back to high energy molecules as current passes through. Since low energy molecules are abundant and high energy molecules are rare when a battery's charge state is really low (0% to 50%), the battery readily accepts charge at that range. As a battery's charge state approaches 100%, the target low energy molecules become rarer and the high energy molecules (unimportant for charging) become more abundant, therefore the battery struggles to accept further charge and eventually becomes unable to accept any more charge at all. Thus, *the closer a battery gets to full charge, the less charging current it can accept*.

3) The Voltage: As a battery nears complete charge, its voltage increases. Charging at too high of a voltage can overheat a battery or cause the electrolyte to split into hydrogen and oxygen (a process called *gassing*). Gassing can destroy sealed lead-acid batteries, and doesn't do anything useful for wet lead-acid batteries. (It does not cause the battery to charge faster; it does make waste heat and hydrogen, which is an explosion hazard.) Therefore, charging voltage needs to be limited. Charging current naturally gets limited when charging voltage is limited.

4) The Storage Capacity: Batteries can be made with electrodes of any size (surface areas). Electrochemical reactions take place in proportion to the surface areas of electrodes, therefore, all other conditions equal, *both* the storage capacities and the maximum allowable charging currents of batteries are directly proportional to the surface areas of their electrode plates and also *directly proportional* to each other. This means that batteries with electrodes of larger surface areas both *store proportionally more charge* and *can accept proportionally higher charging currents* than batteries with electrodes of smaller surface areas. [Example: If a 75 Amp hour capacity battery can be charged with 15 Amps, then under otherwise identical conditions, a 100 Amp hour capacity battery can be charged with 20 Amps.]

5) The Number of Batteries: The maximum charging currents that batteries connected in parallel can accept (at any charge state) are *directly proportional* to the numbers of identical batteries in their systems.

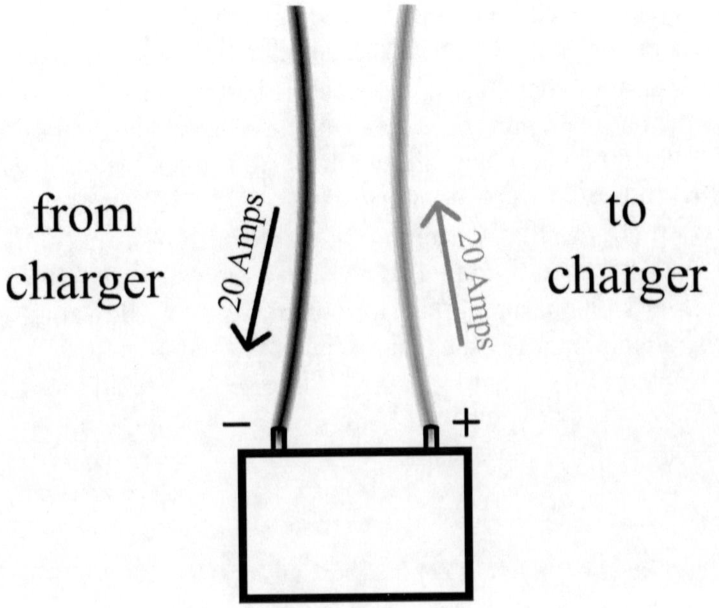

figure 3A

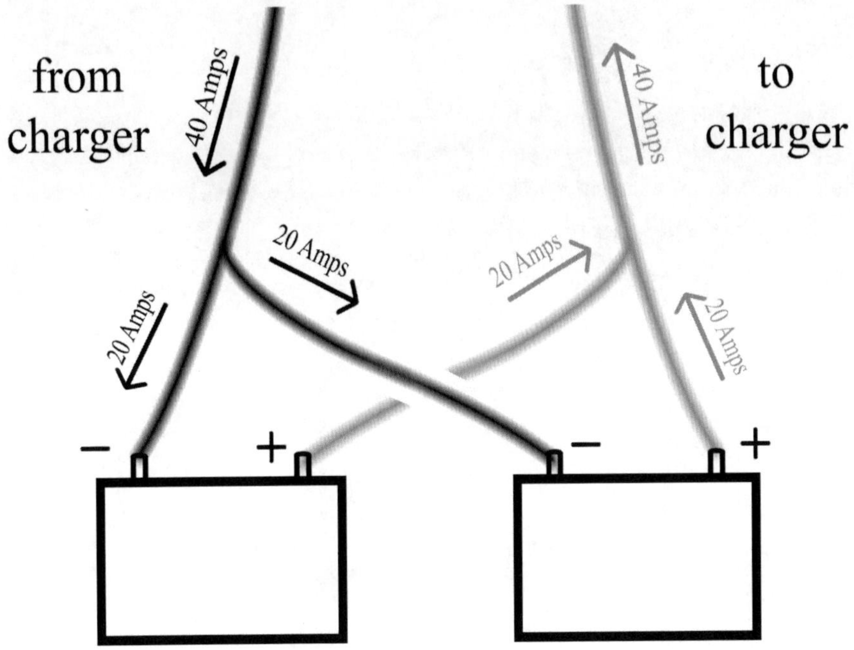

figure 3B

(Two identical batteries, hooked up in parallel, can accept twice the current as one. Three batteries can accept three times the current as one. Etcetera.) To understand why, look at *figures 3A – C*. For this problem, let's assume that 20 Amps is a single battery's maximum allowable charging current. In a single battery setup (*figure 3A*), all the current (20 Amps) goes through that single battery, therefore the maximum allowable current from the charger is the same as the battery (20 Amps). In a two battery setup (*figure 3B*), the current from the charger splits in two, therefore the maximum allowable current from the charger is 40 Amps, which is twice that of each individual battery. In a three battery setup (*figure 3C*), the current from the charger splits in three, therefore the maximum allowable current from the charger is 60 Amps, which is three times that of each individual battery. Notice that the internal charging currents of all the batteries in all three setups are all the same (20 Amps).

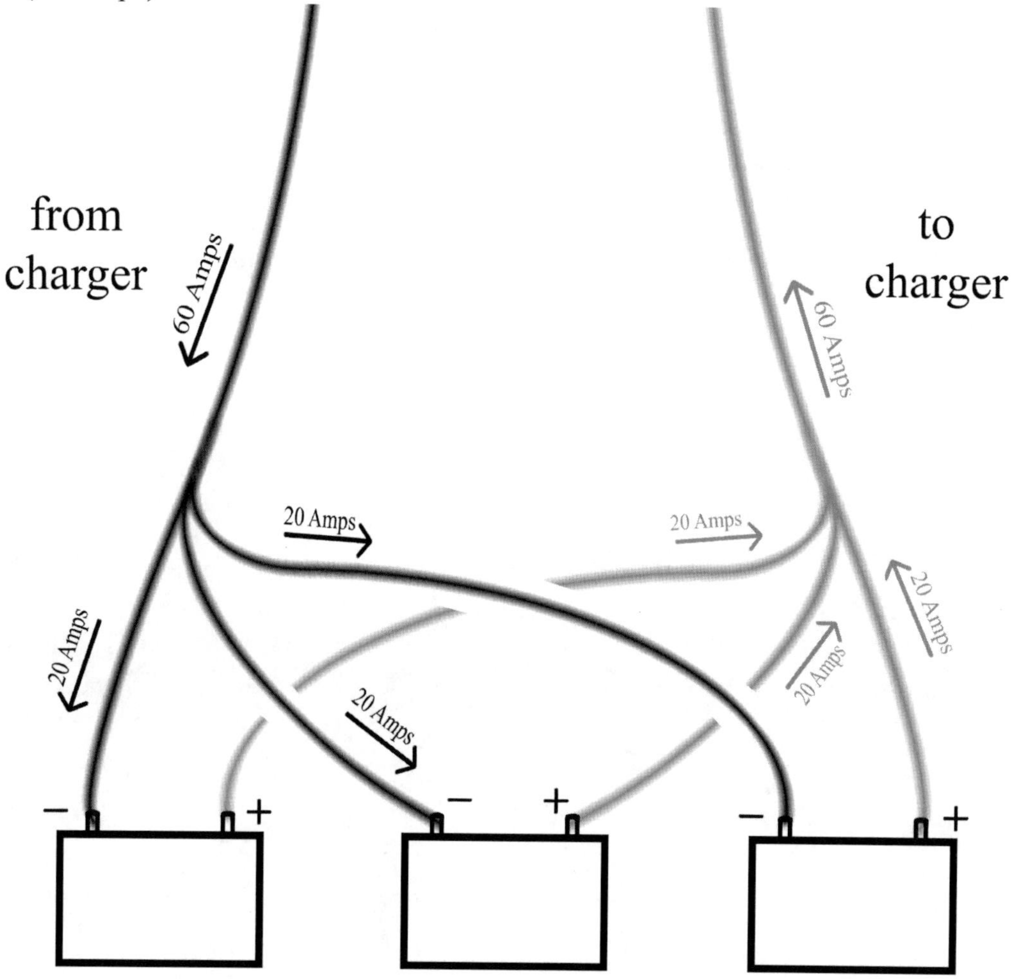

figure 3C

Having more batteries, in parallel, not only helps to store more charge, thus delaying run-down time, it dramatically increases a system's ability to accept charging current.

6) Temperature: Batteries be damaged from overheating during the charging process. The internal temperature of a battery is dependent both upon the heat added by the charging process and the ambient temperature of the world around it. Therefore, if ambient temperatures are hot, the maximum allowable charging current is slightly less than when ambient temperatures are cold. Some advanced chargers have temperature sensors at the batteries, which help them decide how much current to send through those batteries.

B.) Energy Sources that Power the Chargers: Both generators and municipal outlets can provide many times the power needed to charge RV batteries (even multi-battery systems), therefore although they do have power limitations, these limitations are not usually considered variables in the current limiting or charging time equations. Note though that generators are usually operated for only brief time periods during the day. This is considered an important variable. Other charging sources, covered in Chapter 3, do have output limitations that limit charging current (sun and thermoelectric produces limited current, wind is wildly intermittent and unpredictable).

How Many Batteries Could a 3000 Watt Generator Charge?

3000 Watts is a smaller than average output for RV generators, but adequate for most applications.

A decent bulk charging voltage might be 13 Volts. The charger transformer must provide another 1.4 Volts, due to internal voltage drops of the charger (across two diodes), raising the requirement to 14.4 Volts. What is the maximum charging current such a generator could provide?

(Power) = (Voltage) × (Current),
therefore,
(Current) = (Power) / (Voltage),
specifically:
(208 Amps) = (3000 Watts) / (14.4 Volts).

During bulk, a battery can be charged with 20% storage capacity per hour. Since 100 Amp hours is a typical storage capacity for many single RV batteries,

(charging rate) = (20%) × (100 Amp hours per hour) = (20 Amps).

Finally,
(Number of batteries that could be charged from generator) =
(Maximum output current at charging voltage) / (Current needed per one battery) =
(208 Amps) / (20 Amps/battery) = (10.4 batteries).

As you can see, a generator supplies much more power than is needed to charge a battery.

The Four Stages of Battery Charging:

Bulk: This occurs when batteries are below 80% charged. In this stage, a constant hefty current gets used for charging. The voltage is lower than maximum, due to the discharged state of the battery. This is the quickest and most efficient charging stage. In RVs, this can be done with alternators or AC battery chargers powered by generators.

Absorption: This occurs when batteries are between 80% and 100% charged. In this stage, a constant voltage gets applied to the terminals. The current gradually diminishes as the battery nears 100% charge. Because it is not absolutely necessary to charge batteries above 80%, this stage should not be done powered by generators, due to horrible inefficiencies and long generator run times, unless those generators are being run for other simultaneous purposes. This stage can and should be done though if hooked up to municipal electricity.

Float: This is a stage at which a small current gets continuously sent through a fully charged battery, to counteract self-discharge. Since this stage is unnecessary for RVs in active use in wilderness (this stage is necessary for batteries that are in storage for awhile), and only tiny amounts of current are used, generators should never be turned on just for this stage in RVs in wilderness. Float however can and should be done in the wilderness with solar panels, if RV users have them.

Equalization: In my opinion, this is not really a battery *charging* stage; it is more of a battery *conditioning and preservation* stage. In this stage, a large amount of current gets sent through a fully charged battery for a short time. This equalizes the charge levels of the individual cells. (This is because fully charged cells cannot be further charged, but incompletely charged cells can be further charged). It also drives unwanted molecules of sulfate off the electrodes, prolonging the tendency of a battery to lose some of its storage capacity (ability to store charge) as it ages. (Sulfate molecules can harden and become permanently attached to the electrodes, reducing the effective electrode size and hence the storage capacity of the battery.) This stage should be done only occasionally, especially with batteries not regularly brought back to full charge, but should never be done with sealed batteries. In RVs, this stage could be done powered by generators, alternators or municipal outlets.

Determining Charging Currents:

How much current *can* be used to charge a battery? During bulk, substantially more than the 3 Amps rate of many of those wimpy chargers found in RVs. The figure that I have found in some books for wet lead-acid batteries is: (Maximum charge rate) = (20% of the battery's storage capacity per hour), when the battery has less than 80% full charge. This means that a single typical 100 Amp hour RV battery can accept a charging current up to 20 Amps. A 75 Amp hour capacity battery can accept up to 15 Amps. Sealed batteries (gel and absorbed glass mat) can accept charging with twice as much current as wet lead-acid batteries, when below 80% charged. After a battery's charge state exceeds 80%, the maximum allowable charging current drops to a greater and greater extent (to just a few Amps), as the battery approaches full charge.

Should a battery be charged quickly or slowly? There are advantages and disadvantages to charging batteries up either quickly or slowly. Charging quickly with high current gets the job done but shortens the life spans of batteries. Charging slowly with low current is easier on batteries, but to keep up with daily electrical demands, can require too much daily generator run time to be practical or even possible. Charging really quickly (or attempting to do so) can "fry" batteries and become impractical too, because users would need to spend time, money and effort replacing heavy batteries. Different RV experts give different opinions as to how fast batteries should be charged. RV designers seem to favor rather slow charging, presumably to prolong battery life. Focusing lots of attention on pampering batteries may seem like a good idea until one question gets asked: "Which cost more: batteries or generators?". A brand new deep cycle wet lead-acid battery might cost $75 to $100. A brand new generator (installed) might cost $3000. Before choosing a charging plan that requires excessive generator run time, one need to consider that whenever a generator runs, engine parts wear out. After a certain amount of time, a generator's engine parts get thrown out of kilter, and it stops working. Then, it's time for the user to cry all the way to the bank (then the repair shop). Personally, if I were going to use a generator to charge batteries, I would charge with as much current as possible. I would much rather "toast" a few batteries to go easy on a generator, than "toast" a couple generators to go easy on some batteries. Furthermore, generators consume fuel. When the cost of fuel is added to the equation, slow charging makes no sense at all. For example, if an RV could run its generator to charge batteries one hour per day instead of three, two hours of runtime could be saved per day, which would amount to 60 hours of runtime saved per month. If the generator fuel consumption rate were 0.3 gallons/hour (typical), the monthly fuel savings would be 18 gallons. [(18 gallons) = (0.3 gallons/hour) × (60 hours).] If the cost of gas were $4.00/gallon, the monthly money savings would be $72. [($72) = (18 gallons) × ($4/gallon).] If charged one hour per day instead of three, enough money could be saved from unneeded fuel to buy a brand new battery in about a month and a half. Thus, replacing a few "toasted" batteries would actually cost less than the fuel needed to charge batteries optimally with those ridiculously slow chargers. (Note: Generators and 120 Volts AC chargers are often not needed at all for battery charging. All necessary RV battery charging can often take place via the vehicle's alternator, while driving.)

How much current *should* be used to charge a battery?

A) During bulk--- when below 80% charged: Charging is most effective within this range. I would charge with as much current as possible--- 20 Amps for each 100 Amp hour wet lead-acid battery, connected in parallel. (This is almost 7 times the charge rate that the wimpy 3 Amps chargers deliver.) At this charge rate, batteries 40% full could be recharged to 80% full in about two hours. For sealed lead-acid batteries, I would use twice that--- 40 Amps for each 100 Amp hour battery, connected in parallel.

B) During absorption--- when above 80% charged: As batteries approach 100% charge, their abilities to accept current decreases exponentially, so it may be even theoretically impossible for them to attain 100% charge. Therefore, there is some point at which it is advisable to turn the charger (and the generator) off and declare the batteries "charged enough," even though not fully charged. I believe batteries shouldn't even be charged above 80% when using a generator. If they need to be topped off, they should be charged within that range by other methods (like solar, thermoelectric, supercapacitor aided or occasional long drives). "Topping off" batteries by generator requires so much extra runtime (and fuel), at horrible efficiencies, that the process isn't worth it. Unfortunately, wet lead-acid batteries degrade from being left in a discharged state too long and too often, so it is advisable to try to charge them to 80% as often as possible, and to charge them completely one in awhile, if a decent method can be found of doing this. (If in an RV park with electricity or home, batteries should be charged slowly and completely.) Sealed batteries, on the other hand, can withstand being discharged and staying that way many times better that wet batteries, so there may never be a reason to charge them beyond 80%. Since sealed batteries can be destroyed by being charged with too high of a voltage (which might occur when above 80% full), from internal gas bubbles forming which can't escape, special chargers are needed to limit the voltage to 14.1 Volts. Also, sealed batteries cost 3–5 times as much as wet batteries, so if they are chosen, extra care should be taken not to "toast" them---especially when RV owners have multiple batteries in their systems.

> It certainly would be nice if batteries could be charged faster, but batteries are different from other containers that "store" stuff, such as gasoline tanks on automobiles. You just can't "fill" a battery with electricity quickly, like you can fill a gasoline tank in your car with gasoline quickly.

> The Weakest Link
>
> A typical small 3000 Watt RV generator can make over 200 Amps for battery charging.
>
> A typical 100 Amp hour battery can accept 20 Amps of charging current.
>
> A typical standard equipment RV battery charger charges with only 3 Amps.

> What Are 3 Amps Chargers?
>
> 3 Amps chargers are not battery chargers, they're jokes.
>
> 3 Amps chargers are not battery chargers, they're devices which allow RV salespeople to say "this RV has a battery charger," when really it does not.
>
> 3 Amps chargers are what I call: "battery chargers that don't charge."

Charging Efficiency With Multi-Battery High Current Systems

Earlier in this chapter, under the "Inadequate Chargers" defect, the horrible fuel efficiency of charging with a 3 Amps charger was calculated to be .00277 or 1 part in about 360. Charging efficiency (by generator and 120 Volts AC charger) is dramatically increased by having several batteries and charging each with the maximum allowable current Let's consider what happens when three parallel 100 Amp hour batteries get charged with all they can accept, which is 20 Amps of current each, totaling 60 Amps.

It might be tempting to conclude that the efficiency would improve 20 times, since

(60 Amps) = (3 Amps) × 20.

However, generators consume more fuel when loaded. 60 Amps of current @ 15.5 Volts (a possible internal transformer voltage before diode drops--- actual battery charging voltage would be lower) would consume 930 Watts of power, because:

(930 Watts) = (15.5 Volts) × (60 Amps).

A typical generator fully loaded at 4000 Watts might consume .7 gallons of gasoline per hour, whereas a no-load generator might only consume .3 gallons per hour. The general approximate formula for extra fuel consumed per Watt by loading (assuming a linear fuel/power relationship) is:

(Extra fuel consumption rate per Watt) =
(Max. fuel consumption rate – No load fuel consumption rate) / (Max load – No load) =
(.7 gal/hour – .3 gal/hour) / (4000 Watts) =
(.4 gal/hour) / (4000 Watts) = <u>.0001 gallons / (hour × Watt)</u>.

Specific to this situation:
(Extra fuel consumption rate) = (Power) × (Extra fuel consumption rate per Watt) =
(930 Watts) × [.0001 gallon / (hour × Watt)] = <u>.093 gallons per hour</u>.

Loaded fuel consumption rate = (no load fuel consumption rate + extra fuel consumption rate) =
[(.3 + .093) gallons per hour] = .393 gallons per hour.

Thus:
(Efficiency of 60 Amp charging) =
(Efficiency of 3 Amp Charging) × (charging ratio) × (fuel consumption ratio) =
(.00277) × (60 Amps / 3 Amps) × [no load fuel consumption / loaded fuel consumption] =
(.00277) × 20 × [.3 gal per hour / .393 gal per hour] = <u>.04229</u>,
which is about <u>one part in 24</u>.

This is quite a bit better than the 1 part in 360 using a 3 Amps charger.

The Ideal RV AC Battery Charger

The ideal RV AC battery charger should be able to put out hefty current, charge in multiple stages, be adaptable, temporary shut down when electricity is needed elsewhere and be capable of shutting off the generator when the battery reaches 80% full.

Hefty current: I think an RV charger for a midsize RV should be able to deliver 60 Amps or more as a maximum charge rate. (This is based on three parallel 100 Amp hour batteries, receiving 20 Amps each.)

Multi-stage: An AC battery charger should be able to perform all four stages: bulk, absorption, float and equalization. Only bulk and very occasional equalization should be done in the wilderness using a generator, unless the generator gets simultaneously used for other purposes. Float should be done at home, when the RV is in storage and plugged in.

Adaptable: Batteries have different optimal charging characteristics, depending upon their types, their sizes, and the numbers of batteries in the systems. Charging may seem complicated, but it all can be simplified into three variables: *voltages, currents* and *times.* A charger should be programmable to measure and regulate these variables. This could be accomplished, if a computerized charger had a flash drive inserted, with appropriate software individually tailored to the system. It would even be possible for a programmable charger, bought today, to effectively charge a battery type invented five years from now, with new software. Also, a charger needs to somehow sense the source of the AC and differentiate between generator AC and municipal AC. Generator AC is usually available for only brief time periods, therefore chargers should respond by charging with as much current as possible. Municipal AC when available, is almost always continuously available for at least a half a day (often much longer), therefore it is alright for chargers to use lower currents and longer charging times to extend battery life, and also perform absorption and float stages that are not practical in wilderness. (Typical chargers, even expensive multi-stage ones, erroneously assume that generator AC will be available for the next 24 hours, so they make erroneous decisions.)

Automatic temporarily shut down when power is needed elsewhere: The charger should be connected to sensors in the generator or wiring route to the appliances, and be able to grant electrical preference to those appliances that need lots of electricity, and thus temporarily shut down or operate at a lower charge mode until those appliances shut off (then the charger would revert back to normal high-charge mode). However, my recommendation of 60 Amps would only consume just under 1000 Watts of power (@ about 16 Volts upstream of the internal circuitry of the charger; the voltage downstream at the batteries would be lower). This is only about one fourth the power of a typical 4000 Watts generator in a not-too-huge RV.

80% shut off: Because generators should not be used solely for charging batteries beyond 80% in wilderness, chargers should have auto-shut-off optional switches. If a generator gets run solely for the purpose of battery charging, the auto-shut-off switch could be on, which would cause the charger to trip the switch that turns the generator off, when the battery reaches 80%. If a user runs a generator for other purposes besides battery charging, the auto-shut-off switch could be off, allowing the generator to keep running regardless of the battery's charge state.

II.) (solution for moving systems): <u>Have More Batteries, Cut Electrical Waste</u>

Suppose you decided to collect rainwater for drinking, using only a one quart jar. After a rain, you could never have more than one quart of water, even if there was a torrential downpour. You would probably drink that water fairly quickly, then go thirsty for quite awhile. An RV's battery holds electricity like that jar holds water. Suppose an RV has one coach battery that runs down in one day. Suppose also that the RV goes for a good long drive one day per week while relocating, fully charging that battery. The occupants would have electricity for only one day of the six day week (the time while parked), which would be only 17% of the time. Now, suppose they improved their system by having two batteries instead of one. After their ride, they would have electricity for two of the six days, which would be 33% of the parked time. Now, suppose they further improved their system by converting to appliances that consumed only half the electricity. They would then have electricity for four of the six days, which would be 67% of the parked time. Now, suppose they further improved energy conservation, by using appliances that consumed only one third of the electricity of their original system instead of half. They would then have electricity for all six of the days while they're parked, which would be 100% of the time. (They also would have electricity on their one traveling day.) Are you beginning to get how this works? "Charge as you drive" RV electricity is like a TV game show. If you can figure out how to get your electricity to last until your next long ride, you WIN! (You advance to the next round.) If you use your electricity up before that ride, you LOSE! (You go home crying.) Adding more batteries and cutting waste vastly improves "charge as you drive" systems as a whole. Vehicle alternators themselves work quite well and probably don't need much improvement.

> Can the "charge as you drive" game be won? Of course! I've been successfully making all my RV electricity, using this method, for years. I've gotten so good at it, that I no longer even know how long it takes me to run a battery dead, because *I haven't run any batteries dead in years.*

Charging while driving can work really well for primary (or even sole) electricity production for small RVs with modest needs but probably shouldn't be used as the only electricity production method for big RVs. With some modifications though, charging while driving could be a major electricity contributor for everything in big RVs too (except air conditioners), however I believe that big RVs still should have generators and good standing 120 Volts AC charging systems, at least as backups.

<u>Solution #3</u> (to: Generators That Start From "Dead" Batteries):
<u>Start Generators From Other Energy Sources</u>

Generators are internal combustion engine devices. All internal combustion engines can perpetuate their own operation, once set into rotation. Therefore, all that is needed to start a generator is some mechanism to start its rotation.

<u>Possibilities for Starting Generator Rotation</u>:

1.) <u>Hand starting</u>: Small engines, like lawnmowers, weed whackers and chainsaws can be started from the strength and motion of a human body, by means of pull ropes. Larger engines, such as those found in antique automobiles, can also be hand started, though by a slightly different means--- cranks. (Model T Fords started this way.) Despite generators being big and heavy, they too could be designed to hand start by means of pull ropes. Portable generators, with equivalent power of built in RV gensets, often start by pull ropes, however generators designed to be built into RVs start by batteries. RV generators tend to be big and heavy, so hand starting them might require quite a bit of strength, perhaps more than some RV owners might be comfortable delivering, as many are advanced in age. Since very few RVers take along horses, mules or Japanese Sumo Wrestler assistants (to operate pull ropes or cranks), hand starting is perhaps not a good design option for general use, except perhaps for occasional emergency starting when the coach batteries fail.

2.) <u>Spring assisted hand starting</u>: A small amount of torque, applied for a long time, could wind up a spring. Then, the spring mechanism could unwind quickly, releasing its stored energy into the generator to torque it. This would be much more comfortable on the muscles of the person starting the generator, than having to apply the required starting energy all at once. In theory, even a quite feeble person could get a generator started, using this method. Spring assisted hand starting is a basic case of leverage, the same principle that allows weak people to lift big heavy items slowly (like cars and RVs), using jacks. But could such a spring be made small enough for an RV? Could such a spring be made cheaply enough to make installation worthwhile? Also, many RVers would be hesitant to accept a mechanical hand starting method for generators as the only method, as many prefer (and expect) the convenience of "push button starting" inside the RV. Still, this starting method may be worthy of consideration, as a backup for generators that start from coach batteries.

3.) <u>Spring assisted automatic starting</u>: This is the same as above, except that the spring would be wound up during a generator's previous run, either by using an electric motor or a mechanical device that uses the generator's rotation to wind up the spring.

4.) Jump starting: This is the method that *must* be used on RVs of today's common design, in which the generator starts from the coach battery, when that coach battery gets too drained. Some motorhomes automatically connect the coach battery to the automotive battery, upon starting the engine. So if the coach battery is too low to start the generator, do the following: start the automotive (for driving) engine, run that engine for a few minutes to put some charge into the coach battery, start the generator (with the automotive engine still running), turn the automotive engine off. Some RVs use battery isolators, which technically do not connect the automotive and coach batteries. This procedure is unlikely to work on such RVs, unless their coach batteries are only mildly discharged. Some RVs, that do parallel the batteries when the engine runs, use wires not thick enough or too corroded to carry sufficient current to successfully jump start a generator. Sometimes systems, that have solenoid switches that connect coach and chassis batteries together when driving, don't work because those switches fail, getting permanently stuck in the "open" positions. In any of these cases, the coach battery needs to be manually jumped from the automotive battery by means of jumper cables. In motorhomes, this means connecting the positive terminals of both batteries to each other. (Use two cables. Connect one to each positive terminal. Then with the engine running, connect the cables to each other away from the batteries. The final connection causes a spark. This spark should be far from the batteries, to prevent battery explosion.) In trailer RVs, the jumping procedure is fundamentally identical to jumping one car from another--- the positive terminals are connected to each other using one cable, then the negative terminals are connected to each other using a second cable. [On negative grounded systems (most common), make the last negative connection at the car's bumper or engine block, so the final spark is far away from the batteries.] (Note: ALWAYS WEAR EYE PROTECTION WHEN FUSSING WITH BATTERIES! BATTERIES CAN EXPLODE!) Although this method can work for emergencies, jump starting, by means of cables, is messy, bothersome and dangerous. It certainly is no solution for starting generators on a regular frequent basis.

5.) Starting from an entirely separate battery: Suppose an RV had an extra battery for no purpose other than starting the generator. Once the generator started, a small charger, powered by AC made by that generator, would replace the electricity depleted by the starting process (as well as any self discharge). One extra completely separate battery, though heavy and occupying some storage space, could easily fit into an RV. It would ensure generator starting, thus the functionality of the AC electrical system, in the event that the DC electrical system shut down.

6.) Starting from the **automobile** battery, instead of the **coach** (house) battery: Almost all automobiles use the following scheme to start their engines: electricity is "borrowed" from a battery to start the engine, then once started, electricity from the alternator made from the engine's energy is "returned" to that battery, resulting in no net battery drain. This is an extremely effective, well tested and established method of ensuring hundreds of successful engine starts, with little risk of drained batteries. (It works well in your car, doesn't it?) Generators in motorhomes could use the same scheme: the generator could start from electricity in the *automobile* battery, then once started, AC electricity made by that generator could power an auxiliary battery charger, which would replace the electricity borrowed to start that generator. This is a very logical solution, since all motorhomes have an automobile battery, which does not drain while parked, because no live circuits are connected to it. (An alternative to two separate battery chargers, would be a single "smart" charger, which would be programmed to charge the automobile battery first, then switch charging operation to the coach battery, once that automobile battery attained full charge.)

Which Possibility is Best?

Motorhomes: I believe the best generator starting solution is (#6) "starting from the automobile battery." This requires no extra battery, since every motorhome has an automobile battery anyway. It also has an added advantage: An automotive battery in a motorhome parked for a very long time might self discharge to the point of being unable to start the motorhome (especially if the battery is old). Starting and running the generator regularly (or even occasionally) would counteract this self discharge and extend battery life.

Detachable trailers: I believe the best generator starting solution is (#5): "starting from an entirely separate battery." Even though space would be sacrificed and weight would be added, this setup would far outperform the current commonly used setup of starting from the coach battery. A battery specifically for starting a generator could be smaller and weigh less than a standard deep cycle battery for general RV house use.

I believe both of the above strategies would result in extremely reliable generator starting. They would virtually ensure that any defect in the DC electrical system wouldn't also shut down the AC electrical system, leaving users with no electricity nor any standing battery charging options at all.

Solution #4 (to: Energy Hogging Appliances):
Use Energy Saving Appliances

Yes, the solution to energy hogging appliances is simple and contained right there within the above title. Some elaboration is necessary to understand why electricity conservation is so important and the dire consequences of disregarding it.

> Tremendous efficiency savings is not an unrealistic goal.
>
> I've estimated that the lights/furnace combination in a typical electricity wasting system uses 20 times the electricity as the lights/furnace combination in an electricity saving system recommended by me. Efficient lights can run on only 10% of the electricity that energy hogging lights need. Efficient non-electric furnaces can run on 0% of the electricity that energy hogging furnaces need. So if furnace electricity equals lighting electricity (reasonable under moderately cold conditions), efficient versions of these two combined can be run on only 5% or $1/20^{th}$ the electricity as their energy hogging counterparts.

If your electrical system runs dead quickly, what should you do?

Suppose you have an RV with traditional energy wasting appliances. Suppose that your appliances use fifteen times as much electricity as energy conserving appliances in an equal sized RV designed by me. What are your possibilities?

1) Buy a better battery: When RV users notice that their systems run down too quickly, they often ask: "What is the best battery and where can I buy it?" Sorry, buying the best battery won't save the day. The "finest available" battery may cost three to five times as much as the "cheapest discount store special," but it won't store three to five times as many Amp hours--- and if your gadgets gobble lots of electricity, you need the Amp hours. There may be other superior qualities to better batteries, so there may be legitimate reasons for buying them. The "finest available" may have a longer shelf life, a greater ability to withstand discharging cycles without damage, or the ability to accept greater charging current (which allows the battery to be charged faster). But that last characteristic, ability to accept greater charging current, is only useful with a powerful charger. If you don't have a powerful charger or you use a method that naturally charges slowly (like solar panels), this characteristic is superfluous. To run your system as long as mine, you would need a battery that stores fifteen times as many Amp hours as mine. No such battery exists--- at least, not at a similar weight, occupying similar space and available at a comparable cost. Such a battery has not yet been invented, might not be invented for a hundred years, and may even be theoretically impossible. The best you might hope for is a battery with twice the storage capacity. But would it be worth it to pay several times the money for a battery that only stores twice the Amp hours? Would it be worth it to spend three times the money on a battery that can be charged twice as fast, yet only stores the same number of Amp hours?

2) Buy more batteries of the same type: Buying and installing more batteries can help, but this solution only goes so far. If your system uses fifteen times the electricity as mine does, you would need fifteen batteries to make your system perform equally as long as mine does with one. That many batteries would add hundreds of pounds of extra weight to your rig, and cause you to kiss lots of valuable storage space and money goodbye. But things get worse. Your system would need a charger that charges with fifteen times the current that mine does (or you would need to charge the batteries fifteen times as often). Kiss more storage space and money goodbye to accommodate this equipment. What would be the energy source that would power such a heavy duty charger? Things get still worse. Complicated systems (such as yours) tend to be less reliable than simple systems, because complicated systems have a lot more components, any of which can break down. You might need 50% more of everything (batteries, chargers, charger energy sources) to compensate for this unreliability to get your system to work as well as mine. Add still more weight and kiss still more space and money goodbye. Of course, adding a couple thousand pounds of equipment to your rig would decrease fuel efficiency when traveling… Kiss more money goodbye…

3) Convert to energy saving appliances: Yes, this requires effort, some money and patience to get used to such a system. But in the end, it is the easiest remedy. You will save thousands of dollars and have a much more reliable system. This will probably cause you to enjoy RVing more. It may even cause you to forge forward with a trip, instead of turning around and heading for "home." The real reward might be how good you will feel, one night, when somebody sees your well lit RV, notices no generator running, and puzzled by your success, knocks on your door asking you: "What do you do for electricity?" (Yes, this has happened to me.)

Neglecting to consider electricity conservation

Neglecting to consider electricity conservation is in my opinion the most frequently made mistake in the RV world. It is a mistake made repeatedly with many different types of appliances. It is a mistake made repeatedly by many different companies that make both RVs and their components. It is a mistake made repeatedly by RV users and customizers. It is a mistake made repeatedly even by well respected and otherwise intelligent authors on the subject of RV electricity and life off the grid (some of whom give wildly extravagant estimates for daily electrical usage requirements). It is also probably the most serious mistake made by the RV world.

Putting Some Numbers on Electricity Conservation

An electricity wasting lighting system might use 25 Amp hours per day
(using the least efficient type of electric light).
An electricity saving lighting system might use 5 Amp hours per day, yet still
have brighter lights (using the most efficient two types of lights).

An electricity wasting furnace might use 24 Amp hours per day.
(based on 8 hours of 3 Amp forced-air furnace fan use).
A non-electric furnace would use 0 Amp hours per day.

Cooking with a microwave oven might use 20 Amp hours per day
(based on a 1200 Watt model, used 10 minutes each day).
Cooking using only gas would use 0 Amp hours per day.

An electricity wasting refrigerator can use 21 Amp hours per day.
(This is only for its circuit board. Its actual fuel source is propane.)
A non-electric propane refrigerator would use 0 Amp hours per day.

An electricity wasting compression air conditioner uses so much power,
it can't be run by a battery; a generator must be used.
An electricity saving evaporative air conditioner could be run by a battery,
but still might use 20 Amp hours per day.
(One should have solar panels to cut down on this electrical load.)

A water system, using a 12 Volt pump, might use 1 to 2 Amp hours per day.
While some electricity can be saved by using less water or using gravity feed,
water system electricity conservation is usually not worth the effort.

More on how these figures were calculated and
how to set up an electricity conserving system
in Chapter 5.

Final thought: Electricity wasting appliances, hooked up to the finest battery available, will not perform nearly as well as electricity saving appliances, hooked up to the cheapest-discount-store-special battery.

<u>Solution #5</u> (to: The Octopus Effect)**:**
<u>A Computer to Monitor the System</u>

The remedy to the RV electrical & appliance system problem of everything being connected to everything else, is to separate such a system into discrete components.

Decentralizing could be done by cutting the wires and installing multiple separate packs of disposable batteries, with each running just one or two lights/appliances. (At one point, when the electrical system in my mother's RV was shutting down daily, I actually considered doing this.) However, doing this would be self defeating, because such a system would be complicated and difficult to use. Bushels of disposable batteries would be needed, or dozens of individual rechargeable batteries (running down at different times) would need to be recharged, both of which would be a logistic nightmare. Running each appliance from its own separate battery pack is no good long term solution, though it could provide good temporary relief from multiple field shutdowns, because it is simple enough to do in the field. Decentralization of a complex system would probably be best done with a computer.

<u>Computers in RVs</u>

About a century ago, nearly all machines were directly controlled by muscles in the human body. Machines did the work; human brains made the decisions. This is not true anymore. Today, tiny chips of silicon now make many of the decisions that human brains used to make. These computer chips have found their way into nearly every device known to man. The day may come when a person won't even be able to buy a toaster without a computer chip in it. This revolution occurred because computers have gotten smaller, faster, cheaper and capable of storing more information, at a staggering rate.

> Did you know: The average 2011 new car has more computer capability than the Apollo spacecrafts that went to the moon!

Despite these advances in computer technology, many of today's RVs have very little computerization *as a whole*. Usually RVs only have simple gages and fuses to analyze and protect their systems. Gages can tell users how full or empty the freshwater, wastewater, propane tanks or batteries are. (A voltmeter is usually used to estimate the charge state of batteries.) Fuses can prevent fires and melting of wires, by "blowing" (melting then disconnecting the wire), if serious short circuits occur. For water and propane tanks, RV owner's brains, aided by "full-empty" gages, are probably all that is needed to make good decisions as to when to fill or empty tanks. For RV electrical systems, owner's brains and simple gages and fuses are probably not enough to understand and protect these systems, because these systems are just too complex.

The problems caused by both interconnectedness and complication could be greatly reduced by adding a computer to the system. A computer could simplify decision making, quickly identify faulty components, then disconnect those faulty components, keeping the "octopus" alive. Let's consider some specifics.

A Computer For an RV Electrical System Could…

1) Monitor Electron Movements. By taking rapid repeated measurements of current, and storing that information in memory, a computer could keep track of Amp hour movements, both into and out of the battery. (A device that keeps track of Amp hour movements is called an Amp hour meter.) By monitoring individual wiring routes of each fuse/breaker line, it could find approximate locations of problems. It could also tell users what time of day any specific problem occurs, allowing users to correlate that problem to a specific appliance or behavior.

2) Discern the Charge State of the Battery(s). Usually, a battery's charge state is estimated by measuring that battery's voltage, which gradually drops, as the battery discharges. However this is a poor indication of a battery's charge, because it takes time for charged or discharged particles to migrate deep within the battery after charge or discharge. (False readings are common shortly after charge or discharge.) Keeping track of the Amp hours going into and out of a battery, as well as knowing how many Amp hours that battery stores, is a much more accurate way of knowing how "full or empty" a battery is.

3) Detect Problems in the System. A computer could correctly diagnose many problems, that might go unnoticed by anyone who is not an electrical expert. Examples: A) Severe short circuits result in blown fuses or tripped breakers, which protect faulty appliances, faulty wiring routes and batteries. However, short circuits, not severe enough to blow fuses, usually spell disaster. When such a short occurs, current simply flows until the battery goes completely dead--- at which point, every other electrical appliance goes dead too. Usually there aren't any symptoms for awhile, so the owners don't realize there is a problem until it's too late. A computer, could detect unusual phantom currents, such as large current spikes or unexplained current flowing through specific wiring routes for long periods of time. B) A computer could tell if it's time to replace a battery (bank). If a battery's voltage drops too substantially when not very many Amp hours get drained, then the battery no longer stores enough charge. It is too old, been used too much or damaged from being left too discharged for long time periods. C) A computer could determine other battery problems, such as internal intermittent shorts or opens, too much self discharge or inability to hold a charge.

4) Notify RV Owners of Problems and Suggest Solutions. A computer screen, coupled with some distress signal (beeping sound or flashing light), could display messages explaining problems in plain English (or another language). (Examples: "Old battery in need of replacement--- it only stores 40% of the Amp hours that it should." "Sustained excess current flow in wiring route #4--- possible short." "Between 4:00 PM and 5:00 PM, you used quite a bit of electricity. Your batteries drained 10%.") The computer could then suggest solutions. (Examples: "Suggest disconnecting wiring route #4, which possibly has a short." "That gadget you ran from your power inverter at 4:35 PM to 4:55 PM used lots of electricity. Suggest using an alternate gadget .") Locating and correcting small problems before they become big problems could prevent entire systems from shutting down.

5) Help Technicians Quickly Diagnose and Locate Problems. Technicians could fix problems too complex for owners to handle in considerably less time, using the information stored in the computer. Less time in the shop means less inconvenience and lower repair bills for the owners.

6) Obtain Information From RV Users. In order for the computer to make valid decisions, it needs to take into account RVers individual usage patterns and projected future conditions. It could get this information by simply asking the users. Examples: Q#1: "Are you going to be using your furnace tonight? (Y/N.)" (Some furnaces use electricity for their fans.) If yes then, "How cold do you expect it to get? (Select approximate lowest temperature.)" Q#2: "Do you anticipate using your air conditioner tomorrow? (Y/N.)" If yes, "For how many hours?" (When a compression air conditioner runs, the generator also must run. Generators also provide electricity to charge batteries.) Q#3 "Will you be plugging into municipal electricity tomorrow? (Y/N.)" If yes, "For how many hours?" (Plugging in charges batteries.) Q#4: "Will you be taking your motorhome for a drive tomorrow? (Y/N.)" If yes, "Approximately how many miles (or hours) do you expect to drive?" (Driving charges batteries.) Q#5: "Will you be parked at the same location tomorrow as today? (Y/N.) Under similar weather conditions? (Y/N.)" (Future usages of furnace, air conditioner and energy input from solar panels could be estimated for upcoming days, if location, weather and RV usage patterns are similar.) Q#6 (For RVs with solar panels): "Will you be parked in the shade or the open sky tomorrow? (Choose number: 0 = total shade, 10 = completely open sky.)" (Solar panels make electricity from sunshine.) If exposed to the open sky, "What kind of light conditions do you anticipate?" (Sunny/ Mostly Sunny/ Light Overcast/ Heavy Overcast.) Q#7: "Do you anticipate using your generator without running the air conditioner tomorrow? (Y/N.)" If yes, "For how long?"

7) Help RV Owners Make Decisions. Based on internal measurements, projected usages and RV owners input, the computer could make recommendations: "You used your power inverter for quite awhile yesterday. Don't use the power inverter unless you have something plugged into it." "You used quite a bit of electricity this evening. If the temperature isn't too high, try using your gas lights." or "You used 16 Amp hours total yesterday (midnight to midnight)." The real potency of the computer would be revealed when giving advise on battery charging, since this subject stumps many people. Examples: #1: "Since you will be taking a drive tomorrow, and your batteries are about 70% charged, there is no need to run your generator for battery charging." #2: "Run your generator for 45 minutes to charge your batteries." #3: "It's past 10:00 PM and you've got your generator on. You really shouldn't be running a generator at this hour if you have any camping neighbors." #4: "You should run your thermoelectric generator for about 12 hours to charge your batteries. Turn it on? (Y/N.)" (Thermoelectric generators slowly charge batteries with no moving parts, making no noise. Don't confuse them with internal combustion engine generators. See Chapter 3.) Proper RV battery charging involves reading of meters, adjustments of currents and times, and awareness as to when electricity for charging is to be available. It also involves knowledge of electricity, which many RVers do not possess. Despite many variables, the decisions for battery charging are straightforward--- they involve only simple measurements and math. Electrical engineers know how to charge batteries. If they programmed the right parameters into computers, then these computers could make the right decisions, allowing inexperienced RVers to charge batteries too.

8) Do Wonders For Making Complicated RV Electrical Systems User Friendly.

> Most people are used to getting endless amounts of electricity anytime. They rarely think about where it comes from or how much they use. Is it realistic to expect such people to move into an RV and understand consumption rates, battery idiosyncrasies and charging, with absolutely no training?

Conclusion:
(to: Proposed Solutions to Defects)

Many first-time RV buyers realize that their electrical systems don't work well, soon after their initial purchase. They can't get much good advice from other RV users, because all RVs seem to be this way. They can't get much help from RV "experts," because the design mistakes articulated in this chapter are so common, that people in the RV business have begun to believe that such systems are actually OK. People often end up just accepting these systems, but they shouldn't--- I've studied enough electricity to know that these systems are not OK. RV electrical systems ***could*** be designed to work, they just ***aren't***.

Chapter Conclusion: The Biggest Defect of Them All

When I look at RVs, I see many specific defects. I believe that the biggest defects may not be in the RV wiring or hardware itself (which are the tip of the ice burg), the biggest defects may lie within the *attitudes* of those who design and make RVs. (Faulty attitudes are the root cause of all the specific problems.) Ignorance of electricity, lack of concern as to whether or not things work in the field, lack of concern for the people who actually use RVs, are major attitude defects. But in my opinion, the granddaddy of all attitude defects is: *RV designers don't learn from their mistakes, they simply repeat them.* The electrical defects articulated in this chapter, were standard in RVs built 30 years ago, and are still being built into RVs today. And I don't believe the RV industry has plans on fixing or even addressing any of them.

> Idiots make mistakes. Smart people also make mistakes. The difference between these two groups is: *smart people learn from their mistakes, whereas idiots simply repeat their mistakes.*

Chapter 3
Making Electricity & Charging Batteries

For most of the time that I was writing this chapter, I kept debating whether I should name it: "Making Electricity" or "Charging Batteries." Ultimately I decided to include both in the name, because for small to medium off grid systems, these two subjects are almost inseparable. Whenever one discusses making electricity, the subject of battery charging almost always comes up for the following reason: Whenever one makes electricity, there are only two useful things that one can do with it: use it immediately or store it for later. If one stores it for later, one almost always does this by charging batteries. Whenever one discusses battery charging, the subject of making electricity almost always comes up for the following reason: Battery chargers need to get electricity from somewhere in order to do their job. That electricity must be made somehow (either by the battery chargers themselves or by other electricity-making devices). Battery chargers won't run off your good looks or charming personality nor will they charge batteries just to be nice to you.

The Need for Electricity Storage

The need for electricity storage is fundamental because very rarely do the outputs of electricity sources match the required inputs of the appliances they run. For example, generators make lots of electricity but should be left on only for short durations. Many gadgets use small amounts of electricity intermittently but should have that electricity available whenever it is needed (ex: lights), or use even smaller amounts of electricity continuously (ex: circuit boards in refrigerators). In order to avoid massive amounts of waste and noise, generator outputs should be stored in batteries, if used to run low energy consuming appliances. Solar panels have the opposite characteristics; they make small amounts of electricity over long durations. Many gadgets (household vacuum cleaners, microwave ovens) use lots of electricity but only for short durations. In order to power these appliances from solar, panel outputs must be stored in batteries (running them directly from the panels is next to impossible). Also, electrical outputs from many sources often occur at different times from the needed inputs of the appliances they run. For example, a battery may charge while driving but its electricity may be needed while parked. Another example: solar panels produce outputs in bright sunshine during the day, when artificial lights aren't needed, but electric lights are most needed at night, when solar panels produce zero output. Even if the solar panel outputs exactly equaled the required inputs of lights, the panels couldn't run the lights directly. Storing electricity is thus essential.

Harvesting Electricity from Nature

Where Does Electricity Come From?

Consider an analogy. Suppose your RV is parked close to a lake. You have a vinyl floor and you decide you want to wash it. You don't have all that much freshwater in your tank. The nearest clean freshwater source is miles away. You don't feel like driving a really long distance to get some. What might you do? You might grab a bucket, take it to the lake, fill that bucket, bring the water back to your RV, then use it to do your floor washing. If someone asked you where the water came from, you'd probably just say, "from the lake," without analyzing things too deeply. If someone asked you the same question, demanding a more philosophical answer, you might say: "Water already exists somewhere in nature. I don't make it. I just find it, gather it, store it for awhile, then use it later at a time and in a manner as I see fit." Wouldn't it be nice if you could do the same thing with electricity? Imagine just grabbing a "bucket," collecting some electrons, bringing them to your RV, storing them, then using them later to run your TV or something like that! What an appealing fantasy! Too bad it's impossible. Or is it? So where does electricity come from anyway? (Don't be a smarty pants and say "the electric company!") The answer is virtually identical to the answer to the question "Where does water come from?". Electricity comes from energy that already exists somewhere in nature. We don't make this energy. We just kind of find it, collect it, store it, then release and use it later, at a time and in a manner as we see fit. Sure, far more sophisticated costly equipment is needed for collecting and storing electricity than water--- that plastic bucket, bought at some thrift store for $1 or garbage picked for free, simply won't do. And the "storing" step can sometimes be skipped (especially with municipal electricity). Despite these differences, collecting electricity isn't all that fundamentally different from collecting water in a plastic bucket.

How Can Electricity Be Made?

Electricity is made by harvesting flowing energy found in nature. Energy can flow by itself or accompanied by flowing matter. Energy can flow naturally or artificially. The variety of energy-flowing mediums, from which electricity can be made, include flowing water, flowing wind, flowing heat, flowing radiation (sunlight) and flowing chemical reactions. Naturally occurring energy flows (*noun*) need only be located and intercepted with proper equipment to make electricity. Artificial energy flows (*noun*) need to be made by man and intercepted to make electricity. With natural energy flows (*noun*), there is usually some guesswork as to where, when and how strongly the energy flows (*verb*), as nature is somewhat unpredictable. With artificial energy flows (*noun*), man carefully controls when and how strongly the energy flows (*verb*). Artificial energy flows (*noun*) occur when humans do something to cause the release of potential energy somehow locked up in material. Potential energy can be locked up in material by chemical, nuclear, gravitational or by other methods.

Coal or nuclear driven steam power plants are good examples of artificial heat flows (*noun*), used to make electricity. Potential energy, locked up in coal and the oxygen that burns it, or deep within the nuclei of fissionable atoms, becomes heat. The heat then flows to colder areas, boiling water to make steam along the way, driving turbines that make electricity. A dam in a river is a good example of potential energy artificially stored by gravity. Gravity forces water on the high end of the dam to flow through pipes to the low end of the dam, driving turbines that produce electricity.

From What Could Electricity Be Produced Locally at RVs?

With the right equipment, many mediums of flowing energy could potentially produce usable practical electricity in RVs at reasonable prices. These include flowing sunlight, flowing wind, flowing heat and chemical reactions (both inside internal combustion engines and absent of internal combustion engines). Out of all the mediums that carry harvestable flowing energy, only water has no realistic on-site electricity production capability for RVs, as too few campsites have creeks/rivers close enough to make water turbines practical.

Battery Charging / Electricity Making Strategies

I.) Alternators (Charge As You Drive)

Close your eyes for a moment and imagine the perfect RV battery charger. Don't concentrate so much on how it looks, rather focus on how it works. Got an image? Now before opening your eyes, imagine that you are at an RV exposition hearing a salesperson pitching his version of the ideal RV battery charger: "It's free, costs zero to maintain, you don't even need to run it, and it makes lots of electricity reliably for only pennies." How closely does this salesperson's description match your best-of-all-worlds fantasy? Maybe not at all--- I've described something that seems too good to be true, therefore can't exist. Furthermore, how could a salesperson make money on something that costs nothing? But could such a charger exist? Maybe it's not too good to be true. In a way, I am that salesperson pitching just such a charger. I can do this because I make my money selling books, not battery chargers.

In my opinion, the ideal "too-good-to-be true" RV battery charger is the alternator of your vehicle. In a way, it's free because *you've already bought it and you already have it for another purpose*. (If you tow a trailer, you've got one alternator. If you have a motorhome that tows a spare vehicle, you've got two.) In a way, it costs zero to maintain because *if you need to replace it, you probably would have needed to do so anyway*. In a way, you don't even need to turn it on and run it because *you already run it for another purpose (making electricity for your car while driving)*. And it reliably makes lots of electricity very cheaply.

Safety Warning: The following seven illustrations are meant only to show general battery locations; not to imply batteries being stored indoors in unventilated compartments. Storing wet lead-acid batteries inside indoor RV compartments (not ventilated to the outdoors) is dangerous. These batteries make and expel hydrogen gas, especially during charging. Hydrogen is very flammable and can ignite from sparks or flame and cause an explosion. If you do store batteries indoors, make special compartments that are sealed from the inside and ventilate to the outside (or use special battery boxes that do the same). Dry (sealed) batteries are exempt from this requirement, as they are designed not to outgas (but always check with the manufacturer to see if it is safe to store them indoors). If you do choose to store wet lead-acid batteries indoors (which may be like riding in a car without a seatbelt, or riding a bike or horse without a helmet), NEVER store them in compartments that have appliances that spark-ignite or have burning pilot lights (such as furnaces or hot water heaters), and never smoke or use candles or oil lamps in the vicinity of wet lead-acid batteries. YOU ARE RESPONSIBLE FOR YOUR OWN SAFETY.

<u>Charging RV Batteries With Alternators</u> (my recommendations):

Let's start with a tow vehicle and detachable RV trailer combination. If you've got a motorhome that tows a spare vehicle, read on, this applies to you too. (Imagine that the towable trailer is your motorhome and the van is the small car that you tow.) Normally, there are two batteries in the system--- one in the "house" (trailer) and one in the "car" (tow vehicle), as shown in *figure 1A*.

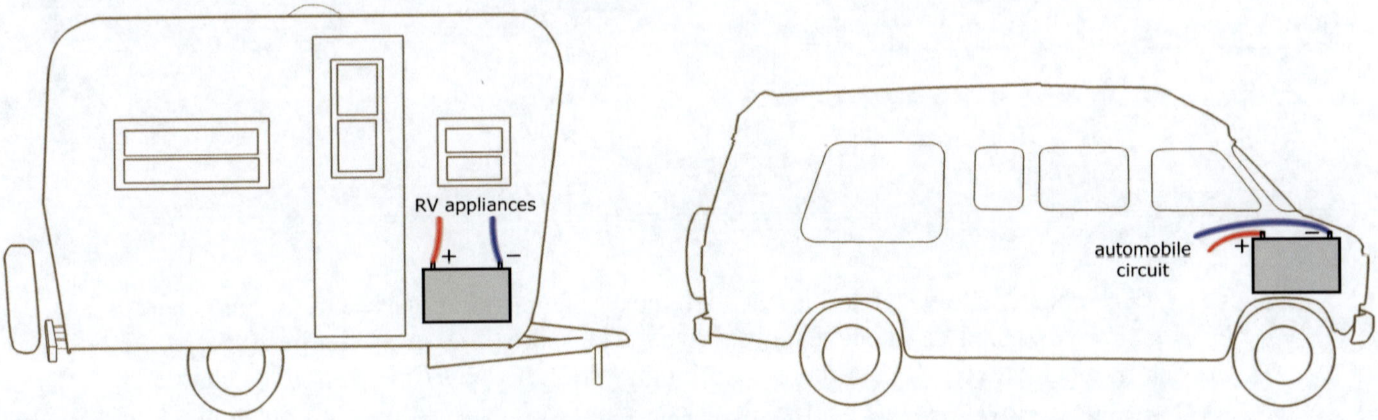

figure 1A
(normal setup)

The problem with the two battery system is: *while detached, the coach (house) battery can't be charged by the alternator while the tow vehicle gets driven, because that battery stays home with the trailer while the car drives away somewhere else.* The only charging that can take place occurs when the tow vehicle and trailer are attached, driven together and their batteries are paralleled (which occurs when relocating to new campsites far away). If detached and parked for long time periods, this type of charging is likely to occur too infrequently to be effective.

Now, let's add a third battery to the system, as shown in *figure 1B*. Battery (A) is the chassis (automobile) battery, located in the car. Battery (C) is the coach (house) battery, located in the trailer. Battery (B), recently added, which I will call the "spare" battery, is located in the car. The coach battery (C) runs all the household electrical appliances in the RV. The spare battery (B) connects (parallels) to the automobile's chassis battery (A) and charging system every time that vehicle takes a ride and disconnects when parked. Connection/disconnection can be done automatically with a battery isolator or solenoid switch, or it can be done manually with a manual switch or just by connecting/disconnecting wires (covered in Chapter 1). Disconnecting when parked protects the chassis battery from being drained too low from the RV's appliances, a condition that would render the automobile unable to start.

[Figures 1B-G Notes: 1) To simplify the drawings, the chassis battery connection to the automobile circuit is not shown. 2) The connection between chassis and spare batteries' negative terminals is shown as a wire. This connection is usually made through the vehicle frame by grounding each to that frame.]

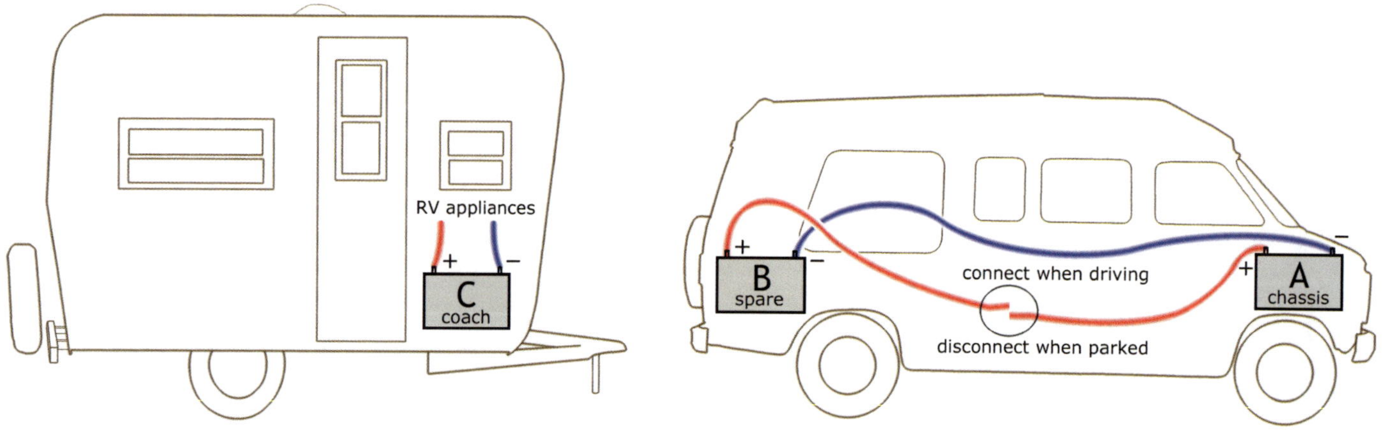

figure 1B
(3 battery setup)

The coach battery (C) drains (through the appliances) but there is nothing to charge it, so as time passes, its charge level goes down. The spare (B) battery charges (every time the automobile takes a ride) but there is nothing to drain it, so as time passes, its charge level goes up. How does one get the electricity from the spare battery into the house so it can run things? There are actually three ways of doing this. I've tried them all. They all work well.

Getting the Electricity From the "Car" Into the "House" ***(3 methods)***

1) Battery Swapping:

Perhaps the most obvious method of getting electricity from the "car" into the "house" is to remove both the spare (B) and the coach (C) batteries, exchange them, then reinstall each in the position formerly occupied by the other. At the "house," the depleted battery gets exchanged for a fully charged one. This is kind of similar to what people do when they exchange empty propane tanks for full ones, except unlike propane tank exchanging, battery exchanging is free and doesn't require one to drive anyplace special, since both batteries are present when users are "home at the ranch" with both their trailers and automobiles. What a great deal! At the "car," the fully charged spare battery gets exchanged for a depleted one. But not to worry--- this battery doesn't power anything. Over the next few rides, its charge level goes up to completion. If all goes well, by the time the new coach battery runs down, necessitating the next swap, the spare battery's charge level should have risen to 100% and probably will have been there for awhile.

Advantages: 1) Batteries don't need to be swapped very often with energy conserving systems. In the early days when I used this strategy, I only needed to swap batteries once per week. 2) Battery swapping doesn't take much time. I did this task many times. It took me about fifteen minutes, including the time it takes to take out and put away the tools.

Disadvantage: Batteries are heavy. Some people might have difficulty repeatedly lifting batteries, which weigh about 40 pounds, or they may just find the task bothersome.

2) **<u>Battery Paralleling</u>:**

In this strategy, the need to remove and reinstall batteries is eliminated by repeatedly electrically connecting and disconnecting the spare (B) and coach (C) batteries.

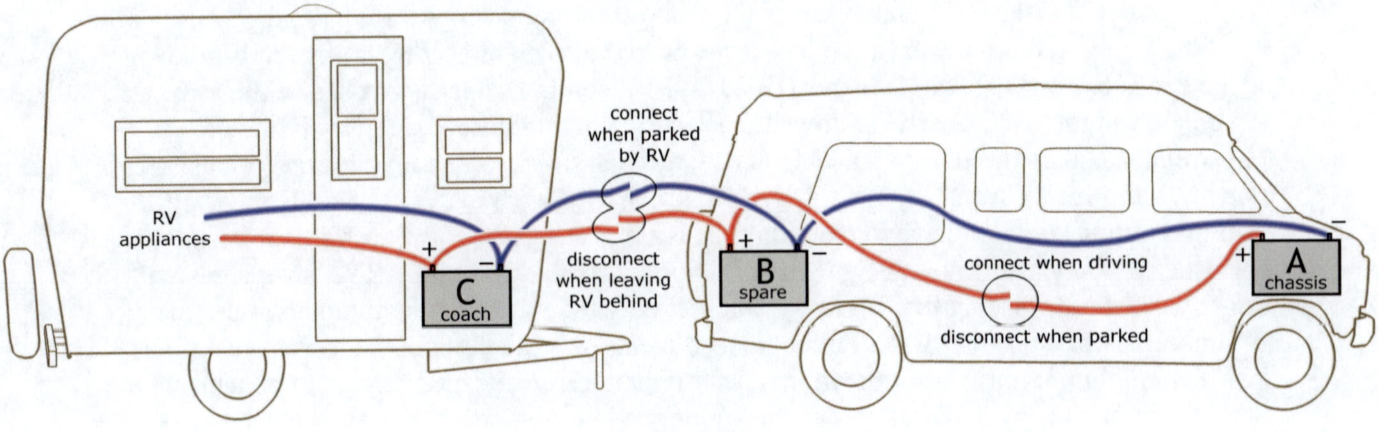

figure 1C
(battery paralleling)

How to parallel batteries: Use wires to connect the like terminals of the coach and spare batteries to each other ("+" to "+" and "–" to "–"). In theory, only two long wires are needed. But use four as shown in *figure 1C*. Having four wires instead of two makes the connection/disconnection process much easier, because batteries may be located in areas where access is difficult. Also, having four wires instead of two makes the connection/disconnection process much safer. This is because the final connection makes a spark, due to a voltage difference between the two batteries. That spark itself is not harmful (it is normal), but if it is made at a battery, it can ignite hydrogen gas released by the battery during charging, causing that battery to EXPLODE and splash SULFURIC ACID at anyone who is near it (which might be you), causing chemical burns and possibly BLINDNESS. With only two wires, the final connection (and that spark) must inevitably be made at one of the batteries, which is extremely dangerous. With four wires, the final connection (and that spark) is made outdoors between the two vehicles far away from both batteries, where the risk of explosion is zero. Therefore, ALWAYS use four wires instead of two and NEVER make the final connection at one of the battery terminals.

What happens when two batteries get paralleled? 1) <u>They electrically unite</u> and essentially become one battery. This new "combo-battery" has the storage capacity of the combined storage capacities of both its component batteries, which if equal, is twice as much. [Example: If two 75 Amp hour batteries get paralleled, they become a 150 Amp hour battery.] 2) <u>Their charge levels equalize</u>. [Example: If when first connected, the spare battery is 100% charged and the coach battery is 70% charged (and of equal storage capacity), current runs through the wires causing the spare battery to donate 15% of its charge to the coach battery which accepts that 15% of charge. They equalize at 85% charge level.] There is however a small loss of energy due to heat in the wires, but this may not be significant. [I haven't been able to notice this efficiency loss.]

When should RV users connect and disconnect? 1) The best time to connect is each night, after the tow vehicle is "home" by the trailer and no longer needs to be driven. Leave the two batteries connected throughout the night and into the next morning while using your electrical appliances. Disconnect only when the tow vehicle is ready to take its next ride. 2) If the tow vehicle leaves the campsite several times during the day, it is advisable to reconnect every time it returns, as frequent reconnecting really tends to keep the coach battery's charge level near 100%. 3) When the trailer is hitched to the tow vehicle and the two are driving down the road together, all three batteries should be paralleled and connected to the vehicle's charging system, so that all batteries can charge.

Important Note: Do Not Drive the Car Away With the Wires Still Connected!!! It may sound stupid to have to mention this, but it is surprising how easy it is to make this mistake. If the car drives away with the wires still connected to the RV, those wires will get yanked out. Doing this could damage not only the wires but anything else near their yanking path. Yanking wires out could also create short circuits in the tow/spare vehicle which might not only shut it down, but destroy batteries or its alternator. Such a mistake could be catastrophic, therefore must never be made even once. At best, such a mistake would cause a lot a crying and/or swearing and necessitate a repair lasting at least several hours.

Here's a trick that I use to prevent such a disaster: *Throw a towel over the steering wheel when you connect the wires.* This reminds you to never drive the car away until the wires get disconnected. To be successful at this "towel" trick, you (and all other drivers in your party) must train yourself to do three things: 1) NEVER drive the car away with the towel still there, 2) NEVER remove the towel until AFTER you are absolutely sure all wires are disconnected and 3) ALWAYS put the towel on the steering wheel BEFORE reconnecting the wires after parking. You don't necessarily have to use a towel as your reminder, you could use something else, such as a box of tissues on the dashboard in front of the steering wheel or some object placed on the driver's seat. Whatever object is used, it should be large and conspicuous enough to not be ignored, it should have no other use that might cause it to be removed by anyone (use an *extra* towel) and it should be always kept in a handy location in the car.

> How well does this "towel over the steering wheel" trick work? Really well. Before devising it, I almost drove away several times with the wires still connected. Fortunately I caught myself before actually doing it. I knew that sooner or later I would make this mistake, so I had to come up with something. After devising and implementing this "towel" trick, I have never driven away with the wires still connected nor even come close to doing so.

Technical Points:

Improper polarization must be avoided! Wires should be clearly color coded or routed through an electrical connecter to make it impossible to accidentally connect the "+" to the "–" and the "–" to the "+." Doing this would cause a current surge upon the final connection, which if left for awhile, would drain the batteries (one would drain completely), probably ruin them, and heat up the wires to an extent that could cause melting or a fire. If such an improper connection is made: A) a great big spark occurs at the final wire connection, B) wires get hot (or warm) and C) appliances simply don't work (because the two battery voltages work against and cancel each other, thus making the overall system voltage very low). If these symptoms occur, disconnect the wires immediately and reconnect them properly.

Thick gauge wire should be used. I recommend at least 10 gauge wire. Some people insist on much thicker wires, but I've had good luck with 10 gauge.

Use some kind of electrical connector. Using an electrical connector, rather than just connecting/disconnecting wires makes the connecting/disconnecting process much easier and quicker. Do not use aluminum trailer connectors. They are wimpy. They have thin wires and the aluminum corrodes, resisting electric current.

Two conditions must be met, in order for this strategy to work: 1) the coach battery must not get too deeply discharged when separated from the spare car battery and 2) on average, the tow vehicle's alternator must be able to add charge to the spare battery faster than the house depletes that charge. For small or medium scale energy conserving RV systems [without air conditioners], these two conditions are not hard to meet.

Deep cycle batteries should be selected for both the spare and coach batteries. Deep cycle batteries are less prone to damage from being deeply discharged [and left that way for awhile] than regular automotive batteries.

Advantages/Disadvantages:

Advantages: 1) No moving of heavy batteries is necessary as is needed with the battery swapping strategy. 2) The coach battery does not drain substantially like it does with the battery swapping strategy. This is because with battery swapping, the coach battery sits for awhile as it gets drained, whereas with battery paralleling, charging takes place often. Keeping batteries at high charge levels as often as possible and for prolonged time spans extends battery life. 3) While connected, the house runs from two batteries rather than one. Two identical parallel batteries store twice as many Amp hours as one. Thus, when connected, there is more electricity available to the house.

Disadvantages: 1) Connecting/disconnecting should take place often [at least once per day]. This is unlike battery swapping, which might take place once per week. 2) The coach battery does not attain full charge with this method. Charging to completion extends battery life. However, when the trailer gets hitched up and driven with the tow vehicle, all batteries get connected to the system and attain full charge, so this may be a moot point. Batteries can age/wear more evenly if the spare and coach battery get swapped once in a while, so occasional swapping may still be a good idea, even though not absolutely necessary. Finally, if a solar panel is installed on the roof of the trailer, the coach battery *can* attain full charge sometimes.

Motorhomes That Tow Cars: These charge-as-you-drive strategies (#2 and #3) also work with motorhome/car combinations. Simply add a spare battery to the car and wire it exactly the same as if it were a tow vehicle (spare battery connects to charging system when driving, disconnects when parked). In the evening, either parallel this spare battery to the motorhome coach battery, exactly as described above, or bypass the coach battery, as described shortly. (Of course, throw a towel over the car's steering wheel.)

Technical points for motorhomes towing cars: 1) A small car with a small alternator may not be able to charge enough to keep pace with consumption, especially if the motorhome is big and has big electrical needs. If this is the case, this strategy can still be used to make *some* electricity, just not as the *only* charging strategy. When connecting the spare battery, the motorhome coach battery should be disconnected from its appliance system [and not paralleled with the spare battery]. Things can be kept running longer by sometimes running appliances from the spare battery and sometimes running them from the motorhome's coach battery. Remember: 2) The coach battery charges when the motorhome drives. Therefore, if the system is weak: 3) Parallel the spare battery in the car with the motorhome coach battery and charging system when the motorhome take a drive. This will make the overall charging system a little stronger.

3) Plugging the "House" Into the "Car":

In this strategy, the RV system of appliances gets plugged directly into the spare battery of the tow/spare vehicle. There are three variations:

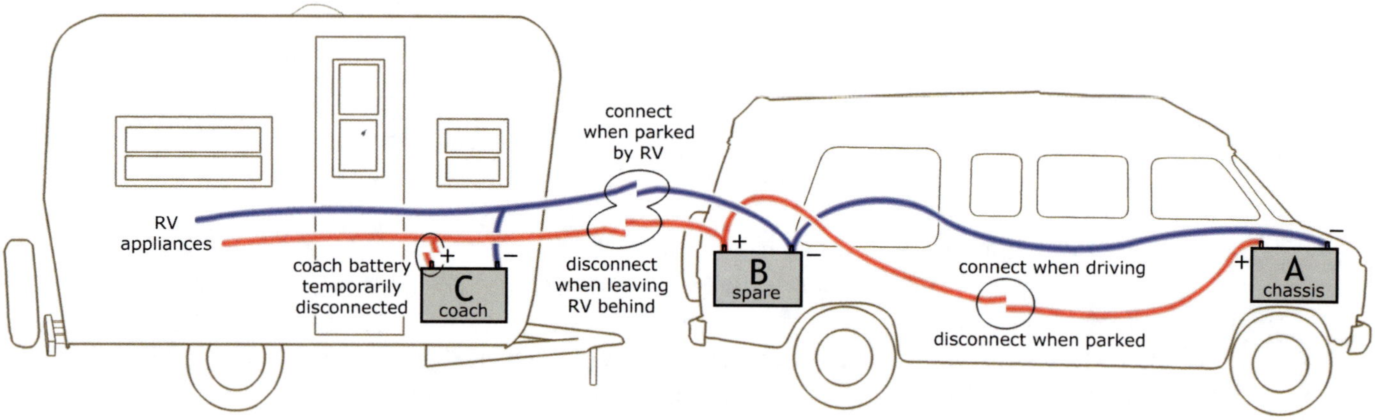

figure 1D
(bypassed coach battery)

Variation #1: *Bypassing the coach battery:* In the setup of *figure 1D*, the coach battery gets disconnected and bypassed whenever the tow/spare vehicle is parked near it, and all the RV appliances get plugged into and powered by the spare battery in that tow/spare vehicle. While the tow/spare vehicle takes a drive and is away from the camp, the coach battery gets reconnected to the RV appliances. While disconnected and bypassed, the coach battery doesn't get drained. If only connected and drained occasionally, it might last quite awhile before running down. This could protect and preserve its electricity reserve, leaving it available for backup power. In both towed RV trailers and motorhomes, the coach battery still does charge when the RV gets driven down the road, so if the battery lasts the duration of its parking period, and charges to completion during an RV drive, this might be a successful approach. Since the coach battery only receives occasional charging by driving, RVers who use this strategy may want to consider having an additional method of charging it, presumably a solar panel.

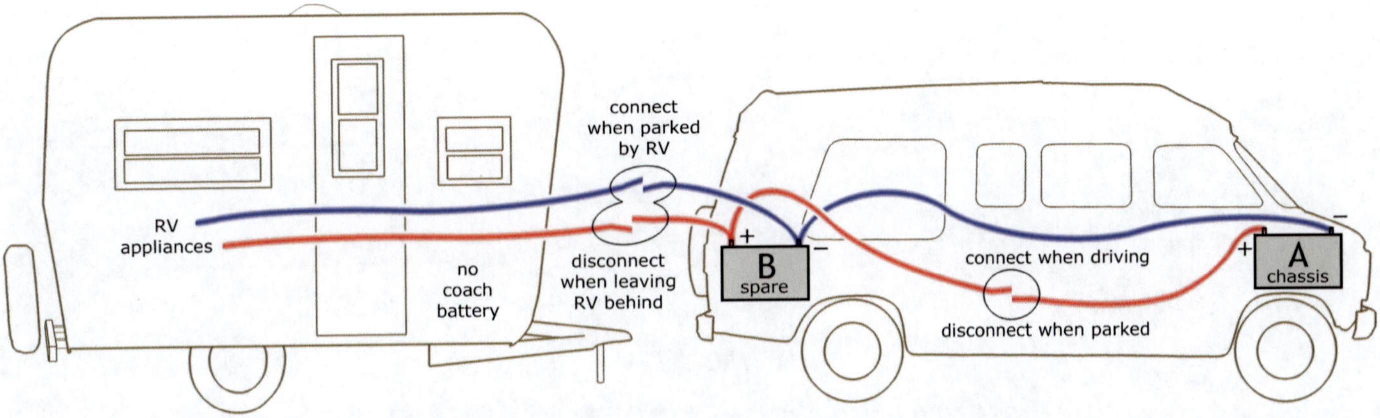

figure 1E
(all batteries in car)

Variation #2: *All batteries are housed in the tow vehicle:* In the above setup of *figure 1E*, there is no coach battery; *all* the batteries that run the RV appliances are permanently housed in the tow vehicle. The above illustration shows only one "spare" battery [which doubles as an externally located coach battery], but it is advisable to use two. ***Advantages:*** 1) This setup makes it easier to recharge deeply discharged batteries. With battery paralleling, a deeply discharged coach battery can only be charged to halfway between its low charge point [when first paralleled] and 100% [if the coach and spare batteries have equal storage capacity]. [Example: A coach battery at 40% full charge can only rise to 70% full charge when paralleled with a fully charged spare battery.] Attaining a charge state near 100% by paralleling batteries requires multiple connection/disconnection cycles and light appliance use, a process that could take days. 2) This setup allows all batteries to be charged back up to 100% when rides are taken. Keeping the batteries as close to 100% charged for as much of the time as possible increases the life spans of those batteries. ***Disadvantages:*** 1) Electricity is only available when the tow vehicle is parked near the trailer and connected. Therefore this strategy only works for RV users who travel alone. If at least one human companion is brought along, there will be times when one person takes the car for a ride, leaving the other(s) home with no electricity at all. 2) This strategy only works if there are no appliances in the RV that need constant electricity. Many propane refrigerators also need some electricity to run their circuit boards. These refrigerators shut off immediately upon disconnection from batteries, so disconnecting them for sustained time lengths renders them useless. 3) If the car needs to spend some time at a repair shop, the RV users stay home with no electricity at all. [Yes, this has happened to me.]

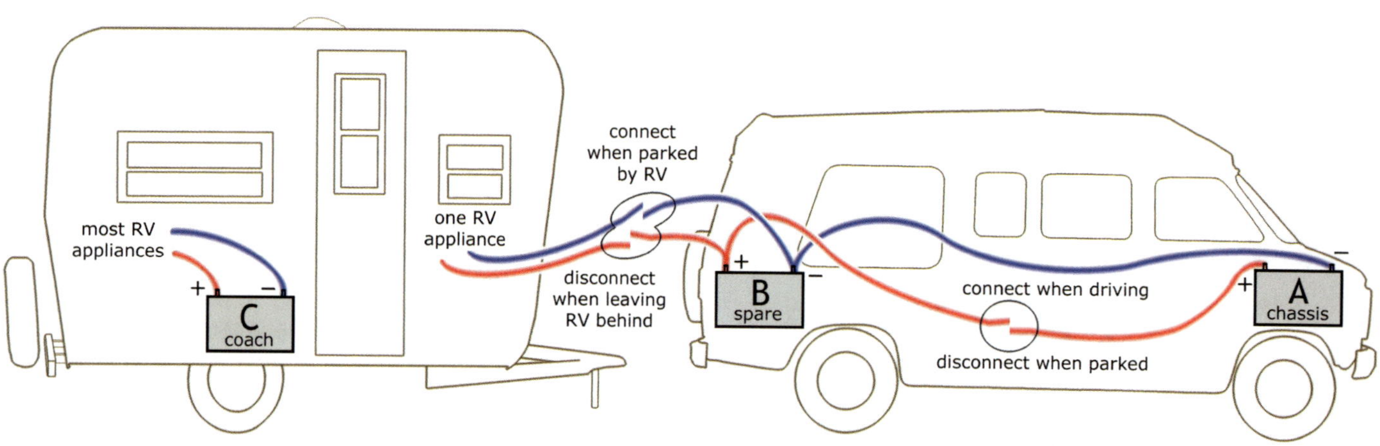

figure 1F
(one appliance run from spare battery)

Variation #3: *Plugging only one or two appliances into the spare battery:* Disconnecting only one or two components from the coach battery and connecting and powering them from the spare battery in the tow/spare vehicle (*figures 1F and 1G*) is a plan worthy of consideration, especially for components that draw lots of electricity. Forced-air furnaces are good candidates for such a strategy, as they are notorious for draining coach batteries, and when they do, they shut down everything else with them. If you run an electricity-hog (like a forced-air furnace) from the spare battery in the tow/spare vehicle, while running everything else from the coach battery, then the rest of your RV electrical & appliance system will be protected from running down due to that electricity-hog. Also, many 120 Volts AC appliances are moderately heavy energy users. You may want to keep your power inverter in the tow/spare vehicle, permanently connected to the spare battery. You may want to refrain from paralleling the spare battery with the coach battery during heavy 120 Volts AC appliance use, in order to preserve electricity in your coach battery and general RV system. It is advisable to parallel a second set of wires between the rogue appliance and the coach battery containing an on/off switch (*figure 1G*). This allows the RVers the choice of powering that appliance from either the coach or spare battery. [Turning switch on while car is away runs the appliance from the coach battery. Turning the switch off and connecting to the car runs the appliance from the spare battery.]

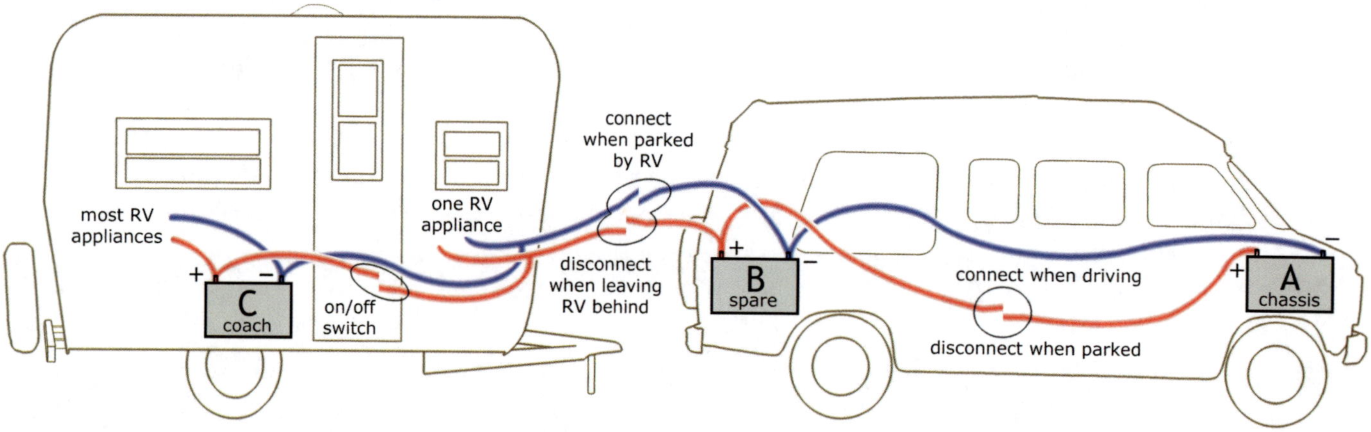

figure 1G
(one appliance run from choice of batteries)

Universal Advantages/Disadvantages/Considerations of three battery systems:

Advantages: 1) These systems can provide electricity indefinitely for both trailer RVs and motorhomes, without ever needing to plug into municipal electricity nor using [or even having] a generator. 2) These systems are extremely cheap to install. All that is needed is wire, terminals, a cheap [$6] switch, a connector [or one could just connect and disconnect wires manually], some hardware to mount the spare battery and the battery itself [maybe $100]. However, one may choose to get a bit more elaborate and use a battery isolator or solenoid switch to connect the spare battery to the automotive charging system. These are covered in Chapter 1, so I won't repeat them here. 3) These systems are fundamentally simple. One does not need advanced electrical training or knowledge to wire and install a three battery system.

Disadvantages: None, as far as I can discern. But-- Do Alternators Really Make Good Battery Chargers? I have read a comment by an author of another well respected book on RV electricity stating that alternators make lousy battery chargers. This seems to contradict my own opinion and experiences. I don't want to bash my book-writing colleagues; there may be a reason for this discrepancy. I think this author may have a habit of draining the batteries too substantially and too quickly, probably due to the use of energy-hogging appliances. True, a very deeply discharged battery might take a long time to recharge with an alternator. I suspect that my electrical use habits just don't drain the batteries deeply enough to cause alternator charging to struggle to keep pace with consumption or take too long.

> Do alternators really take a long time to charge batteries?
>
> Consider this: I had a small (old tube type) TV in my conversion van, which was powered by the chassis battery. (Not a great plan--- it was in there when I bought the van.) Once, I watched a movie in the evening. After the movie was over, I shut everything off and went to sleep--- well, at least I *thought* I shut everything off. After the DVD player was shut down, the TV screen went dark, but the TV was technically still "on," using power. The next morning, the van wouldn't start. I measured my battery's voltage at about 7 Volts. I got the van started using jumper cables and a cooperative fellow camper. I drove to a destination about forty minutes away then parked. When I tried to start the engine again, it started up right away with no problem.

Technical considerations: 1) Long wiring routes with multiple connection points along the way may be needed between batteries, especially if batteries are located some distances from one another. Because the net alternator output voltage is less than 2 Volts above the battery voltage (a very low net voltage), the system is very sensitive to electrical resistances along the way. If there is the slightest bit of corrosion at connection points, or wires too thin are used, then electricity has difficulty flowing into the spare and house batteries. Resistances do not stop the spare/house batteries from charging, but they do limit current and cause charging to take place much more slowly, sometimes unacceptably so (charging does not keep pace with consumption). Using a digital voltmeter allows one to check the system. When the house battery is properly connected to the spare battery, its voltage should start to go up immediately (and reach about 12.7 Volts). Also, the voltage measured between the spare battery terminals should be greater than 13 Volts when that spare battery is connected to the running car. 2) The coach batteries are shown indoors in these diagrams. Wet lead acid batteries should be either mounted outdoors or if indoors, they should be in compartments or special battery boxes sealed from indoor air and vented to the outdoors, to prevent explosive hydrogen from entering indoor living space and compartments. Dry (sealed) batteries do not have this requirement; they may be mounted indoors.

Fuel Consumption and Efficiency of Charge-As-You-Drive Electricity:

Since all battery charging by this strategy takes place during driving trips that would occur anyway, it is tempting to conclude that no fuel gets consumed to charge spare battery(s), therefore efficiency is infinite. However this is not quite true. Whenever electrical energy gets drawn from any generator (including an alternator) for any reason (including battery charging), a magnetic torque gets created that acts to slow that generator down. In this case, the alternator doesn't actually slow down, instead an extra torque gets applied by the fan belt (to counteract that slowing down), which in turn transfers that extra negative torque to the engine. The engine also does not slow down, instead it burns just a bit more fuel to counteract that negative torque. How much extra fuel must an engine burn to charge an RV battery? What is the efficiency of energy transfer from the fuel to the battery? How much does it cost to charge an RV battery? The following three text boxes attempt to answer these questions.

How much gasoline would be needed
to charge a 100 Amp hour battery from 50% to 100% full,
at 100% efficiency?

Technical Data:
Energy Density of gasoline = 130,000,000 Joules per gallon
Charging Voltage is assumed to be 13.5 Volts

The general formula for energy is:
(Energy) = (Voltage) × (Charge),
therefore,

(Energy to charge from 50% to 100%) = (Charging Voltage) × (number of Amp hours) =
(13.5 Volts) × (50 Amp hours) = (675 Watt hours),
which converted to Joules =
(675 Watt hours) × (3600 Joules / Watt hour) = 2,430,000 Joules.

(Fuel consumed) = (Energy) / (Energy Density of fuel) =
(2,430,000 Joules) / (130,000,000 Joules per gallon) = .0187 gallons
or 1/53rd of a gallon.

If gasoline cost $3.00/gallon, this gasoline would cost about 6 cents.

Alternator Charging Efficiency

Assumptions:

1) All charging takes place during trips that would occur anyway.

2) Everything spinning spins at the same speed regardless of whether or not a battery is charging, therefore the mechanical friction is zero. [Only magnetic torque will be considered.]

Much of the energy in gasoline gets wasted, even in the best engines. Some of the gasoline does not burn or burns incompletely, therefore that chemical energy gets wasted. Some of the energy that does get released just turns into useless heat, which either goes out the exhaust pipe or dissipates through the radiator. Thankfully some of the energy does turn into useful mechanical motion. The efficiency figures that I get from the internet are a bit hard to decipher, because there are so many places in the automobile combustion/locomotion process where energy can be (and is) lost. Also, the figures may be different when autos cruise down highways at high speeds than they are when autos stand still idling at red traffic lights or in gridlocked traffic. Like MPG ratings, these figures are only extreme approximations. The figures I get are:

Efficiency =
Efficiency of conversion of fuel energy to mechanical energy =
(Useful Mechanical Energy)/(Total Energy in fuel consumed) =
.15-.25 or 15%-25%,
which is one part in 4-6.

This is quite a bit better than the 1 part in 360 of charging with a typical 3 Amps charger powered by a generator (about 60 to 90 times as efficient), using equipment likely to be supplied in a standard RV.

This is also quite a bit better than the 1 part in 24 of my proposed improvement of charging 3 parallel batteries with a total of 60 Amps of current powered by a generator (about 4 to 6 times as efficient).

Estimated Cost of Charging While Driving

Cost of Charging a 100 Amp hour Battery from 50%-100% =
(Fuel needed at 100% efficiency) × (price per gallon of fuel) /
(Useful Mechanical Energy from fuel / Total Energy in fuel) =
(.0187 gallons) × ($3.00/gallon) / (.15 to .25) = 22 to 37 cents.

How I (the author) Learned to Charge Batteries (while driving)

Most of this book is either informational, problem-solving or exposé. I've decided to include this section of my personal electrical odyssey, because many of you readers may be wondering how I (the author) charge my own batteries in the field and how I managed to acquire this skill. This should serve as a good "motivational speech" to many of you RVers who are struggling to figure out how to charge your batteries. You are not alone in your struggling. When I first got started in RVing, I was just as much of a clueless goofball about electricity as anybody else. I learned how to charge my batteries and so can you.

Lesson #1: Conservation:

After buying an old dilapidated travel trailer and spending six months fixing it, it was ready for its maiden camping voyage in December 1985. Back then, the only electrical equipment I had (other than appliances) was two batteries (one in the car and a second in the trailer) and an electric battery charger. I had absolutely no plan for charging the trailer battery except to remove it, find someone with an electrical outlet who was willing to let me use it, and plug in for a half-a-day or so.

The voyage: I began the trip, taking along my old high school friend Jeff Mullen. We arrived at our destination, Florida. Early the very first full day we camped, we went swimming. After we were done, each of us changed into dry cloths in a small closet in the trailer deemed the "changing room"--- first me, then Jeff. This "changing room" had no windows and was therefore dark, but there was an electric light on the ceiling. Before Jeff changed, I told him: "Be sure to turn that light off when you're done. We are running on battery power and I don't want to run that battery down." Dry, we left the campground and went shopping and sightseeing all day long. We arrived back at the trailer--- I opened the closet door--- Oh no! The light was on and had been left on all day. I was not happy with Jeff. I let him know so. The lights were struggling. I told Jeff that since he ran the battery down, it was his job to remove it, take it to the ranger station, and see if they would let him plug in and charge it there overnight. Jeff did so. The rangers obliged this request. (I don't think they would do so today anymore. Things were different three decades ago.) That evening, with no battery, we cooked and ate supper in a "blacked-out" trailer under the dim light of a malfunctioning gas light (since been replaced) and flashlights. The next day, Jeff retrieved and reinstalled the recharged battery and we were back in business. Yes, I learned about electricity conservation *on the very first day I ever camped in my trailer.*

<u>Lesson #2: Battery Swapping</u>:

We had electricity for over a week after this first mishap had been resolved. But even with diligent conservation, the battery started to get low. (I didn't use a voltmeter back then. The only way I knew the battery was too low was by noticing that the lights would get dimmer and be slow to turn on.)

Jeff and I started talking about battery charging. We decided that there must be a better way to get a battery charged than removing it and imposing on rangers' or somebody else's electrical outlets. Talk focused on the automobile, specifically the alternator. Neither one of us were auto mechanics, but we knew a little. We knew that alternators made electricity from rotating engines while cars were running, and cars needed electricity to keep their engines running, so we wondered if a running car engine would continue to run, if its battery were to be removed after being started. If so, a car could be started with a fully charged battery, that battery could be removed, then a discharged battery could be connected in its place. After the car drove awhile, its new low charged battery would get recharged. There was only one way to find out if this scheme would work. We needed to experiment. We started the car. I disconnected the battery, and low and behold, the engine kept running, though it made a different sound upon disconnection. I connected and installed the low battery, and upon reconnection, once again the engine made a different sound. The only unresolved issue at that point was whether or not the low battery would get recharged enough during its ride to start the car after stopping. (For this scheme to work, it wasn't necessary for the battery to become *fully* charged during the first drive, only *charged enough to start the car*. After several drives, for sure that battery would become fully charged.) We took our ride to town, which we were going to take anyway. Before we did, as a precaution, we put the newly removed fully charged battery and the toolbox in the trunk, in case the ride didn't charge the low battery enough to start the car. (If this occurred, we would have needed to put the fully charged battery back in the car.) After our first drive, the car restarted with no problems. Success! That evening, we put the fully charged battery in the trailer and once again enjoyed electricity for about another week. I think this "swapping batteries with the car running" idea was mostly Jeff's, so I must give him credit. Although Jeff had "tested" both the electrical system and me initially, this time, Jeff saved the day.

During this maiden voyage, which lasted about two and a half weeks (including travel time), we performed this battery exchange procedure twice. Six months later, Jeff and I took a cross country camping trip lasting 50 days (all the way to the coast of California and back from New York). During that trip, we made all of our electricity using this battery swapping procedure multiple times. We never ran black again. Two years later in 1988, I took a solo trip to Florida lasting 95 days. Once again, I made all my electricity using this battery swapping procedure, never plugging into municipal electricity once. I never ran black. I returned home from that trip with my batteries as fully charged as they were the day I left.

Although deep cycle batteries are recommended for RVs, we used regular automotive batteries. I never trashed any of my batteries through misuse; they all lasted several years. My electricity system was simple back then. My only appliances were incandescent lights (and one fluorescent), a small "boom box" (which I had wired to run off the coach battery) and an electric musical keyboard resembling a piano.

Now, let me give you some advice on this procedure: **DON'T DO THIS!!!** There are two reasons: 1) Sparks made at the battery terminals upon connection and disconnection can ignite hydrogen gas made by the batteries and cause them to EXPLODE in your face. 2) This procedure can substantially damage today's automobiles. More recent cars/trucks have computerized controllers in them. These controllers can be damaged by spikes of high voltage. Batteries in vehicles act as filters that remove extra electrons, preventing such spikes of voltage from occurring. If a battery is disconnected while a vehicle is running, voltage spikes can occur and "toast" its controller. Replacing a controller costs several hundred dollars. In addition, the vehicle becomes inoperable upon losing its controller, so it will also need to be towed to the repair shop. Although this procedure doesn't work with newer cars, it did work on older cars which had less computerization (such as my 1976 and 1982 cars).

Lesson #3: Three Battery Rotation:

I knew I needed to improve my battery charging strategy. Believe it or not, I was always aware that batteries could explode in my face during swapping. I liked electricity and was addicted to it to such an extent that I foolishly took this risk. I did take precautions though. I always wore safety goggles. I always had a container of water nearby (in case I needed to wash sulfuric acid off my face--- what a lovely thought). I also fanned the battery by waving a sheet of cardboard just prior to connection and disconnection, to blow away any possible hydrogen gas. I was damn lucky none of my batteries ever exploded.

In 1989, shortly after buying my second car (a 1982 Buick Regal Estate Wagon), I saw an opportunity for just such an improvement. I noticed that there was room under the hood of my recently acquired car to install a second battery, so I devised and used the following three battery system (two batteries in the car, one in the camper):

Normally, a car has one battery connected to the alternator, the starter motor (through a solenoid) and the rest of the car's circuitry as shown in *figure 2A*.

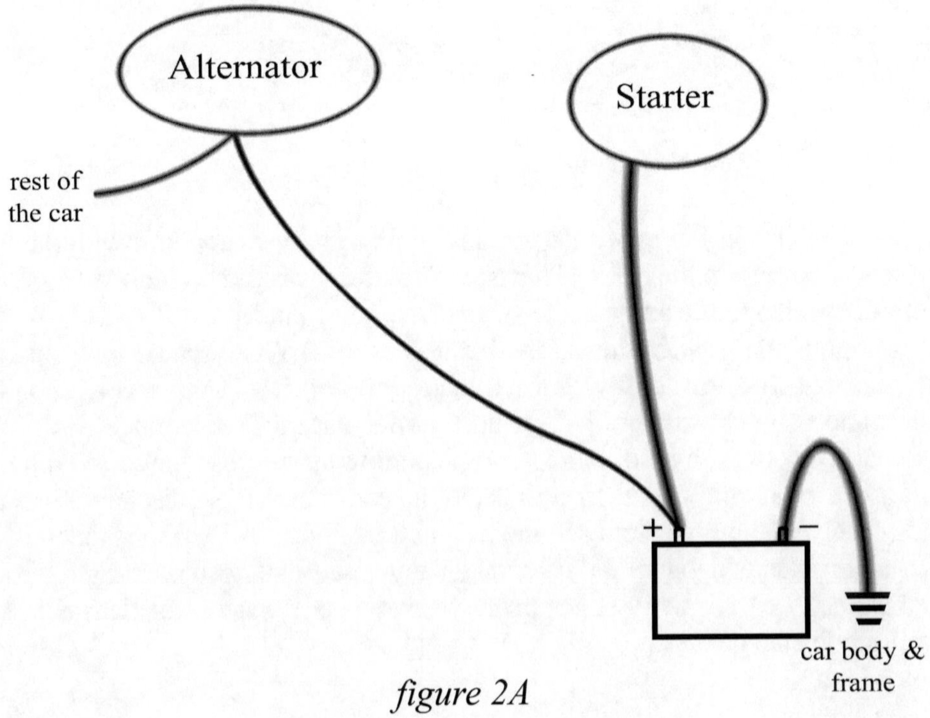

figure 2A
wiring diagram of normal car

I installed a second battery under the hood, so the vehicle's operation could be shifted from one battery to the other *electronically* instead of *physically*, by means of two high current on/off switches called "master disconnects," as shown in *figure 2B*. (I refer to this electronic shifting from one battery to the other as *changeover*.)

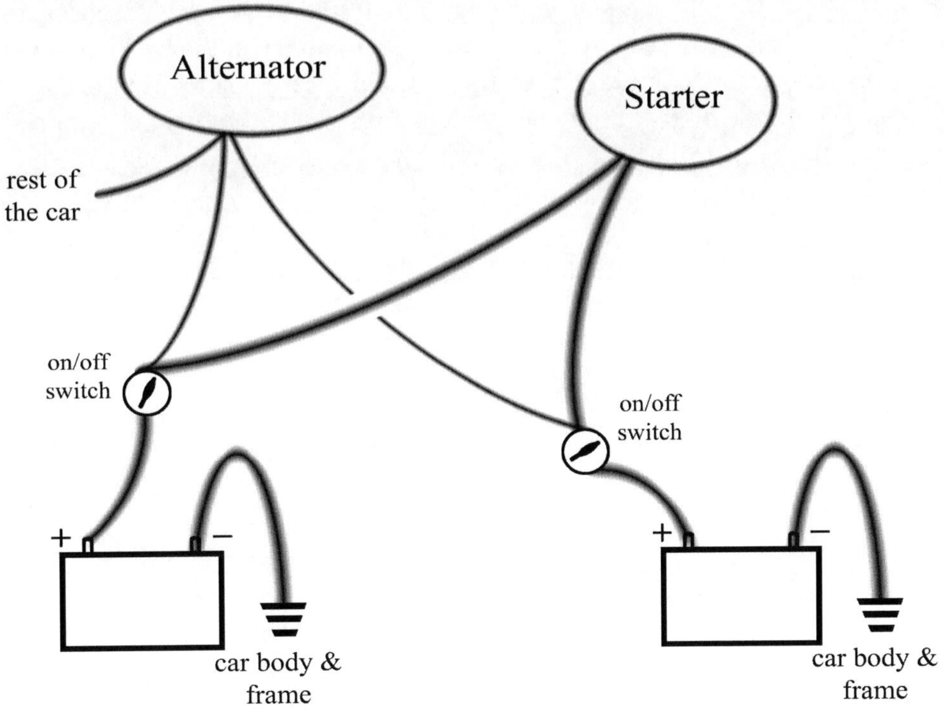

figure 2B
wiring diagram of car with two batteries

Here's how I used the system:

1) When the coach battery in my trailer got low, I physically swapped it with the fully charged "spare" battery in the car. [The "spare" battery was the one in the car not electrically connected to the automobile--- its master disconnect switch was "off."] This physical swap took place when the car engine was off. 2) When ready for a long ride, I started the car from the other [fully charged] battery [the battery whose corresponding master disconnect switch was "on"]. 3) Shortly after starting, I disconnected the car from that fully charged battery, by moving its corresponding master disconnect switch from the "on" to the "off" position. 4) I then connected the car to the other [discharged] battery by moving its corresponding master disconnect switch from the "off" to the "on" position. 5) I then took my ride which I would have taken anyway. Doing so recharged the recently connected battery. 6) The newly disconnected spare battery was fully charged. It sat in the car waiting for the next swap.

Battery role sequence:
(Note that all three batteries go through the same sequence of roles, though obviously not at the same time.)

1) A fully charged battery located in the house. With use it becomes: 2) A discharged battery located in the house. At this point it gets swapped with the fully charged "spare" battery in the car. At this point it is: 3) A discharged "spare" battery located in the car. The next time the car gets started, electronic changeover takes place, after which it becomes: 4) The automotive battery in the car (connected to the car's electrical system). After the next physical swap [which involves the *other* battery in the car] and another electronic changeover, it becomes: 5) A fully charged "spare" battery in the car. It stays there for awhile until the battery in the house gets low. Then the two get swapped, and it once again becomes: 6) [role (1) again] A fully charged battery located in the house.

Advantages/ Disadvantages/ Technical Consideration of Three Battery Rotation:

Advantages: 1) Very low risk of explosion. This is because during changeover: A) Disconnection and reconnection takes place away from the battery terminals, where there isn't much probability of hydrogen gas. And: B) The sparks occur *inside* the master disconnect switches. Since the insides of those switches are almost completely sealed from the outside air, there is almost no chance of them having hydrogen in them. A spark [thus an explosion] still could take place during physical swapping, if a short circuit occurs through the metal wrench or ratchet, so be careful not to let this happen.
2) Physical swapping takes place with the car engine off. If done right, no sparks get made during physical swapping. There are no moving engine parts to accidentally put one's hands into. Also, hydrogen gas gets released during charging, so if the car has been sitting for awhile when batteries get swapped [likely], there is much less hydrogen around to become ignited and explode if an accidental spark does get made. 3) It's easy to change back to the other battery, if first ride doesn't charge the low battery enough to start the car after taking that ride. All one needs to do is turn one switch off and the other one on.

However, in all the years that I charged batteries by swapping them [both by this and the preceding method], the first ride never failed to charge the recently connected low battery enough to start the car after parking and turning the engine off.

Disadvantages: 1) Deep cycle batteries cannot be used. RV experts recommend deep cycle batteries instead of regular automotive batteries for household use. Regular automotive batteries must be used for this strategy because each battery takes a turn as the regular car battery and must be able to start the car. Regular automotive batteries can provide the high current that's needed for starting, but they are more prone to damage from being repeatedly and deeply discharged [than deep cycle batteries]. 2) Swapping heavy batteries is a bother. If one uses lots of electricity, one might have to do this often. 3) This procedure cannot be done as described with today's computerized cars, which have controllers in them, because disconnecting all batteries can burn controllers out.

Technical consideration for today's cars: Three battery rotation might still work on today's cars if a switching technique called "make before brake" gets used. Instead of disconnecting the fully charged battery *first*, then connecting the low battery *soon thereafter* (a technique called "break before make"), do these steps in reverse order: connect the low battery *first*, then disconnect the fully charged battery *soon thereafter*. In "break before make," there is no time when both batteries are connected to the car, and a brief potentially destructive time period when neither batteries are connected to the car. In "make before break," there is no potentially destructive time period when neither batteries are connected to the car, and a brief time period when both batteries are connected to the car. During the time when both batteries are connected, there will be a current surge from the charged battery to the low battery, because in addition to both being connected to the car, the two batteries are also connected to each other. This current surge might heat wires and cause melting, but probably won't, since this time period is likely to last no more than a few seconds.

How well does this strategy work?

I used this three battery rotation system just described for about seven years (1989-1996). During that time period, I took a two and a half month trip, a three month trip, and several shorter trips, without plugging into municipal electricity. My batteries charged faster than I could deplete them. Once again, I never ran black.

Lesson #4: Battery Paralleling:

Just when I had my battery charging strategy down to a science, another RVer with a towable trailer shattered my nearly perfect universe by telling me his very simple electricity making strategy: "Every evening when I'm parked for the night, I parallel my coach battery with the battery in my truck, by connecting them with long wires. Electricity seeps from my truck battery into my coach battery and charges that coach battery. I don't disconnect until the next morning when I'm ready to drive away. As long as I do this every night, enough electricity seeps into that coach battery to keep it charged indefinitely without ever running dead." (RVers love to share insights and technical tricks. It is uncanny how much I have learned from these people.)

Wow. There is a way to charge batteries that doesn't involve physically removing, moving and reconnecting them. I immediately saw a problem with this procedure though: If too much electricity got used during the night, or the coach battery was too low prior to connection, too much electricity would be drained from the car/truck battery, and the car/truck would be unable to start the next morning. This especially might happen if the car/truck were parked for more than one day without moving. Also, the mere fear of running both batteries down might cause people to neurotically conserve on electricity when they don't really need to. If such a rundown occurred, RVers would need to find and impose upon a cooperative stranger willing to jump start the car/truck from that stranger's vehicle.

I knew the solution immediately: Use a three battery system (two in the car, one in the house) similar to the one I was already using. When parking for the night, the chassis battery would get disconnected from the spare battery in the "car," before the spare battery would get paralleled with the coach battery. Doing this would ensure that the chassis battery would never be able to run down, so the "car" would always be able to start. It would also foster peace of mind in the RVers.

This battery-paralleling-three-battery-system strategy I began using in 1996. Though I have also tried storing both deep cycle batteries in my van paralleled (I have two such batteries), I have returned to storing one deep cycle battery in the trailer and one in the van and paralleling them nightly just as described, and use this strategy to this day. It works very well. Twice I have gone over four months without plugging into municipal electricity. I've also used it on numerous shorter trips (almost too many to count) for extended periods of time. This strategy can make electricity until the RV user gets tired of traveling and concludes the trip, the RV user runs out of money or the world runs out of oil.

II.) Solar Panels

What Are Solar Panels? How do They Work?

What are solar panels? Solar panels (also known as photovoltaics) are flat devices that make electricity from sunlight. They are composed of individual light collecting units, called solar cells, electrically connected to each other in both series and parallel configurations, encased in plastic. The encasing holds all the components (cells and wires) together and prevents the configurations from breaking apart from handling and weather damage. The side of each panel that faces the sun is transparent, allowing light to enter. Multiple solar cells, packaged together in single casings are called *solar modules.* Multiple solar modules, packaged and arranged together, are called *solar arrays.*

How do solar panels work? When light strikes a solar cell, its energy causes electrons to be removed from the wire where they are less crowded ("+") and deposited in the wire where they are more crowded ("–"), which is not the direction electrons normally want to flow, so they are essentially moved "uphill." In solar modules and arrays, individual cells work together to do the same thing on a grander scale. As the electrons move, a voltage builds. If used to run gadgets directly, some of those electrons are allowed to leave the cell/module/array and run through the circuit from "–" back to "+." If connected to a battery, once the cell/panel/array voltage builds to equal the battery's charging voltage, any extra electrons move through that battery, charging it.

A bit more technical explanation: Each solar cell is a diode, which is very flat, to allow maximum collection of sunlight. Within every diode is a p-n junction, which is a region that permanently holds two opposite charges (–) and (+). When light of sufficient energy strikes electrons within this charged region, it knocks them loose. The electrostatic attractions, within this charged region, cause the loose electrons to migrate to the (–) wire, while their removal causes electrons in the (+) wire to move into the charged region, to replace them.

Advantages of Solar Panels

1) Solar panels are ***passive*** *devices, which means that they have* ***no moving parts****.* Because they have no moving parts, when they are used: A) they do not wear out and B) they make no noise. Since no wear and tear nor objectionable noise results from their making electricity, C) they can be left on continuously (or for very long durations). They are "on" whenever strong light strikes them and they are connected to batteries capable of accepting their electricity. They only turn "off" when insufficient light is present (mainly at night, but also in heavy shade/overcast) or when they are not needed (because all batteries connected to them have attained full charge).

2) Solar panels have very long life spans. If you buy solar panels, and they are not damaged by shock (being cracked somehow), they probably will outlive you. I have read somewhere that the earliest solar panels manufactured are still making electricity today.

3) Solar panels can charge batteries slowly and completely. This is the way batteries "like" to be charged. Charging slowly and completely extends battery life and greatly extends the usable storage capacities of batteries.

4) Once the panels and supporting equipment are purchased, the electricity that they make is free.

Disadvantages of Solar Panels

1) The steady electrical output of solar panels is small (especially when compared to generators). Solar panels are usually used to charge batteries because their output is usually not enough to run most gadgets directly. [Even two 60 Watt incandescent light bulbs would require a considerable size panel to run them directly. However some gadgets using very small amounts of electricity (such as calculators) can be run directly from solar cells/panels/arrays.] Running gadgets directly would pose the following disadvantages: A) huge panels would be required which would be very costly, B) when no gadgets were on, all electricity made by the panels would be wasted and C) no electricity would be available at night. This is why batteries, to store the electricity that panels make, are an almost inevitable complimentary component of any self contained solar electricity collection station. Recall the formula from Chapter 1:

$$(\text{Charge}) = (\text{Current}) \times (\text{Time}).$$

Even though currents produced by solar panels are small values, the time spans that they are on collecting energy are usually big values (maybe all day). This allows quite a bit of charge to build up in the batteries, which can be released in moderate to large amounts for short times when needed, including at night.

2) Solar panels are not reliable because they need strong sunlight to make significant amounts of electricity. Sunshine is not always available, as some heavy overcast days and shade are almost inevitable. One strategy of increasing their reliability is to have enough panels to make more electricity than needed on sunny days and enough batteries to store that excess to sustain the users though several sunless days, while users wait and hope for the sun's return. Still, even the best solar electricity collecting/storage systems could be shaded out under prolonged dark weather conditions (especially possible in RVs). So…

3) Backup systems of making electricity are highly recommended. Backup systems add to the overall system cost, and if the backup systems are too effective, they might even render the solar panels redundant. In RVs, the backup systems are often generator powered 120 Volts AC battery chargers or "charge as you drive" electricity.

4) Solar panels cannot power some of the extreme energy hungry appliances often found in RVs (especially compression air conditioners). Big RVs with big electrical requirements still need generators. RV users with solar panels still need to spend nearly the same money to purchase and maintain their generators as RV users without solar panels, both for backup charging and running energy-gobblers. However, a really good solar panel system could allow an RV user to get by with a less expensive generator, or use the generator only occasionally.

5) Solar panels add to overall system cost. Supporting equipment must be purchased besides. However, adding solar panels to an RV could save the owner more money than their cost in the long run because: A) they may reduce or eliminate the need to stay in expensive RV parks with electricity and B) RVs with solar panels are likely to keep their batteries both more fully charged and charged a greater percentage of the time than RVs without solar panels, thus extending battery life and reducing battery replacement costs. Also, anyone who has the money to buy an RV, probably also has the money to add solar panels.

Differences Between Solar Panels and Generators

Solar panels and (internal combustion engine) generators are so different, both in the way they work and in their physical construction, that comparing them may be more like comparing apples with jet airplanes, than comparing night with day. About the only thing solar panels and generators have in common is: *they both make electricity.*

*1) Solar panels are **passive** devices whereas generators are **active** devices.* By definition, passive devices (like solar panels) have no moving parts, hence they make no noise and use does not wear them out. By definition, active devices (like generators) do have moving parts, hence they do make noise and use does eventually wear them out.

2) Solar panel/battery systems are more ideally suited for light-duty long term applications, whereas generators are more ideally suited for heavy-duty short term applications. Solar panels are ideal for light-duty applications because they make small amounts of steady current. Because they make this current for long durations, substantial amounts of charge can build up in the batteries, so solar panels can also be used for medium-duty applications, especially if average daily electrical usages aren't terribly high. However, running extreme energy hogs (like microwave ovens) from power inverters, even briefly, causes substantial battery drain, requiring hours of sunshine to replace the electricity used. Also, deeply discharging batteries before recharging them (and doing this often) shortens battery life spans. If many energy-hogging gadgets are attempted to be used, or if even one is attempted to be used for a substantial time length, then batteries may run down, because on average, solar panels must be able to make electricity faster than gadgets use it, in order to be successful. Generators, on the other hand, make huge amounts of electricity when compared with solar panels (perhaps a hundred times as much). Because of this, they are ideal for running energy hogs (like microwave ovens), especially if those devices are used only briefly. However, generators are extremely fuel inefficient when used for light-duty operations, including battery charging. This is because a minimum amount of fuel must be burned to start and keep them rotating. When energy gobblers are connected, more fuel makes usable electricity, making the use/waste ratio not too bad. When only energy misers are connected, the use/waste ratio is terrible, because only a tiny portion of the fuel burned makes that electricity--- most of the energy from burning fuel is used to overcome friction to keep them rotating (or wasted as heat).

3) Solar panels are usually used to charge batteries, whereas generators are usually used to run appliances directly (skipping electricity storage). Solar panels are ideally suited for battery charging, because batteries can only accept modest amounts of current through them when charging, and sometimes need long charging periods.

Generators are not great at battery charging, because in order to do so effectively, they need to be running for long time periods, which is inadvisable due to their annoying noise, their wear and tear during use, their consumption of fuel and their horrible light-load fuel efficiencies.

4) Solar panels are powered by sunshine whereas generators are powered by fuel. Sunshine is free but not always available. Fuel (gasoline, diesel or propane) costs money and requires a small effort to fill fuel tanks, but once in those tanks, is available for burning anytime.

5) Solar panels make DC whereas generators make AC. DC is the kind of electricity that batteries store, so solar panel outputs can charge batteries directly. AC is the kind of electricity that is delivered to household wall outlets, so generator outputs can power household 120 Volts AC appliances directly. Each electricity type can be converted to the other one, using converters or inverters (covered in Chapter 1), and the efficiencies of such transformers are pretty good, so this may be a moot point only of interest to electrical engineers.

6) Solar panels are flat and lightweight, whereas generators voluminous and heavy. Solar panels can fit nicely on an RV's roof, whereas generators require quite a bit of an RV's internal storage space.

Practical Considerations for RV Solar Panel Use

1) Equal size solar panels do not produce as much electricity for RVers as they do for people with permanent houses/cottages in the same parts of the country. Why? People with permanent properties don't need to move around, so they only need to position their panels once. Thus, they can take lots of time to do it right. They can find that "perfect spot," with no shade from large rocks or foliage. They can adjust their panels to the perfect angle for maximum output. They can perch panels high on big rocks, trim foliage, remove trees, or even construct special towers upon which to mount the panels above the trees, to ensure that when the sun shines, those panels will receive maximum sunshine. RVers are not so lucky. Many times, typical RVers find themselves under heavy leaf cover, in deep ravines (where the sun hardly shines) or immediately adjacent to high rock embankments. [If parked west or east of high rock embankments, RVs receive sunshine for only half the day; if parked north of high rock embankments (in the northern hemisphere), RVs receive almost no direct sunshine at all.] RVers need to be realistic in understanding that they won't always park in places that are perfect for maximizing sunshine, so their solar panels won't work all the time.

2) Maximum allowable size and weight are limited for RVers. Permanent property owners don't have this problem. If permanent property owners need more electricity, they always have room for more/bigger panels and more batteries. The extent of their electricity production is limited only by their needs/desires and their willingness to spend money. Special shacks can be constructed for batteries. And if their batteries weigh thousands of pounds, so be it--- each battery only needs to be brought to and removed from the site once during its useful lifetime. RVers are not so lucky. Only so much free area exists on their roofs. Only so many big heavy batteries can be fit in their relatively small RVs. Thus, RV solar power plants can only grow to be so big.

3) Many RVs can't stand being in the sun! This is because many RV shells (roofs and sides) are made of plastic, rubber or fiberglass. These materials tend to absorb heat rather than reflecting it. If RVs with heat absorbing shells are also poorly insulated (often the case), when parked directly in the sun, they absorb massive amounts of heat and dump that heat into their interiors, often making them unbearably hot. How does a heat-absorbing shell effect RV solar panel performance under various conditions?

Scenario #1: *The cold day in the sun:* (60° F or lower.) The solar panels will work just fine. Any extra interior heat from the sunshine is likely to be welcome.

Scenario #2: *The moderate day in the sun:* (About 70° F.) RV owners may still be able to inhabit their RV, without using their air conditioner, by opening all their windows, to get rid of some or most of the unwanted sun-generated heat, especially if some wind is present. If the air-conditioner isn't needed, the solar panels will work fine and be useful.

Scenario #3: *The warm day:* (Maybe 75° F or higher.) Suppose that the RVers arrive at a campground, to discover that they have a choice between a sunny site and a shaded site. A) What happens if the RVers park in the sunny site? If they vacate their RV during the day (for shopping or sightseeing), and don't have any pets they need to leave in there, they won't need to run their air conditioner, so their solar panels will work fine and be useful. (If the RV interior gets hot when nobody is in it, so be it.) If however they inhabit their RV during the day, they will need to run their air conditioner (perhaps all day long) to remove the massive amount of unwanted extra heat from the interior. In order to power their air conditioner, they will also need to run their generator. When the generator runs, it not only makes electricity for the air-conditioner, it also makes plenty of extra electricity for battery charging. Good 120 Volts AC battery chargers are capable of charging batteries considerably faster than moderately sized solar panels. Charging batteries with a good AC charger and solar panels simultaneously can be done, but it is a bit like filling a swimming pool with water using both a large hose hooked up to a fire hydrant and a small garden hose hooked up to a regular house faucet simultaneously. The fire hydrant hose fills the swimming pool so much better and faster than the garden hose, so if both are used simultaneously, the hydrant hose overshadows the garden hose's usefulness, rendering the garden hose unimportant and unneeded. Likewise, a good 120 Volts AC charger charges batteries considerably faster than solar panels, so if both are used simultaneously, the AC charger overshadows the output of the solar panels, rendering them unnecessary. In other words, generator powered 120 Volts AC battery charging doesn't stop solar panels from working, but it does make them redundant. B) What happens if the RVers park in the shaded site instead? The inside of their RV is likely to be nice and cool, so the air-conditioner won't need to run, therefore the generator won't need to run, but in the shade the solar panels just won't work. Even so, parking in the shade might be a better option than parking in the sun.

Scenario #4: *The really hot day in the sun:* (Maybe 85° of higher.) Both the air conditioner and the generator will need to run almost all day long, rendering the solar panels redundant.

RVs with skins made out of sheet metal have much better abilities to reflect away heat from sunlight (than those with plastic, rubber and fiberglass skins), therefore metal skinned RVs need less air conditioning (especially if well insulated), so they are much better suited for solar panels.

4) Getting extra panels and extra batteries, doesn't work very well for RVs. Although this strategy for dealing with solar panel unreliability works quite well for off-grid stationary houses/cottages, there are two issues particular to RVs: A) Many campsites are almost entirely shaded out. Based on my camping experience, I estimate that solar panels may only be useful about 30% of the time. Thus living off of stored electricity, while waiting for sunny days at such sites, just doesn't work, especially at campsites where the sun seldom or never shines. B) Having extra batteries greatly improves overall electrical system performance even without the inclusion of solar panels. This is because extra batteries can also be charged while driving the tow/spare vehicle or motorhome. One could try solar charging batteries at one location, then relocating, but the long drive between campsites charges batteries anyway, thus rendering the solar panels redundant [for charging at one location, then relocating].

Technical Points

1) RV solar panels can be used while flat on the roof or tilted toward the sun. Tilting them toward the sun increases their output. If tilted toward the sun, they can be positioned just once per parking location or moved several times per day, to track the sun. If used flat, output is lower but zero RV user effort is required, other than trying to find sunny locations to park. If tilted once per parking location, facing south at such an angle as to be perpendicular to the sun at celestial noon, output can be increased in the range of 15%–30%, depending upon the latitude and the time of year. [In Northern Hemisphere: The further north, the better the efficiency improvement. The closer to the winter solstice (December 21), the better the efficiency improvement.] Not much effort is required for tilting, because they only need to be tilted once (shortly after parking) and flattened once (shortly before driving away) per location. If moved an infinite number of times to track the sun perfectly, with twelve hours of perfect sunshine, a panel would make $\pi/2$ times the electricity as it would if stationary tilted south at the perfect angle for celestial noon, a 57% increase (on top of that obtained from just tilting them south). Of course, nobody is going to move their solar panels an infinite number of times, but if one is willing to take the time to position them several times per day to "sort of" track the sun, some output improvement can be realistically achieved. Also, few people want to use mathematics to determine the perfect tilting angles, but this is not a problem--- one can just look for the sun's location and "wing it." If tilting solar panels: A) be sure they are anchored to the roof so the wind does not carry them away or knock them down and B) always return them to the flat position for travel.

2) High temperatures slightly reduce panel electrical output. This difference is only about 15% (between hot and cold days), a small enough quantity such that RV owners might not notice it. There is nothing that RVers can do to make hot panels as productive as cold ones, short of dousing hot panels with cold water (assuming cold water exists nearby), a feat requiring so much effort and needing to be repeated so often, that it is not worth considering.

3) Some solar cell types are more efficient than others at converting sunlight to electricity. Efficient panels make the same electricity with smaller surface areas. This greater efficiency is achieved within the individual solar cells by a process called multilayering. In this process, two or more layers, placed on top of one another, convert light *of different frequencies* to electricity. (Light not used by one layer might be used by another.) Since roof areas on RVs are limited, smaller more expensive high output panels may be a venture worth considering, if high electrical outputs are desired.

4) Additional circuitry is required for solar panel battery charging. While it is possible to connect a solar panel/array directly to a battery to charge it, this practice is not recommended for two reasons: A) During the day, when the sun shines on the panels, such a setup will charge the battery(s), however at night, when there is no sun, electricity will flow in the opposite direction through the panels and discharge the battery(s). The simplest solution is to have a blocking diode in the wiring. (A diode is a crystal which allow electricity to flow in only one direction.) This allows charging during the day but prevents useless discharging at night. B) Overcharging needs to be prevented. With larger panels, overcharging is a very real possibility. A computerized controller is needed. This device temporarily electronically disconnects the panels from the battery(s), when battery(s) attain full charge. Smaller panels are sometimes self-regulating. This means that they produce a voltage that is equal (and opposite) to the battery at full charge, therefore when the battery is at full charge, no current flows.

Conclusion: A few decades ago, solar panels were considered exotic technology. Today they are mainstream. They have in fact now become the choice method of electrical production for many people who live in places where municipal electricity is unavailable. The bad news about solar is: *it is unlikely that the RVers will **always** be parked in the sun, so sometimes the solar panels **won't** work.* The good news about solar is: *it is also unlikely that the RVers will **never** be parked in the sun, so sometimes the solar panels **will** work.* While solar panels alone can't solve all RV electrical problems, they definitely hold an important place in the design and optimization of RV electrical systems.

III.) Wind Turbines

What Are Wind Turbines? How do They Work?

What are wind turbines? Wind turbines are devices that make electricity from wind. Each wind turbine contains: 1) a rotating component with a set of blades to catch moving air, 2) a stationary component upon which the rotating component is mounted and 3) coils of wire through which current can flow that make or capture magnetic fields and thus energy. It may also contain permanent magnets, diodes (which only allow current to pass in one direction), brushes (which allow current to exchange between rotating and stationary components) and complicated regulation circuitry, depending upon its design type.

How do wind turbines work? Wind turbines are essentially electric fans operating in reverse. In a fan, a flow of electrons through wires moves a set of rotating blades to make wind. In a wind turbine, wind moves a set of rotating blades to send a flow of electrons through wires. In a fan, electricity sent through the coils torques the rotating blades, speeding up the wind. In a wind turbine, electricity removed from the coils torques the rotating blades in the opposite direction, slowing down the wind. Wind turbines make electricity by the same mechanism that water turbines, stream and internal combustion engine generators use--- through interacting magnetic fields occurring between coils of wire and magnets where the rotating and stationary components meet. (No doubt, you've noticed that magnets near other magnets can push or pull each other, depending upon their orientation. If done right, this can torque a motor or make electricity.) The magnetic fields can be stored in permanent magnets or they can made by sending current through coils of wire.

How many designs of wind turbines exist? There are literally thousands of individual wind turbine designs. Most are probably not appropriate for RVs. To work for an RV, a wind turbine must be small and lightweight enough to transport, easy enough to set up and dismantle multiple times, and not easily damaged by the shaking of road travel.

Advantages of Wind Turbines

1) *Wind turbines can make lots of electricity* (compared to solar panels).
2) *Once initial equipment is purchased, the electricity that they make is free.*

Disadvantages of Wind Turbines

1) Wind turbines are active devices. This means that they have moving parts, so they do make noise and do wear out.

2) Wind turbines can actually be damaged by very high winds. Thus, they need to have some sort of braking systems, to stop them from rotating too quickly. Under extremely high wind conditions, they should be temporarily taken down.

3) Wind turbines don't produce all the time; they only produce when the wind blows (which usually isn't all that often).

4) Wind is even more unreliable and unpredictable than the sun. So…

5) Backup electricity producing systems are needed. It is unlikely that too many RVers would choose a wind turbine instead of solar panels, rather they might choose a wind turbine in addition to solar panels. Of course, the more equipment one buys, the greater the cost of an overall system.

Practical Considerations for RV Wind Turbine Use

1) Wind turbines need to be set up and taken down, at every campsite in which they are used. One of the harsh realities of the RV lifestyle, is that considerable time and effort is needed to set up and dismantle campsites. (I know this for sure, from over three years of full time camping.) Setting up campsites consists of simple tasks (such as putting tablecloths on picnic tables, taking out folding chairs, taking down bicycles), complex tasks (such as setting up screen tents, awnings, unhitching, leveling) and lots of little tedious tasks (mainly removing items from their "riding" places and putting them into more convenient "house" places). Each task alone doesn't take too much time, but all tasks together require considerable time and effort. Campsite dismantling takes about the same amount of time as setting up a campsite, so this doubles the time and effort required for each move. If too many setting up/ dismantling tasks are required, then RVing becomes impractical or impossible. Wind turbines might work rather well on stationary properties, where they only need to be set up once, or for RV campground hosts who stay at the same locations for months, but for highly mobile traveling RVs, the time and effort required for multiple setting ups and dismantlings probably makes them not worth the effort.

2) Wind turbines are not useful at campsites when RVs only stay for short time periods (like four days or less). This is because: A) setting up/ dismantling times are the same for both short and long stays, but for short stays, potential benefits diminish. And: B) Driving an RV charges house batteries. After a long ride, an RV parks with fully charged batteries. It probably has enough stored electricity to last for a short stay with no stationary battery charging whatsoever. If the batteries get low by the end of the stay, this doesn't matter, because the batteries recharge after departure.

3) Wind turbines take up lots of storage space. Space is always an important consideration to RVers. Wind turbines are large enough such that they probably need to be stored on the outside of an RV (possibly on the tow or second vehicle), rather than on the inside.

4) Outside mounted equipment poses certain problems. A) During travel, outside mounted equipment increases air drag, which can lower gas mileage and make a vehicle less stable in high winds (which can occur from nature, passing trucks, or from an RV's own motion through the air). B) During travel, outside mounted equipment can be damaged from getting subjected to high winds. If not secured properly, pieces of equipment can rattle around or break free and leave the vehicle entirely. Ejected equipment is lost equipment. Ejected equipment also poses an extreme hazard to other motorists, both while airborne and while laying in the street. C) Air hitting such equipment at high road speeds can cause annoying "whistling." D) Equipment, mounted on a roof of an already high RV, raises the effective height of that RV. This would make the RV unable to pass under some low bridges and tree limbs (in campsites), that it might otherwise be able to pass under, if it didn't have such equipment fastened to its roof. "Clipping" (colliding equipment with a low bridge/treelimb) would be disastrous. E) Large equipment fastened to the outside of a vehicle can pose "head bumping" or "obstacle course" safety hazards. F) Large mounted equipment can block access to other equipment, storage compartments or access doors, or prevent windows or roof vents from opening. And: G) Large equipment fastened to the outside of a vehicle can just be an ugly eyesore.

5) Wind is not available at all campsites. There are places where and times when setting up a wind turbinc for an RV would be pointless.

Wind Verses Sun

Wind turbines need to compete with solar panels, for RV user's electricity gathering dollars. For many RVers, buying additional solar panels may be a better option than buying a wind turbine in addition to solar panels. Consider:

1) Unlike wind turbines, roof mounted solar panels, used in the flat position, require no setting up or taking down.

2) Unlike solar panels, wind turbines make some noise and have moving parts to wear out.

3) Wind turbines and solar panels often succeed (or fail) in the same places. Whenever two separate systems are delegated for a single task, it is best for those systems to succeed under different conditions. This way, the two systems complement each other. If both systems simultaneously fail under the some conditions, then they both become unusable when those conditions occur. If both systems simultaneously succeed under some other conditions, then one system becomes redundant when those conditions occur. Sun and wind do however often occur at different *times* at a single location, so this makes the two systems somewhat complimentary.

<u>Finding Wind</u>

I have personally never used a wind turbine. However, when I'm out living in my trailer, I pay quite a bit of attention to weather patterns. Since I've camped in 32 different states (so far), I am probably in a good position to give some really good insights as to where the wind blows and where it doesn't.

Forests block both sun and wind. Overhead leaf cover obscures sunlight, tree trunks and branches block wind. You are very unlikely to harvest any electricity from nature, when parked deep in some forest. If you do get good wind in a forest, it might be advisable to get both you and your gear out of that forest, because tree limbs may come falling on your rig.

Flat lands (primarily prairies and deserts), free from tall vegetation and obstructions (such as boulders or houses), are good places for wind. However, flat lands do not guarantee wind; atmospheric conditions must still produce it. Prairies and deserts (especially) are also great places to find sun.

Boulders, hills and mountains can block both sun and wind, especially if they are large. However, boulders hills and mountains usually don't actually stop the wind, they redirect it around them, providing certain locations where wind is absent and other locations where wind is intensified (*figure 3*).

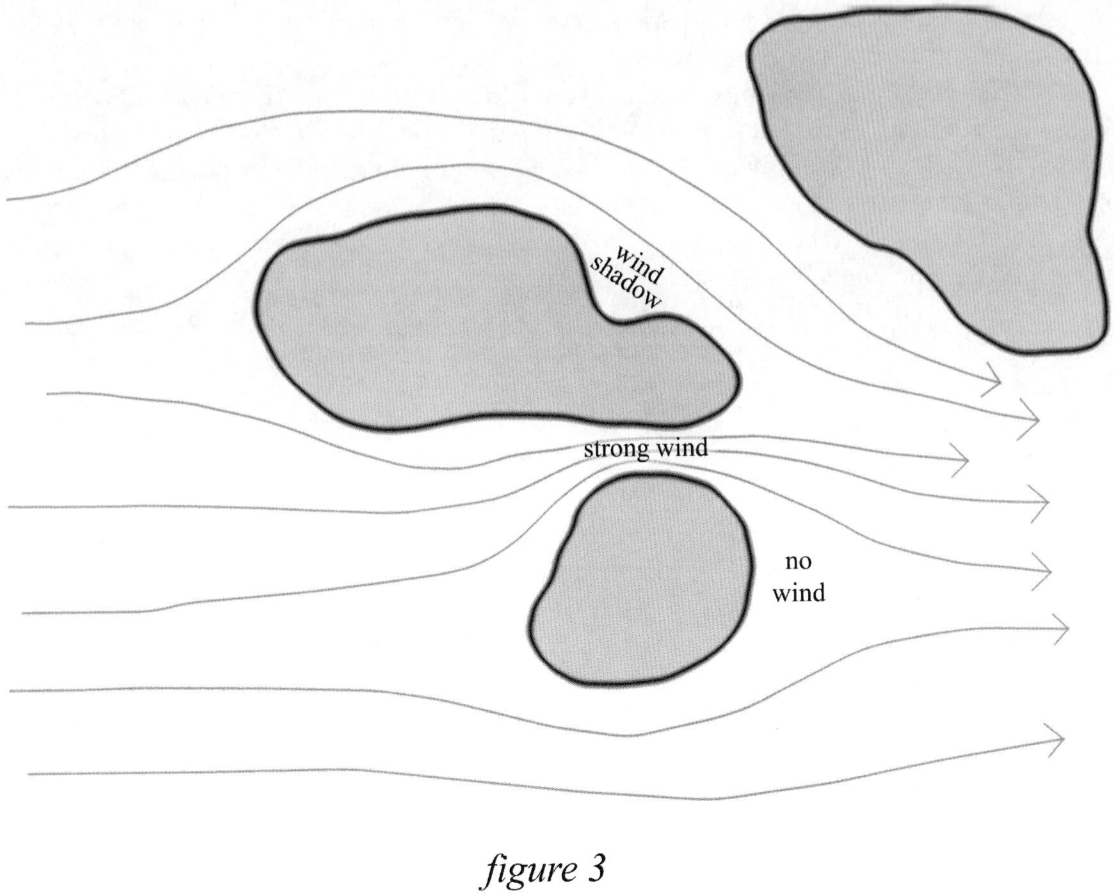

figure 3
(wind going around rocks)

Recessed areas in boulders/mountains tend to create "wind shadows" (areas with no wind at all) in them (also in *figure 3*). However, strong wind may be very close by, so check general areas carefully, if looking for wind. I have personally observed "wind shadows," by large rock embankments only about a dozen feet from places where the wind was strong.

Valleys can act the same way as horizontal recessed areas in mountains, because valleys are vertical recesses. A valley can create a wind shadow in it, especially if it is not very long, has at least one end closed (to block wind), and (especially) if the wind blows perpendicular to its long axis. However, a valley can instead intensify wind, if both ends are open (or it is very long) and especially if the wind blows parallel to its long axis. (Once, I walked down into the Grand Canyon. The North-South descending valley was short and perpendicular to the direction the wind blows, with the upper end blocked. Yes, there probably was lots of wind *over* the valley, but *in* the valley, the air was motionless---you couldn't even *hear* wind off in the distance. After I descended into the main East-West portion of the canyon, which was extremely long, unblocked and parallel to the direction the wind blows, the wind was strong and constant.)

Altitude makes a difference. Wind near the ground tends to be substantially weaker than wind high above the ground. This is because low-to-the-ground wind tends to be blocked and slowed down by low vegetation and objects on the ground, whereas high-off-the-ground wind doesn't get blocked and slowed. This is why people with portable wind turbines try to mount their rotors as high as possible, which on the RV size scale is ten to fifteen feet off the ground.

Cause of wind: Wind is caused by uneven heating of air at the Earth's surface. When air is heated, its pressure rises and it expands. Air that is heated more expands more, so it blows into areas where the heating/expansion/pressure is less. The same thing happens in reverse when air cools. When air cools, its pressure lowers and it contracts. Air that loses more heat contracts more, so it tends to suck in higher pressure air from areas where the cooling/contraction is less. Hence…

Land areas near large bodies of water are great places to find wind. This is because: 1) Water needs to absorb considerably greater energy in order to heat up than land does. Consequently, when the early morning sun hits, giving equal energy to all, the water (and the air over it) warms much more slowly than the land (and the air over it), so the warmer air over the land expands more and blows onto the water. When the sun goes down, the reverse takes place. When energy flows out of material at an equal rate, the land (and the air over it) cools more quickly than the water (and the air over it), so the air over the land contracts more and sucks up air from over the water. And: 2) The surfaces of bodies of water are almost perfectly flat (except for waves). Lack of obstructions, for tens or hundreds of miles, allows wind speeds to build up unimpeded over water. (It is known that hurricanes coming from out in the ocean start to lose their speed once they make landfall. It is also known that buildings close to the seashore are more vulnerable to hurricane winds than buildings far from the seashore.)

Time of day: By water, I have frequently observed strong wind in the evening and for a couple hours after sunset, dead air stillness deep in the night, then strong wind again at and for a couple hours after sunrise, and rather still air at midday. This is logical, since after sunset, the Earth cools, and after sunrise, the Earth warms, whereas around midnight and midday, temperatures neither rise nor fall.

Direction: Wind most often comes from the West and heads East.

Cardinal boulder directions (which sides face North, South, West and East): When wind hits a circular boulder on an otherwise flat land (usually from the West), the wind moves around that boulder, providing very poor wind on the two sides of the boulder that face and trail the wind, and very good intensified wind on the two sides that face perpendicular to the wind. Usually both the West and East sides of the boulder get very poor wind, whereas both the North and South sides get very good wind (*figure 4*).

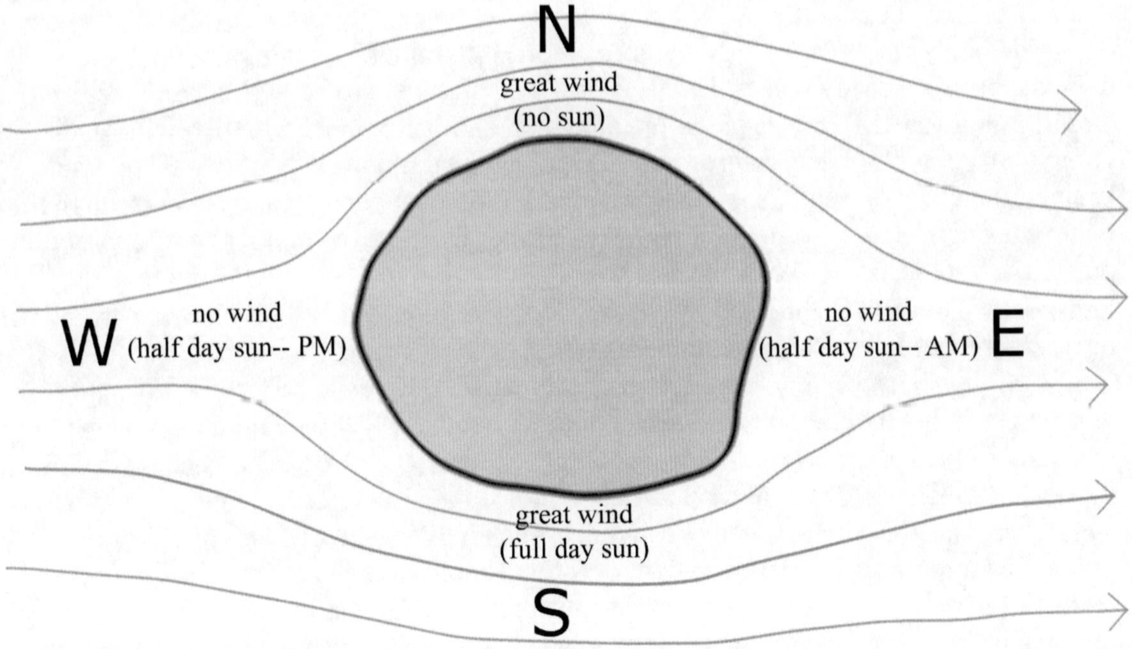

figure 4
(typical wind & sun expectations for each direction)

Conclusion: Wind turbines can make lots of electricity, but only when the wind blows. Wind turbines can be good choices for people with permanent stationary cabins (especially near water or on prairies), where space is not an issue and where equipment only needs to be set up only once. But for highly mobile RVs, wind turbines are not a good option, because they take up lots of storage space, require too much time for setting up and dismantling multiple times, and are unreliable.

IV.) Thermoelectric Generators

What Are Thermoelectric Generators? How do They Work?

What are thermoelectric generators? Thermoelectric generators are devices that passively make electricity from the *flow of heat*, using electrical components called *thermopiles*. Thermoelectric generators are also called thermogenerators, by their acronym TEGs, or Seebeck generators (named after the man who discovered the principle that makes them possible).

How do thermoelectric generators work? If you look at a thermopile, the component that actually makes the electricity, you are likely to be initially unimpressed. You'll usually see a flat rectangular metallic plate, not very thick, with two wire leads coming off of it--- not very interesting. But if you heat up one side of that plate, while keeping the other side cold, something very interesting happens. Heat flows through this plate, causing electrons to flow through the thermopile and become more concentrated in one of the wire leads. (The electrons flow "uphill" from the "+" wire lead where they are less crowded to the "–" wire lead where they are more crowded.) This electric current can be used to either run devices or charge batteries.

A bit more technical explanation: All material strives to be electrically neutral, which means that the conglomeration of atoms tries to have the same number of negatively charged ("–") electrons as positively charged ("+") protons. However, because different metals have different physical properties and molecular structures, when two different metals are placed in contact with one another, one tends to hog a few extra electrons (compared to protons), giving it a slight negative charge over the other, while the other tends to surrender a few electrons (compared to protons), giving it a slight positive charge over the other. In *figure 5A*, which shows two junctions between metals A and B at the same temperature, the relative electron density pressure difference (ΔV) at the first junction going from metal A to metal B, is equal and opposite to the relative electron density pressure difference (ΔV) at the second junction going from metal B to metal A, so the voltage (ΔV) between the two endpoints of the circuit P1 and P2 is zero. (Electron pressures are like heights on a stairway. If you take a step 10 inches up on a stairway, then take another step 10 inches down, you are at the same height that you started.) Stated mathematically:

$$\Delta V(\text{from A to B}) \;=\; -\Delta V(\text{from B to A}), \quad \text{so}$$
$$\Delta V(\text{from P1 to P2}) = 0$$

However, at higher temperatures (*figure 5B*), this tendency for one metal to hog electrons (and the other to donate them) is exaggerated, so if the two junctions are at different temperatures, the two opposing ΔVs (from A to B and from B to A) are not equal, therefore there exists a voltage difference (ΔV) within the same metal (A) between the two endpoints of the circuit P1 and P2 that is not zero. (It's like taking a step 12 inches up on a stairway, then taking another step 8 inches down. You end at a different height than you started.) Stated mathematically:

$$\Delta V(\text{from A to B}) \neq -\Delta V(\text{from B to A}), \text{ so}$$
$$\Delta V(\text{from P1 to P2}) \neq 0.$$

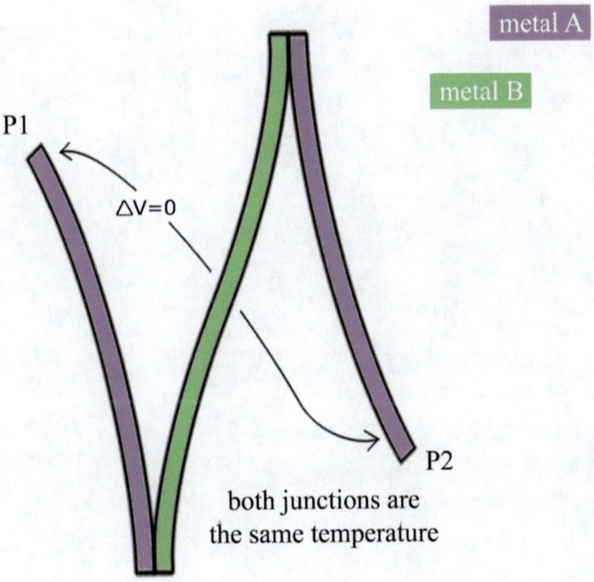

figure 5A
(thermocouple)
(not being used)

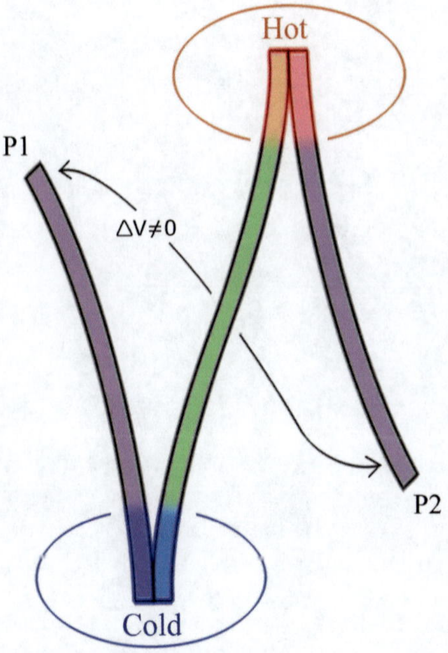

figure 5B
(thermocouple in action)

A pair of junctions between two different metals, with both ends being the same metal (*figures 5A and 5B*), is called a *thermocouple*. A single thermocouple (with one junction hot, the other cold) cannot produce significant voltage. But multiple junctions, connected in series (end to end) can produce significant voltage between the two endpoints of the circuit. A chain of thermocouples, connected in series, is called a *thermopile* (*figure 6*).

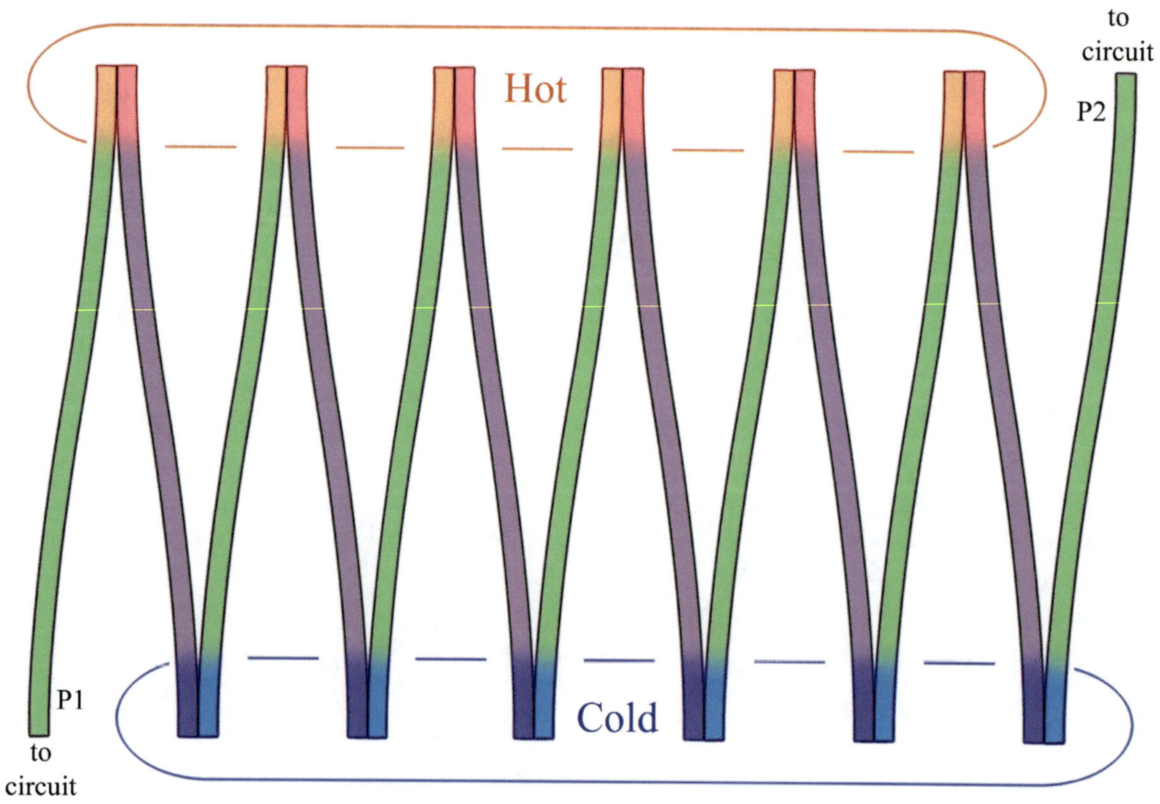

figure 6
(thermopile)
(made of wires)

Today's functional thermopiles do not use wires, instead they use alternating flat plates of p and n doped semiconductors, electrically insulated (thus separated) from each other, but electrically connected at their endpoints by conductors (*figure 7*). Using p and n doped semiconductors (instead of non-doped metals) greatly improves thermopile electrical outputs. Older models used non-doped metals. They didn't work as well.

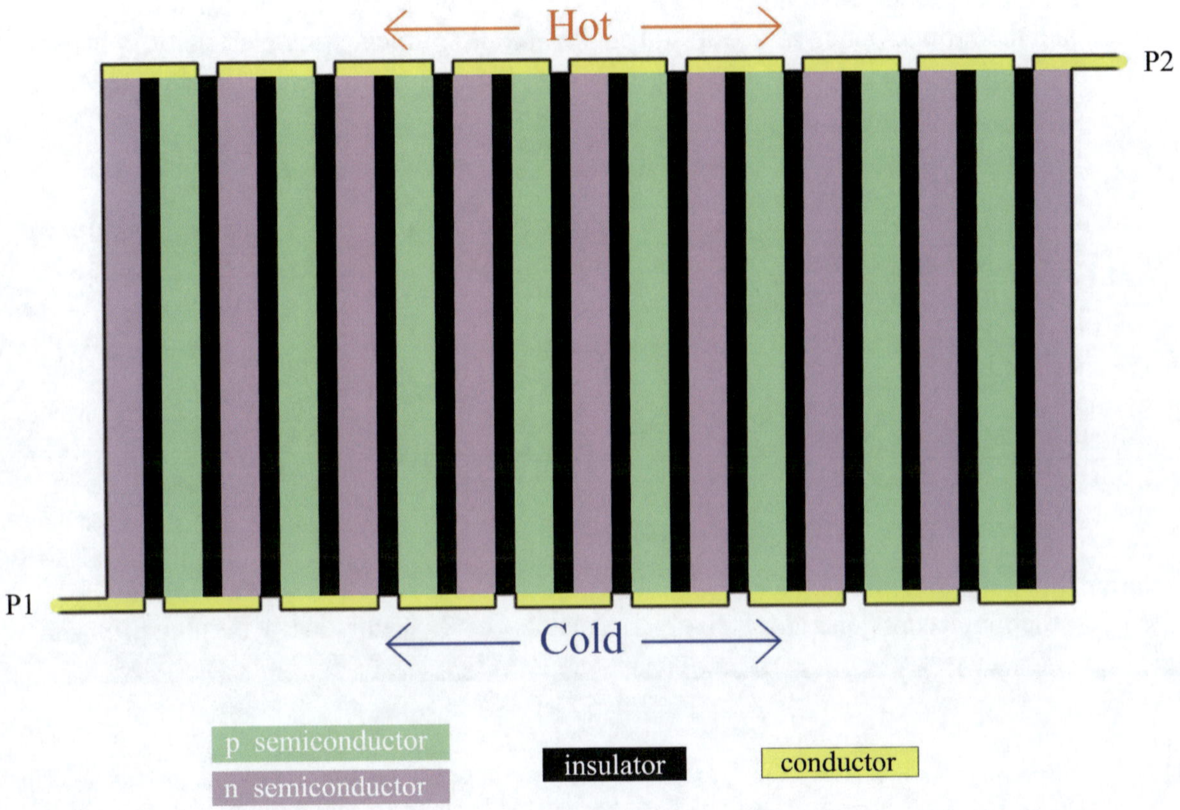

figure 7
(functional thermopile)
(cross sectional view)

Getting a thermoelectric generator to work

What needs to be done: One side of the thermopile must be kept hot while the other side must be kept cold. In an open circuit situation (nothing connected), the voltage difference (ΔV) between the two wire leads is directly proportional to the temperature difference (ΔT) between the two sides of the thermopile plate. Stated mathematically:

$$\Delta V \ \alpha \ \Delta T.$$

In a circuit where current gets drawn, heat flow ($\Delta Q/\Delta t$) is proportional to current (I). (Stated another way: *heat flow is proportional to electron flow.*) Stated mathematically:

$$(\Delta Q/\Delta t) \ \alpha \ I.$$

(However, this formula assumes 100% efficiency. In reality, whenever heat flows through a thermopile, much of it leaks through without producing any electricity.) Heat flow through the plate is approximately proportional to the temperature difference between the two sides of that plate, thus it is advantageous to keep that temperature difference as great as possible, without damaging that plate. If heat gets continuously applied to one side, while the other side gets continuously cooled, that temperature difference gets maintained indefinitely, so heat flows through the plate indefinitely, hence the thermopile makes current indefinitely.

The driving mechanism for a thermoelectric generator: A temperature gradient must exist (meaning something "hot" in close proximity to something "cold"), in order for the "hot" and "cold" to be directed to opposite sides of the thermopile. Temperature gradients can be found in nature or made artificially. While thermoelectric generation is possible from natural temperature gradients (such as warm rocks in the sun by cold water from a river, or geothermal springs by cool air), most often, the "hot" is made artificially through combustion, while the "cold" is just ambient stuff that's around, such as water or air. It is almost always easier to make something hot than it is to make something cold, so for artificial temperature gradients, it is almost always the "hot" that is made artificially. An RV thermoelectric generator is likely to make its heat by catalytically burning propane.

Actual Uses of Thermoelectric Generators

1) Natural gas companies sometimes use them to power remote sensing equipment along gas lines. This makes a lot of sense, since natural gas, needed for burning to make the hot spots, is always available along gas lines.
2) NASA has been using thermoelectric generators aboard space probes to power equipment for decades. (Voyagers used them to photograph Jupiter, Saturn, Uranus, Neptune & their moons.) These devices use lumps of radioactive material as their heat sources, so they are called radioisotopic thermoelectric generators (acronym RTGs). Radioactive material, by its very nature, produces heat reliably and steadily for decades, centuries or millennia, depending upon isotopes used. Since many parts of outer space are quite cold, conditions are ideal for heat to flow.
3) Thermoelectric generators are sometimes used by the military in combat settings, to charge batteries, since the noise of generators would give away locations and prevent troops from hearing approaching enemies.
4) Thermoelectric generators are sometimes placed around chimney pipes of or upon woodburning or pellet furnaces, to charge batteries (for off grid homes) or to make electricity to sell back to the grid (for homes connected to the grid).
5) One company used the heat difference, between a live human body and the outside air, to drive a tiny thermoelectric generator to make electricity to power a wristwatch.
6) In the Soviet Union, battery free thermoelectric generator radios were common in the 1920s.

Proposed Uses of Thermoelectric Generators

1) Some have suggested that thermopiles be placed on the backs of solar cells, to add efficiency to electricity production, since heat, from sun shining on solar cells, is just wasted.
2) Some have suggested placing thermopiles around the exhaust manifolds of automobiles. If most or all of an automobile's electrical needs (including battery charging) could be met by this method, the alternator could be shut off sometimes or eliminated entirely. This would reduce the need of the engine to torque the alternator, which could increase the automobile's overall fuel efficiency by 5% to 15%.

Technical Points

1) Thermoelectric generators have other components besides just the thermopile. A TEG may have a catalytic burner as its heat source (propane or natural gas is often used as fuel), an automatic igniter for that burner, an active cooling system (to keep one side of the thermopile as cold as possible), computer control (to shut down operation when the battery attains full charge) and a diode (to prevent electricity from flowing the wrong way and discharging the battery, when the heat source isn't on).

2) Thermopiles can be damaged by high heat. One needs to be careful, if trying to run one by a poorly controlled heat source, such as a campfire or woodstove. Generally, overheating is not a problem for TEGs that have carefully controlled burning units.

3) Some TEGs are cooled by actively circulating liquid (water or antifreeze) and/or heated by actively circulating oil. On the "cold" side, the liquid is pumped between the thermopile and a radiator by a low energy electric pump. The liquid absorbs heat at the surface of the thermopile and loses that heat to the environment at the radiator (which is often an automotive transmission cooler). On the "hot" side, oil may move heat from the heat source to the thermopile. The pumps required to do such jobs are small, inexpensive, quiet (about as noisy as household aquarium pumps) and readily replaceable. Of course, the electricity needed to run such pumps is considerably less than the electricity produced by the thermopile. Also the extra electricity gained by having pumped systems (as opposed to just passively cooling with air) is more than the pumps use, otherwise such pumps would not be justified. Batteries are not used for the pumps; they run directly from electricity made by the thermopile.

4) Thermopiles can be reversed. This is because a thermopile couples the flow of heat to the flow of electrons. So while it is true that a flow of heat causes a flow of electrons, it is also true that a flow of electrons causes a flow of heat. (Sending a DC electric current through a thermopile causes one surface to become cold and the other surface to become warm.) A thermoelectric generator, used in reverse, is called a *thermoelectric cooler*.

Thermoelectric Coolers

Where can thermoelectric coolers be found? You might even have a thermoelectric cooler and not even realize it. Do you by any chance have a small cooler, that plugs into the cigarette lighter receptacle of a car, that keeps food cold while out on a road trip? (Even if you don't have or use one, odds are, you know somebody else who does.) That's a thermoelectric cooler.

Can thermoelectric coolers be used in reverse to make electricity? Apparently they can. I've read several internet accounts of people who wondered the same thing, then attempted to answer this question experimentally. They heated one side of a cooler's thermopile, and found that it *does* make electricity. If you removed all diodes, the fan, and all other circuitry from your thermoelectric cooler, then created a heat gradient greater than that which the battery's voltage can produce (by filling your thermoelectric cooler with dry ice and submerging its outside in hot water, for example) then plugged it into your car, your thermoelectric cooler would start *charging* your car battery, rather than *draining* it!

5) Charging batteries with TEGs is not a new idea. I found an internet photo of an antique TEG from the first decade of 1900s that connected to a woodstove to charge a house battery. Back in the early 1900s, house batteries may have been used for little more than ringing doorbells--- few people knew what to do with electricity in those days. The principle that makes thermoelectric generation possible, called the *thermoelectric effect*, isn't new either. It was discovered way back in 1811.

6) Steam turbines also make electricity from the flow of heat. But they are completely different, because unlike TEGs, steam driven turbines: A) are active devices (engines with lots of moving parts), B) make electricity by a totally different mechanism (interaction of moving magnetic fields), C) usually make more electricity (than TEGs) and D) are much more commonly used and known about than TEGs (nuclear and coal burning power plants use steam driven turbines to produce municipal electricity).

Thermoelectric Generators Verses Solar Panels

Thermoelectric generators have very similar properties to solar panels, but they do have two major differences.

Similarities: Like solar panels, thermoelectric generators: 1) make DC, 2) in small amounts, 3) with no moving parts [to wear out or annoy someone], so 4) they can be left on for long periods of time, 5) allowing substantial charge to build up in batteries. Like solar panels, TEGs: 6) are ideally suited for light-duty long term applications, and not well suited for heavy-duty short term applications, and 7) are ideally suited for battery charging, and not so good at running appliances directly.

Differences: 1) Thermoelectric generators [that make their heat by burning fuel] are reliable. They can be used anywhere anytime, as long as propane fuel to run them is available. Solar panels are not reliable, because in order to produce, they need sunshine, which is not always available. 2) Thermoelectric generators that burn propane have some operational cost, since propane is not free. Solar panels make electricity that is free.

My Own Proposal for an RV Thermoelectric Generator

I believe that a propane fueled catalytic burning thermoelectric generator would be an ideal way to charge RV batteries because: 1) propane is inexpensive, 2) propane is readily available (even in rather remote locations), 3) all RVs already have propane storage tanks and 4) all RV users are already accustomed to buying, storing and using propane. The only needed modification, besides making room for the unit and installing it, would be tapping an extra gas line into the propane system (and perhaps having larger propane tanks to accommodate increased consumption). Also, a thermoelectric generator would be the only method for wilderness battery charging that would be both reliable and quiet. However, there is one issue: *such a unit would also make heat.*

Heat is both an advantage and a disadvantage for RVers, depending upon the circumstances. On cold days/nights, heat dumped into an RV is a gift from the heavens. On hot days (especially when parked in the sun), heat dumped into an RV is a curse from the devil himself. It would thus be great if the heat from an RV thermogenerator's burner could be directed indoors on cold days/nights, and outdoors on hot days. This would make it usable under all outdoor temperature conditions. My proposed solution: *mount the TEG on a small slideout*. When outdoor temperatures were cold, the unit would be used indoors, for simultaneous battery charging and interior RV heating. When outdoor temperatures were hot, the unit would be slid to the outside and used outdoors, for battery charging alone. When used indoors, the exhaust from the burner could either be directly discharged indoors (acceptable because it is catalytic) or hooked up to a quick connecting thin metal chimney (I'm imagining a connector that looks a bit like a flexible vacuum cleaner hose), where the exhaust would slowly work its way through the chimney from some low point near the floor up to and out of the roof, heating the metal chimney along the way, which would in turn heat the RV interior. (The chimney would be physically located inside the RV but would vent its exhaust to the outside of the RV.) Note that the thermogenerator would not be designed to be used primarily as a furnace, however its waste heat might as well be dumped indoors on cold nights/days, where that heat would be useful. Since the primary furnace of the RV is temperature sensitive, it would work less aggressively at times when the thermogenerator was also heating the indoors, conserving propane. When used outdoors, the chimney would be disconnected; all exhaust and heat would be directly discharged outside.

There is still one more possibility for indoor RV thermogenerator use, on days with temperatures that were warm but not ridiculously hot. Use the unit indoors, vent its exhaust outdoors through the RV's side (instead of through the indoor vertical chimney) and insulate the unit well in a compartment, using flat removable thermal barriers, resembling cupboard doors or shelving. (Special flat compartments would be needed, to store the thermal barriers, when not in use.)

On very hot days, if a compression air conditioner were needed and operated, the thermogenerator would not be useful. This is because, in order to power the air conditioner, an internal combustion engine generator would need to run simultaneously. Internal combustion engine generators make lots of electricity, so besides running the air conditioner, the (internal combustion engine) generator would make plenty of extra electricity to operate a 120 Volts AC battery charger, capable of pushing as much current through the batteries as they were able to accept. When the (internal combustion engine) generator and the 120 Volts AC battery charger were on, it would be pointless to also turn on the thermogenerator, because the thermogenerator would be redundant.

Advantages of Thermoelectric Generators

1) Thermogenerators are reliable. They are the only electricity-making devices that are reliable (compared to all other electricity-making equipment/procedures), provided that propane to run them is continuously supplied. This is true because...

2) Thermogenerators work under all weather conditions. This is not true for two other popular electricity-making options: solar panels and wind turbines.

3) Thermogenerators are quiet. Except for a water pump, for an optional cooling system (which makes about as much sound as a household aquarium), and possibly an oil pump, a thermogenerator is noise free. Therefore...

4) Thermogenerators can be operated at night. Other electricity-making equipment can't or shouldn't be operated at night. Don't expect your solar panels to produce at night. Also, please don't turn on an internal combustion engine generator at night, anywhere where you've got neighbors.

5) Thermogenerators can operate at low costs. I found one unit on the internet that claims to consume 0.8 gallons (3.29 pounds) of propane per day (24 hour operation). That works out to only $2 to $3 per day (based upon the 2014 price of filling a 20 pound propane tank of $12 to $18). That's quite a bit cheaper than staying in private RV parks with electricity for $25 or more per day.

6) Thermogenerators can produce considerable outputs. The same unit described above claims an output of 1.8 Amps at 12 Volts. After considering charging inefficiency (batteries release about 5/6 of the electrons put into them), that works out to an effective charging rate of 1.5 Amps. This would result in 36 usable Amp hours per day, which could easily power everything (except a compression air conditioner) in even a big RV with energy-conserving appliances, even if no other battery charging sources were used.

7) Thermogenerators could be used as cogenerators of both electricity and comfort heat. I think the best time to use them would be during cold nights, while the RV occupants were sleeping. Since the burning propane would be used for two purposes, the actual cost of producing electricity would be further diminished.

8) Thermogenerators can charge batteries slowly and completely. This extends battery life and greatly extends the usable storage capacities of batteries.

9) Thermogenerators make excellent backup electricity, when all other systems fail. Remember that a thermogenerator would not need to be turned on during times when enough electricity were made by other methods, such as by charge-while-you-drive, or from solar or wind.

10) Thermogenerators are here today. They have been used to charge batteries and power gadgets for over 100 years. The technology that makes them possible was discovered over 200 years ago. Two of the other strategies proposed in this book (fuel cells and supercapacitor aided charging) rely on futuristic technology that, although already here, has numerous technical difficulties that need to be worked out. Also, futuristic technologies may not be available at reasonable prices, even if and when they do become commercially available.

Disadvantages of Thermoelectric Generators

1) Nobody apparently manufactures thermogenerators specifically for charging RV batteries. If you do internet searches, you'll find plenty on RVs, and plenty on thermoelectric generators. But I seem to be the only one thus far who links these two ideas together, advocating thermogenerators for RV battery charging. And as for mounting such a unit on a small slideout, so it could be used in both warm and cold weather, well to me, this seems to be about as obvious of a thing to do as eating food when one is hungry. Other people however don't find this so obvious. Why? Don't they read books on electricity? Are they so busy trying to copy other people's designs that they just have blinded themselves to the obvious? Thus…

2) If you want a thermogenerator in your RV, you will probably need to do some parts matching and install it yourself. Many thermogenerators do not come with heat sources; it is the responsibility of the buyer to provide hot woodburning stoves or gas burners to make them work. This is not too much of a problem, as propane catalytic heaters are cheap and readily available. However, if the burner isn't hot enough, electrical output may be too low; if the burner is too hot, it may damage/destroy the unit. Some TEGs require even more assembling and parts matching--- one needs to hook up pipes and a radiator and fill them with liquid for cooling and heating purposes, install water/oil pumps and heat sinks. Then there are gas duct/vent/chimney parts and thermal insulation (preferably flat sheets) that need positioning and installing. Making a slideout would be tricky and take some time and thought (tracks and wheels needed), but it is doable with parts that are readily available at hardware stores. The problem is: parts may need to be acquired from several sources, and since those parts are not specifically made to go together, much experimentation would be needed to get such a unit to work as a whole. Not everyone is well suited for do-it-yourself designing. It would be nice if someone sold a complete computerized TEG unit, specifically for an RV, requiring only hookups to a propane gas line, two electrical wires (to the battery) and a chimney/duct, but believe it or not, I've just described something that doesn't seem to exist anywhere in the present day commercial market.

> One does not need to be a rocket scientist in order to heat one side of a flat plate while keeping the other side cool. I saw a video in which a user in wilderness heated one side of a thermopile by facing it toward a campfire (about one foot away). The other side was part of the inside of a container, which he kept cool by filling it with cold water from a creek, changing water every 40 minutes. He made electricity!

3) Thermogenerators cost money. Everything does. Before adding significant cost to an RV, one needs to question whether one's existing electricity-producing system makes enough to comfortably accommodate one's existing electrical needs.

4) Propane-fueled thermoelectric generators aren't all that energy-efficient. Their efficiency is only about 2%, which is 7 times better than charging with the 3 Amps 120 Volts AC charger provided by many RVs (powered by an internal combustion engine generator), but only about half that of charging three batteries with 60 Amps of current (powered by an internal combustion engine generator).

Thermoelectric Generator Charging Efficiency
(Propane Fueled)

Technical Data:

Energy density of propane = 98,000,000 Joules/gallon
Consumption per day = 0.8 gallons per day
Charge rate = 1.8 Amps
In/out battery charging efficiency = 5/6

Total energy per day = Total energy in propane burned per day = (0.8 gallons/day) × (98,000,000 Joules/gallon) = 78,400,000 Joules/day.

Usable energy transferred to battery per day =
(Power) × (time) / day =
(Effective charging rate in Watts) × (24 hours)/day =
(Voltage) × (charging rate in Amps) × (in/out efficiency) × (24 hours)/day =
(12.5 Volts) × (1.8 Amps) × (5/6) × (24 hours)/day = (450 Watt hours)/day,
when converted to Joules =
(450 Watt hours)/day × (3600 Joules / Watt hour) = 1,620,000 Joules/day.

Efficiency = (Usable energy transferred to battery per day) / (Total energy per day) = (1,620,000 Joules) / (78,400,000 Joules) = .0207 or 1 part in 48.

Conclusion: Thermoelectric generators are perhaps the least well known and least frequently used method for making RV electricity. Most people don't even know what they are. Most people have never heard of them. But that doesn't mean they can't or shouldn't play an important role for RVs. Propane fueled thermoelectric generators are probably the best and most reliable standing method for charging RV batteries in the wilderness.

V.) Fuel Cells

What Are Fuel Cells? How do They Work?

What are fuel cells? Fuel cells are devices which catalytically combine fuel with oxygen, in such a manner as to move and concentrate electrons. Multiple individual fuel cells connected together in series/parallel are called *fuel cell stacks.*

How do fuel cells work? Fuel (such as gasoline, propane, hydrogen) contains chemical energy, which gets released when that fuel burns and combines with oxygen. Under ordinary circumstances, the burning of fuel is not an *electrochemical reaction*, which means that it does not move electrons in such a manner as to concentrate them someplace and rarefy them someplace else, therefore the burning of fuels does not ordinarily directly make electricity. In electrochemical reactions, which occur in batteries, electrons do get moved in such a manner as to concentrate them someplace and rarify them someplace else (the opposite poles of a battery). A fuel cell is essentially a marriage between these two types of reactions: combustion and electrochemical reactions. Since combustion releases energy, it stands to reason that carefully controlled combustion in the presence of the right catalysts at the right temperatures could move electrons "uphill" (from the "+" region where they are less crowded to the "–" region where they are more crowded). This is exactly what happens in fuel cells.

What types of fuel do fuel cells use? There are many different types of fuel cells. Most are probably yet undiscovered. The type of fuel used depends upon the type of fuel cell used. Hydrogen is the most common fuel (one common variety is called a *proton-exchange-membrane* fuel cell), but natural gas, propane, methanol, diesel and other fuels are also sometimes used. There is even speculation that sugar could run small fuel cells for portable gadgets like cell phones and laptops.

How Do Fuel Cells Differ From Batteries?

Batteries and fuel cells have one thing in common: within them, chemical reactions take place which concentrate electrons someplace and rarefy them someplace else. Thus, it might be tempting to consider batteries and fuel cells to be the same types of devices. But there is one fundamental difference. In batteries, the electron moving chemicals come from the *insides* of the batteries and their used up byproducts stay inside those batteries. Batteries are thus closed systems. In fuel cells, although the reactions occur inside the fuel cells, the electron moving chemicals are fed into them from the *outside*, and their byproducts must also exhaust to the outside. Fuel cells are thus open systems. This difference causes batteries and fuel cells to have some very different properties.

Inside verses outside chemicals:

In batteries, the electron-moving chemicals are stored inside the batteries permanently. Those chemicals are put there at the time and place of the battery's manufacture. With the exception of small amounts of water that are added to some electrolytes, to replace water that is split into hydrogen and oxygen during the charging process, no chemicals are ever added nor removed from batteries during their useful lives.

In fuel cells, the electron-moving chemicals are fed into the fuel cells and the exhaust is allowed to escape. One exterior tank is needed to store the flammable fuel (unless the system is running from natural gas lines). The oxygen, which is also needed, may be fed into the fuel cells from the Earth's atmosphere or stored compressed in a second tank. The exhaust from the chemical reactions is usually water vapor and sometimes carbon dioxide.

Discharging & Recharging:

Discharging: *While discharging, batteries weaken gradually whereas fuel cells maintain steady electrical outputs.* As a battery's high energy electron-moving chemicals get used up, they become progressively rarer. Since the chemicals in a battery are quite evenly mixed, this causes the concentration of high energy molecules to diminish very gradually, hence a battery loses its ability to pump electrons very gradually. Since a fuel cell is continuously fed the electron-moving chemicals from the outside, it maintains full electrical output until it runs out of fuel, at which time its output quickly drops to zero.

> Have you ever noticed how batteries lose their strength very gradually? You probably throw away flashlight batteries when the lights get dim. The batteries still have energy at that point, just not much. This is very different from the way a car runs out of gas. A car's engine has as much power moments before the gas tank goes empty as it does when the gas tank is full. When it runs out, the engine sputters a bit and goes from full power to zero power briefly. Fuel cells are likely to behave more like a car running from a gas tank, than like batteries, maintaining full power until only moments before they run out of fuel.

Recharging: *Method:* Batteries get recharged by sending DC electric current through them whereas fuel cells get "recharged" by refilling (or replacing) the fuel tanks. Therefore: *Recharging Time:* Batteries take long times to recharge (because they can only accept limited currents), whereas fuel cells can be "recharged" in minutes (because filling or replacing fuel tanks doesn't take very long).

Advantages of Fuel Cells

1) Fuel has a very high energy density (compared to batteries). By definition:

(Energy density) = (usable energy amount) / (volume of space required to store that energy).

This means that an equal amount of energy, to that which can be stored in a lead-acid battery, can be stored in fuel occupying a much smaller space. Fuel also has a very high *specific energy* (compared to batteries). Specific energy is a similar concept to energy density, except specific energy compares the usable energy to the *mass (often called weight)* required to store it. By definition:

(Specific energy) = (usable energy amount) / (mass required to store that energy).

This means that an equal amount of energy, to that which can be stored in lead-acid batteries, can be stored in fuel weighing much less. In practical terms, this means: *you can store much more energy in the same space and weight, if you store that energy in fuel instead of in batteries.* Since available space and allowable weight are limited in RVs (and all vehicles), storing energy in fuel, if it could be done, would be a great advantage.

Sample comparisons:

It would take over 75 times as much space to store in batteries the same energy that can be stored in propane fuel. Do you have room to store 75 propane tanks in your RV? I sure don't!

It would take over 700 times as much weight to store in batteries the same energy that can be stored in hydrogen.

Source: Wikipedia--- "Energy Density"

The extremely high energy density of fuel, compared to batteries, is a fundamental reason why electric cars have never been put into widespread production. Imagine if all the energy of fuel combustion (or a good portion thereof) could be used to move electrons instead of simply being burned--- that would make a lot of electricity.

2) Fuel cells could potentially make enormous amounts of electricity in RVs. Fuel cells could out-produce all other electricity making/gathering methods many times over. This could fundamentally change the very nature of RV electrical/appliance systems.

3) Fuel cells are mostly passive technology. They do not have big heavy noisy moving parts, like internal combustion engine generators do. They might however require small motors to precisely feed them their fuel.

Disadvantages of Fuel Cells

1) Fuel cells are kind of futuristic technology. The fundamental discovery that made them possible occurred way back in 1839, when electrolysis was successfully reversed. ("Electrolysis" is the *splitting* of water into hydrogen and oxygen, *using* an electric current. "Reverse electrolysis" is the *combining* of hydrogen and oxygen into water, *making* an electric current.) Despite this, fuel cells are loaded with problems that have yet to be effectively solved. Among them…

A) Fuel cells need high operating temperatures. They cannot work at normal room temperatures, the way batteries can, so their insides must be heated prior to their use. Users must therefore either wait for they to warm up, or keep them perpetually warm, in order for electricity to be ready upon demand. Proton-exchange-membrane hydrogen fuel cells (the most likely type to be used in RVs and electric cars) require a temperature of about 80° C, which is slightly lower than the boiling point of water. Other hydrogen fuel cell types require higher temperatures in the range of 150° – 200° C, which is hot, but not ridiculously hot. Some high temperature fuel cells, that consume natural gas or propane, require temperatures of 800° C – 1000° C, which is really hot.

B) Hydrogen, required for most fuel cells types, is not yet widely available.

C) Hydrogen is difficult to store. Hydrogen cannot be liquefied by compressing it, the way propane can be, at normal ambient Earth temperatures. It must thus be stored in tanks as a gas under extremely high pressure. These tanks must be super strong, in order to withstand those pressures without bursting. Or, if hydrogen isn't stored…

D) Hydrogen must be made (extracted) from hydrocarbon fuels using an external device called a ***reformer***. In higher temperature fuel cells, reformers are not needed, because those high interior temperatures decompose the fuels, and hydrogen is one of those decomposition products.

E) The fuel and oxygen must be fed into the cells in very precise quantities. The waste product, water, must be evaporated from the cells in very precise quantities. Otherwise, the fuel cells could be damaged or destroyed.

F) Internal temperatures must be carefully regulated.

G) Fuel cells are extremely sensitive to contamination. They can be poisoned by carbon and rendered useless.

2) Fuel cells have not yet been put into widespread commercial production. This is due to the many unsolved problems they present. I have yet to see fuel cells available at Wal-Mart.

How Might Fuel Cells Be Used In RVs?

Because fuel cells need high internal temperatures and careful regulation, most likely, they would be turned on only occasionally, to charge batteries or to power high energy consuming equipment, such as compression air conditioners.

> **Houston, we've had a problem...**
>
> The mishap, which crippled the Apollo 13 space mission to the Moon, and almost cost its three astronaut crew their lives, involved fuel cells. They stirred a tank of hydrogen (used for the fuel cells). Faulty exposed wiring in an oxygen tank (also used for the fuel cells) started a fire, which resulted in an explosion, which blasted a big hole in the spacecraft and vented precious oxygen into space. Because oxygen was required for fuel cell use, they had to conserve it, by shutting down most of their fuel cells. They had to really conserve electricity, in order to survive.

Conclusion: Fuel cells are not yet well developed enough to be available to power today's RVs. But they can theoretically make so much electricity, that if their technical problems could be solved, and they were to be put into widespread commercial production, in the future, they could make so much electricity for RVs, that they could render nearly everything written in this book obsolete.

VI.) Generators Powering 120 Volts AC Chargers

Since internal combustion engine generators have been covered extensively in both Chapters 1 and 2, I'll be brief. (I use the term "internal combustion engine generators" to mean the traditional engine driven generators often used by RVs. Not to be confused with "thermoelectric generators," which are a completely different type of device.)

Generators Versus Other Electricity-Production Methods

Major Advantage: Generator outputs can be used to immediately run things. With only a few light-duty exceptions, all other electricity-producing methods almost always store their outputs in batteries, for later use.

Other Advantages: 1) Generators can make lots of electricity, 2) generators can be started upon demand, 3) generators take only seconds to get started, 4) generators don't use much fuel if run for only short time periods, 5) generators can run for long total time periods without breaking down and 6) cheap portable models cost only a few hundred dollars.

Major Uses: Because of their high outputs and ability to start upon demand, generators are great for occasional short term usc of high energy gobblers, like household vacuum cleaners, microwave ovens and compression air-conditioners (which absolutely need them), all of which can be found in the RV world. In the non-RV world, generators are especially useful for emergency power, if/when municipal electricity goes out, where power outages are infrequent enough such that it doesn't make sense to have and learn to use alternative electricity-conserving or non-electric machinery. Generators are especially useful for essential facilities/services such as hospitals, police stations or emergency disaster relief.

Disadvantages: 1) Generator noise annoys people. 2) Generators make way more electricity than is usually needed, so most of the time, they just grind away accomplishing nothing. 3) There are better ways to charge batteries in wilderness.

> Have you ever heard of a person who weighs 517 pounds, never bothers to exercise, and eats only deep fried greasy and highly processed foods and no more than 2.6 ounces of fresh vegetables per year, who then takes anti-cholesterol drugs, insulin, and swallows at least 8 other medications like they're candy? What is the role of his medications? To offset the effects of a very unhealthy lifestyle. Does he need his medications? If he continues his poor lifestyle, yes--- it's a better alternative than dying--- but what he really needs to do is fundamentally change his habits. Are his medications likely to work? They may keep him alive a little longer, still, he's heading for an early grave. If you ask an average RV user what the role of a generator is, he might answer from a utilitarian perspective and say, "to run lights," "to run my electric coffeepot and toaster" or "to keep my batteries from going dead." If you ask me the same thing, I will answer from a more philosophical perspective and say: "The role of generators is to rescue poorly designed RV electrical & appliance systems. They are a quick-fix solution and not a very effective one." The poorly designed RV is like that 517 pound man. The generator is like the man's medications. Does the RV need a generator? If design improvements aren't made, yes--- it's better than being in the dark--- but what the RV really needs is energy-saving appliances, more batteries and a quiet effective charging system. Is the generator likely to make the RV system work well? It may keep the system alive a little longer, still, the RV is likely to struggle, go dark and head for the junkyard quicker than it probably should.

VII.) Charging From Municipal Electricity

Yes, this is cheating, because it technically is not a true off-grid strategy, but it can work. It might be especially appealing as a temporary solution for those new to RVing who need to come up with a battery charging strategy quickly.

Variations:

1) **Have multiple batteries and charge them either at an RV park or at one's regular home:** Besides the batteries themselves, all that is needed for simultaneous charging is one cheap 5-6 Amp battery charger for each battery and enough parking time to complete the task. (24 hours is preferred, but even 12 hours can do substantial charging.) If one has enough batteries to last a duration of a couple weeks, and the entire trip length is shorter than that duration, all batteries can be charged at one's regular "home" prior to embarking, then recharged upon returning. For longer trips, RVers could just decide to stay in an RV park with electricity for one day/night every two weeks, do their charging there, then head back into the wilderness for the rest of the time.

Advantages: 1) This strategy is fundamentally simple, both to do and to understand. 2) Batteries can be charged from a choice of multiple locations using the same equipment (home, RV parks, or at gas stations when batteries are removed).

Disadvantages: 1) This is costly. One might need half a dozen batteries and chargers to sustain a two week interval, using an RV with traditional energy wasting appliances. This could cost almost $1000. Of course, one could have less batteries & chargers and go to RV parks more often, but this defeats the purpose and isn't true off-grid living. 2) Batteries are heavy. Half a dozen batteries and chargers might weigh about 250 pounds. Still, carrying around this added weight isn't an impossibility, especially in big RVs. 3) Multiple batteries & chargers take up storage space. I've already covered this point elsewhere, so I won't hammer it again here. 4) RV parks are costly. However, if one limits patronage to one day/night each two weeks, the cost becomes manageable. 5) Moving the RV to and from an RV park is a major bother. Packing, relocating, unpacking, staying the day/night, repacking, re-relocating, will probably take about 48 hours and lots of effort, and could take longer if the RV park is far away. This is certainly not a task that I would want to do more than once every two weeks. However, if one carefully coordinates one's movements, in such a manner that one hits the RV parks while in-transit between off-grid camping destinations, the time and effort needed to do this could become manageable. 6) There are better easier ways to charge batteries. Charge-as-you-drive works very well, is quite simple to do (but does require placing and installing wires), and works without requiring multiple big heavy batteries, so why resort to this multiple battery multiple charger municipal electricity charging strategy that is more complicated to do?

2) Remove batteries and charge them elsewhere: For this strategy to work, one needs two house batteries, a cheap 120 Volts AC charger, and a nearby automotive service station or somebody else with a municipal electrical outlet willing to let you borrow it. One battery stays at the RV powering the appliances, while the other battery and the charger spends the night plugged in at the service station charging. Be sure to give the repair shop a few bucks for the privilege of allowing you to charge your batteries.

Advantages: 1) This strategy is fundamentally simple. 2) Only two house batteries are needed. Even with energy wasting appliances, it is easy to get a single battery to last 24 hours, which is the time that it would take to fully charge the other battery at the service station. 3) No modification of any existing equipment or installing is necessary. This is so easy to do, one could buy the extra battery and charger and be good to go in less than an hour, provided one is near civilization. 4) This is inexpensive. A single extra deep cycle battery and charger costs about $150 [2016 prices].

Disadvantages: 1) One must impose upon someone else's electrical outlets. Many people might not mind RVers charging batteries at their facilities once in awhile, but they might not want RVers doing this on a regular basis. Lead-acid batteries are unappealing because they are dirty, they have acid, and they are explosive. 2) Swapping and moving heavy batteries frequently is an inconvenience, especially if it must be done often. So too can driving out of then back into the wilderness with the batteries be an inconvenience, especially if the commute is of a considerable distance.

Conclusion: Although many businesses currently are reluctant to allow RV battery charging at their facilities, who knows, maybe this will change. RV/camping stores sell all kinds of goodies that campers need, such as propane lanterns, LED flashlights, warm sleeping bags, propane (both in small portable canisters and as tank fillings), charcoal, matches, insect repellant, etcetera. It is surprising how much stuff is available near areas where people camp, even in areas that are remote. Maybe someday battery charging service will catch on and become commonplace enough such that RV users could survive alright with no charging equipment or strategy of their own. But because other strategies work better (especially charge-as-you-drive), charging by municipal electricity is primarily for RVers who are new to RVing, take limited short trips, so seriously pressed for time that they haven't considered other options, haven't done substantial reading, and need a quick fix (and possibly temporary) solution right away.

Supercapacitors

What are Supercapacitors?

Supercapacitors are an "alternative to batteries" class of devices that store electrical charge, for the purpose of powering electrical gadgetry. The term "supercapacitor" essentially means "a really big capacitor"--- big enough to be capable of directly powering electrical stuff for reasonable time periods. Supercapacitors differ from smaller capacitors in that besides the size difference, smaller capacitors exist in circuits for reasons other than powering things.

What is a capacitor? A capacitor is a double "dead end" circuit, where both dead ends are separated by an electrical insulator (*figure 8A*). (An insulator is material which does not permit the flow of electrons from one plate to the other.) A capacitor is to electrons a bit like what a dead end street is to cars. Cars can enter a dead end street but there is nowhere for them to go, so if many of them enter, they accumulate and the street becomes crowded. The cars could however enter the street in significant numbers, park there for awhile, then leave later. Likewise, electrons can enter one end of a capacitor, accumulate in significant numbers, then leave later. Capacitors differ from the "dead end street" analogy in one fundamental respect: unlike the street, a capacitor actually has two "dead ends" of opposite polarity and in close proximity. In one end, extra electrons accumulate, giving that end a negative charge. In the other end, electrons that should be there get sucked out and removed, leaving behind a partial electron vacuum, giving that end a positive charge. (Absent electrons, that should be in a material for electrical neutrality but aren't, are referred to as "holes.") Because the plates are close, and because electrons repel one another, when an electron heads into one end of a capacitor and gets stuck there, it causes an electron in the other end to dislodge, leave the capacitor and head into the circuit (*figure 8B,* a time lapse illustration). This can give the illusion that electrons are passing "through" a capacitor when in fact they are not, so such electrons and currents are said to pass "across" capacitors rather than "through" them. Capacitors often have a rolled up structure--- like two long sheets of aluminum foil, separated by wax paper, all rolled up on the same cardboard roll. Capacitors are often electrolytic. In an electrolytic capacitor: 1) there is some chemical change that exists when it is in a charged state [as compared to when it is not in a charged state] and 2) it is polarized [meaning that one side must have the positive (+) charge and the other side must have the negative (–) charge--- else it fries].

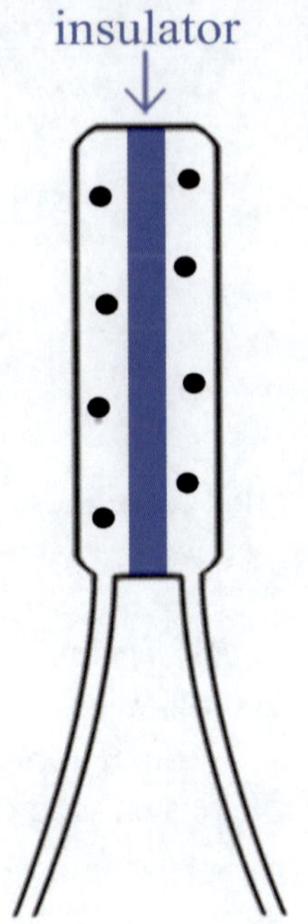

figure 8A (capacitor)

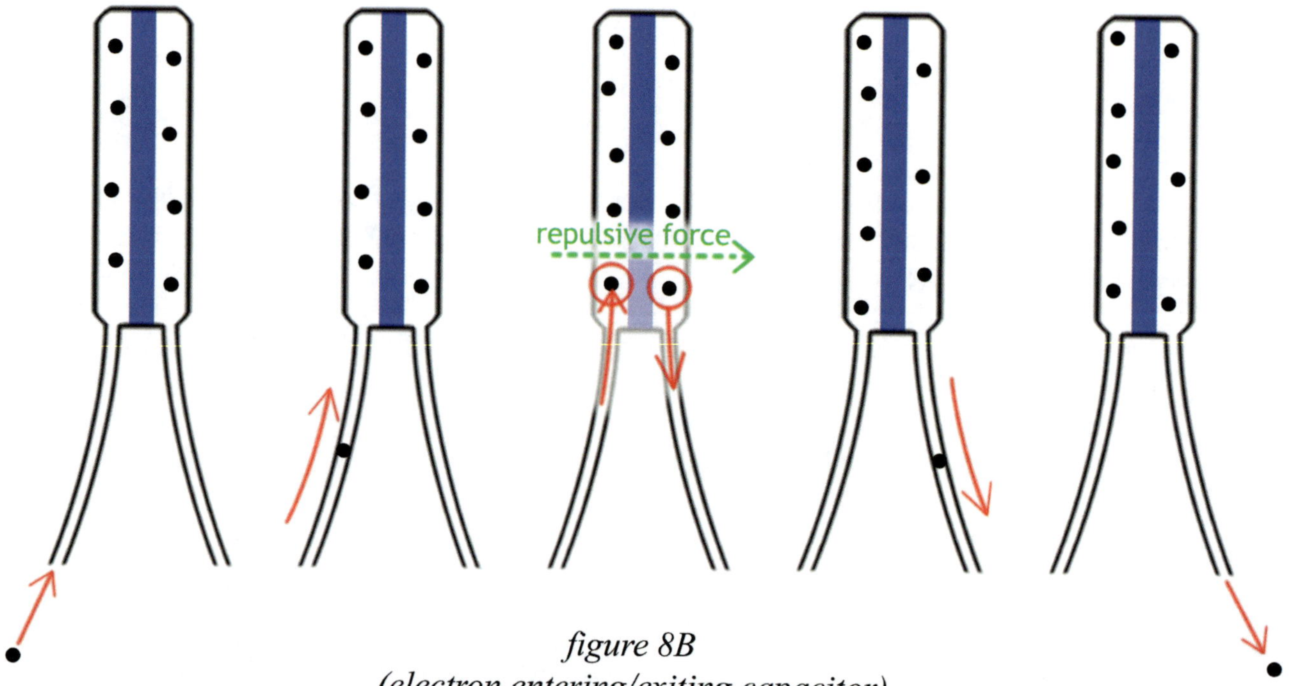

figure 8B
(electron entering/exiting capacitor)

Note that the electron that leaves the capacitor down the right wire lead (illustration far right) is not the same electron that enters the left wire lead (illustration far left).

Where can capacitors be found? Small capacitors are an integral part of almost all electronic devices. They are almost as abundant in electrical circuits as trees are in forests. Capacitors can be found in radios, TVs, computers and in some motors. They can exist as small etchings in tiny integrated circuits, or as large discrete components connected (usually soldered) to other things. They are usually used to store small numbers of electrons for very brief time periods or to block DC while permitting AC to pass across them.

Differences Between Batteries and Supercapacitors

Supercapacitors store electrons very differently from the way batteries do. Batteries are essentially "chemical demand pumps" of electrons, behaving much the way that RV demand water pumps do. Supercapacitors are essentially "tanks of compressed electrons," behaving much the way tanks of compressed air do. Hence, batteries and supercapacitors have very different properties.

1) Batteries charge slow whereas supercapacitors charge fast. In a battery, chemical reactions must take place, both during charging and discharge. During charging, electrons and electrolyte molecules must "find each other," then the charged molecules must migrate away from the electrode surfaces, while uncharged molecules must migrate toward the electrode surfaces. Migrations take time, and electrons and molecules finding each other is a matter of chance, so what this means is: a battery has a limited ability to accept current during charging. Also, when a battery nears full charge, it can't accept much charging current, because the target uncharged molecules become increasingly rare. In a supercapacitor, chemical migrations don't take place (or travel only tiny distances) and electrons don't need to find anything. The electrons are simply withdrawn from one end creating a rarity ("+" charge), and jam packed into the other end creating a surplus ("–" charge). Hence, unlike a battery, a supercapacitor can be charged up very quickly (minutes or even seconds)--- electrons can be added/withdrawn almost as fast as the wires can carry them or the charger can supply them. Also, unlike a battery, a supercapacitor can charge with the same current when it nears 100% charge as it can when its charge state is lower.

2) Batteries discharge slow whereas supercapacitors can discharge fast. This is for the same reasons as above. (Capacitors are often used to zap things with a lot of energy quickly, such as the electronic flashes of cameras. Batteries can't do this.)

3) During discharge, supercapacitor voltage drops rapidly whereas battery voltage drops slowly. A battery is essentially a chemical demand pump, behaving much like an RV demand water pump. Just as an RV demand pump turns on and pumps when water gets lost through faucets, chemical reactions take place in a battery whenever electrons get lost through the circuit. Therefore, when a battery gives up electrons, its voltage drops very little. A supercapacitor has no such chemical voltage stabilization; its only mechanism of storage is the jam packing of electrons themselves. A supercapacitor is essentially a tank of compressed electrons, behaving much like a tank of compressed air. Just as the pressure in a tank of compressed air drops whenever air is allowed to escape from that tank to do something useful, whenever electrons get removed from a supercapacitor and flow through a circuit, the voltage starts to lower immediately. This causes the supercapacitor's voltage to vary drastically during its use. Most electronic devices can only function within a narrow voltage range, so devices designed to work when a supercapacitor is at full charge need to be shut off early during the discharge cycle, rendering much of the supercapacitor's stored energy unusable. One way of getting around this voltage lowering problem is to simply accept it, by only operating the supercapacitor within a small portion of its peak voltage range (like maybe from 4 Volts to 3.5 Volts). But this is wasteful--- it makes the usable range of the "tank of electrons" considerably smaller than its actual storage capacity. Another way of getting around this voltage lowering problem is to use a DC to DC power inverter (which transforms the variable voltage into a stable voltage). But DC to DC power inverters suffer from the same problems that DC to AC power inverters (covered in Chapter 1) do: A) Power inverters are not completely efficient--- there is a 10%–20% energy loss. B) Power inverters add to the cost of the system. And: C) Power inverters use electricity even when no gadgets are turned on, so they should only be on when something is actually being used.

4) Supercapacitors are not degraded by cycling, like batteries are. The number of times that supercapacitors can be charged and discharged is almost limitless. Batteries degrade from being used, even if done so under ideal conditions. Also, supercapacitors are not damaged at all by being completely discharged and left that way indefinitely. Lead-acid batteries can quickly become trash, if they are discharged too deeply and left sitting in that discharged state for too long.

5) Supercapacitors are much easier to destroy by overcharging than batteries are. If a supercapacitor's voltage gets too high, it shorts out and quickly becomes trash. If a (wet lead-acid) battery's charging voltage gets too high, often the only result is boiled away electrolyte, which can be replaced by adding distilled water. Electronic voltage regulation is needed to prevent overcharging of supercapacitors (possibly zener diodes).

6) Supercapacitors have much longer lifespans than batteries do. During the lifespans of most rechargeable battery powered devices, batteries need to be replaced several times, because the devices often live longer than the batteries do. In supercapacitor powered devices, the supercapacitors often outlive the devices they run.

7) Supercapacitors self discharge at considerably greater rates than batteries do. Hence, supercapacitors are not suitable for long term storage of electricity. (But they can hold their charges for a little while.)

8) Supercapacitors cost much more than batteries (of equal energy storage). Right now, they can't even be considered as possible substitutes for RV batteries.

9) Supercapacitors have very poor energy density, compared to batteries (which already have low energy density). The energy density (space required to store a given amount of energy) of supercapacitors is only about one tenth that of batteries.

Energy Storage and Voltage Drop

Energy Storage:

The formula for the energy stored in a supercapacitor/capacitor is:

$$W = \frac{1}{2}V^2C$$

where W is the energy (work) measured in joules, V is the voltage measured in volts and C is the capacitance measured in farads. The term "capacitance" refers to the *size* of the capacitor, which if you think of a capacitor as being a "storage tank of electrons," this value would be the *size* of that tank. (A one farad capacitor is defined as the size of capacitor that stores one coulomb of electrons, when charged to a voltage of one volt.) The capacitor's rating in farads is directly proportional to both the number of electrons it stores and the energy it stores at any particular voltage. If a supercapacitor is not used over its entire voltage range (almost always the case), its usable energy is given by the formula:

$$W = \frac{1}{2}(V_{max}^2 - V_{min}^2)C$$

where V_{max} is the maximum voltage that the supercapacitor is charged to (the voltage at the beginning of its discharge cycle) and V_{min} is the minimum voltage at which it is used before turning the circuit off (the voltage at the end of its discharge cycle). Please note that since a capacitor's voltage varies during its use, the number of Amp hours it stores is not really a meaningful quantity, since more Amp hours are needed to produce the same energy when a lower voltage is used. The stored energy (W) is a much more meaningful quantity. This value (in joules--- which are Watt seconds) can be converted into Watt hours by dividing the W (in joules) by 3600, and this value can be converted to equivalent Amp hours at 12 Volts by diving again by 12.

Like batteries, supercapacitors can be arranged in series or parallel configurations or combinations of both. When identical capacitors are arranged in parallel, the total capacitance of the group is proportional to the number of individual capacitors. (Two parallel 12,000 farad capacitors equal one 24,000 farad capacitor, three equal one 36,000 farad capacitor, etcetera.) When arranged in series, the capacitance of the group is inversely proportional to the number of that group. (Two series 12,000 farad capacitors equal one 6000 farad capacitor, three equal one 4000 farad capacitor, etcetera.) Like batteries, the energy stored by a group of identical equally charged capacitors is proportional to the number of capacitors in that group, regardless of whether they are arranged in series or in parallel.

Voltage Drop:

If you look back at *figure 8B*, you should note that before the electron enters the left side of the capacitor (illustration far left), both plates have an equal number of electrons (four on each side). After the electron enters and gets stuck in the left plate and dislodges and pushes out an electron in the right plate (illustration far right), this is no longer so--- there are more electrons in the left plate than there are in the right plate (five in the left, three in the right). This is an example of the basic law by which capacitors work: *Whenever electrons enter/exit a capacitor, the voltage changes*, and *whenever the voltage changes, electrons enter/exit the capacitor*. (If electrons head into the side where electrons are more crowded, the voltage rises. If the electrons leave the side where electrons are more crowded, the voltage drops.) The formula relating these is:

$$\frac{\Delta V}{\Delta t} = \frac{I}{C}$$

where ΔV is the change in voltage, Δt is the change in time in seconds (also known as the time the current flows), I is the current in Amps, and C is the capacitance in farads. [Example: If 1 Amp leaves a 1000 farad capacitor initially charged to 4 Volts for 500 seconds, its voltage drops by 0.5 Volts, so it would finish at 3.5 Volts.]

Supercapacitor Aided Battery Charging

Right now, supercapacitors should not be considered as possible *substitutes* for RV batteries. Their extremely high cost, low energy density, variable voltages during their discharge cycles and tendencies to self discharge, are all unappealing characteristics. However, a hybrid system containing supercapacitors *in addition to* batteries may be worth considering.

The case for a supercapacitor aided system:

Since both generators and alternators can make lots of electricity, but usually only for brief time periods, and batteries are best charged by small amounts of electricity over long time periods, a supercapacitor could substantially improve the performance of a generator/alternator based charging system, by storing electricity intermediately. If a generator or alternator were to charge a supercapacitor (bank) instead of a battery (bank), a large amount of electrical energy could be stored quickly and temporarily (while the generator or alternator were on), then that energy could be slowly and steadily released later to run a battery charger (while the generator or alternator were off). To work, the supercapacitor bank would have to be the right size. Too many supercapacitors would cost too much, require too much storage space and defeat the purpose of the batteries. Too few supercapacitors would render such a system ineffective. I suggest a supercapacitor bank capable of storing about 10% of the energy that the batteries store.

Such a system could allow batteries to be charged by running the generator for several brief periods of only a few minutes each, instead of for long time periods involving mind-numbing noise at horrible efficiencies. If the supercapacitor bank were to be located in the car, even short drives could allow the alternator to build up substantial charge to transfer to the RV batteries. Best of all, the supercapacitor bank could be designed to be charged by either the generator or the alternator, so such a system would be doubly effective. And when "house" and tow/spare vehicle traveled hitched together, the supercapacitor bank would charge, enabling battery charging to continue to take place when parked for short rests, such as while eating, napping or engaging in "in transit" shopping sprees.

Finally, because supercapacitors are not damaged by multiple charge/discharge cycles, the supercapacitor bank could be used to release large amounts of electrical energy briefly for other purposes, such as for the microwave oven or some type of big heavy motor/machine that needs to be active for only short time periods.

8-Minutes of Generator Run-Time
(for charging)

A supercapacitor bank, capable of storing 10% of the energy of a 300 Amp hour storage capacity battery system, could be charged in only about 8 minutes, using a 4000 Watt generator, if most of its power could be directed into the supercapacitor.

A battery bank discharged to 50% could be fully charged with only five 8-minute generator sessions, each occurring every couple of hours.

If 30 Amp hours were enough for one day, then only one single 8-minute generator session would be needed per day (assuming no other charging sources get used).

Calculation

300 Amp hours would be stored in three 100 Amp hour batteries wired in parallel.
10% of this in the supercapacitor bank would be 30 Amp hours.

300 Amp hours × 10% = 30 Amp hours.

Since: Energy = Voltage × Charge,
the amount of energy needed would be 360 Watt hours, because:
(360 Watt hours) = (12 Volts) × (30 Amp hours).

Since: Energy = Power × time,
and: time = Energy/Power,
the time needed for a 4000 Watt generator to charge at 100% efficiency would be .09 hours, or 5.4 minutes, because:

(.09 hours) = (360 Watt hours) / (4000 Watts)
and
(.09 hours) × (60 minutes/hour) = (5.4 minutes).

If the efficiency were 2/3, this would be about 8 minutes:

Actual time = (time at 100% efficiency) / (efficiency factor) =
(5.4 minutes) / (2/3) = 8.1 minutes.

The case against a supercapacitor aided system:

1) Right now, supercapacitors capable of storing this much energy are cost prohibitive--- such a system, and the necessary hardware (like the chargers), might add thousands of dollars to an RV. 2) Buying and installing more batteries is still a much cheaper method of enhancing an RV electrical system, and a much better method, even though batteries are big and heavy and charge slowly. 3) There are other much cheaper ways of slowly charging batteries in wilderness, namely solar and thermoelectric. And: 4) Charging ordinary batteries while driving is extremely effective and cheap, especially if one's electrical needs aren't off the charts, so why spend money enhancing a system that already works quite well?

Conclusion: Supercapacitor aided battery charging may not be practical with today's technology. But there are some very intelligent minds working to develop both better batteries and supercapacitors for use in electric cars. All advances made in the field of electricity storage, for use in electric cars, are likely to also be very useful for RV "home" electricity systems. If supercapacitors get better and cheaper, they might be a viable RV option sometime in the future.

Comparing Battery Charging / Electricity Making Strategies

Advantages(+) / Disadvantages(–) Summery

Alternators (Charge As You Drive):

+ extremely inexpensive to install, as you already have the basic equipment
+ extremely low operational cost
+ fundamentally simple
+ makes lots of electricity
+ no noisy additional engine idling needed (in campground or anywhere else)
+ reliable, provided the vehicle is reliable and gets routinely driven

– may not make enough electricity for high energy consuming systems

Solar Panels:

+ quiet
+ no moving parts
+ not subject to mechanical breakdown
+ can be operated for long time spans, building substantial charge in batteries
+ good for keeping batteries charged while RV is in storage
+ extremely long life spans
+ occupies very small storage space
+ makes free electricity

– not reliable, as sunshine isn't always available
– small output, therefore multiple panels might be needed
– may not make enough electricity for high energy consuming systems

Wind Turbines:

+ makes lots of electricity (compared to solar), when the wind blows
+ makes free electricity

– unreliable, as wind is only occasionally available
– occupies considerable storage space
– requires setting up and dismantling, which might take considerable time
– somewhat noisy when operating

Thermoelectric Generators (propane fueled):

+ quiet (but may require tiny motors, which make very small noise)
+ extremely reliable
+ can work under all weather conditions
+ can be operated at night
+ low operational cost
+ could also heat RV in cold weather, while simultaneously making electricity
+ excellent backup when all other methods fail

– no known complete models specifically made and tested for RVs, therefore
– must assemble a complete unit from several unrelated components
– overheating can destroy thermopiles
– low steady output (but can do considerable battery charging over time)

Fuel Cells:

+ flammable fuel has very high energy density, therefore
+ fuel cells could potentially make enormous amounts of electricity

– rather futuristic technology
– many technical problems associated with fuel cells
– not yet commercially available

Generators Powering 120 Volts AC Chargers:

A) using 3 amps charger provided by many RVs:

+ no advantages (as far as I can discern)
– far too weak and slow to be practical

B) sending 60 Amps through 3 parallel batteries (20 Amps each)

+ relatively good efficiency
+ reliable (but only use during daytime and at places where generators allowed)
+ good backup when all other methods fail

– generator noise annoys people
– charging range limited to less than 80%* full battery
– there are better ways to charge batteries

* Not all of a battery's charge range is useful. On the low end, batteries struggle if discharged below 50%. On the high end, batteries can only accept low charging currents above 80%, so charging powered with generators is inadvisable above this range. The total usable charge range with this method is only 30% (between 50%-80% full), whereas with all other methods the usable charge range is 50% (between 50%-100% full).

Charging From Municipal Electricity:

+ simple to understand
+ can be done very soon with minimal or no equipment purchase and modification

– may cause RV park dependence, which is very expensive
– a single battery runs down very quickly in wilderness
– multiple batteries & chargers are heavy, take up space and could be expensive
– moving heavy batteries frequently to other locations for charging is a royal pain
– may have difficulty finding folks willing to charge batteries

Supercapacitor Aided Charging:

+ unlike batteries, supercapacitors can be charged incredibly quickly
+ generators and alternators can deliver very large currents, therefore a supercapacitor bank could "catch" large current surges, then use that stored energy to charge batteries slowly and completely

– supercapacitors are expensive
– supercapacitors have a much lower energy density than batteries
– largely a futuristic experimental technology, therefore
– not yet widely commercially available
– not a quiet charging strategy

Efficiencies

(ranked from best to worst)

Solar Panels:	infinite[1]		
Wind Turbines:	infinite[1]		
Alternators (Charge As You Drive):	15%-25% [2]	or	1 part in 4-6
Generators Powering 60 Amps Chargers:	4.2% [3]	or	1 part in 24
Thermoelectric Generators (propane fueled):	2.1% [4]	or	1 part in 48
Generators Powering 3 Amps Chargers:	0.3% [5]	or	1 part in 360
Fuel Cells:	unknown[6]		
Supercapacitor Aided:	unknown[7]		

Efficiency = (energy transferred to battery) / (energy in fuel consumed)

Notes:

[1] Both solar panels and wind turbines consume zero fuel, but they do harvest energy to charge batteries. Since they consume no fuel, their fuel efficiency is taken to be infinite.

[2] Calculated earlier in this chapter. Since vehicles drive vastly different distances under different conditions, this number is only a rough guess.

[3] Calculated in Chapter 2. [Solution #2 (to: Inadequate Chargers)]

[4] Calculated earlier in this chapter.

[5] Calculated in Chapter 2. [Defect #2: Inadequate Chargers]

[6] There are too many variables to calculate this accurately, such as type of fuel cells, size of the stack and how long they operate (as energy is needed to heat the stack to operating temperature).

[7] There are too many variables to calculate this accurately. Such a system must contain both batteries and supercapacitors, both primary and secondary chargers and a DC to DC power inverter. Furthermore, battery charging characteristics vary greatly as the battery bank charge level climbs from 80% to 100%.

Costs* per KiloWatt Hour

Municipal electricity delivered to regular houses (average US):	$ 0.13 [1]
Charging an RV battery at regular "home" before/after a trip:	$ 0.17 [2]
Solar Panels or Wind Turbines:	Free ($ 0.00)
Alternators (Charge As You Drive):	$ 0.33 to $ 0.55 [3]
Generators Powering 60 Amps Chargers:	$ 1.96 [3]
Thermoelectric Generators (propane fueled):	$ 5.19 [4]
Generators Powering 3 Amps Chargers:	$ 30.00 [3]
Charging a battery at an RV park:	can't be calculated [5]

Notes:

* These are operational costs only. The cost of buying or servicing the equipment is not included. Except for municipal electricity delivered to regular houses, these costs are for electricity that is stored in batteries then used later.

[1] Rounded off from 12.7 cents. Source: EIA Electricity Data--- cost for July 2016.

[2] Average municipal electricity cost multiplied by 1.2 (in/out inefficiency factor for battery charging), then again multiplied by 1.1, to account for 1.4 Volt drop across diodes of bridge rectifier within the battery charger. Rounded off from 16.8 cents.

[3] For gasoline powered systems. These figures are obtained by dividing the volume of gasoline containing 1 kWh of energy by the respective efficiency (expressed as a decimal), then by multiplying by the price of gasoline (using $3.00 per gallon). The volume of gasoline that stores 1 kWh of energy is .0277 gallons.

[4] This is the worst case efficiency, in which fuel is consumed solely for making electricity. If heat from the catalytic burner also warms the RV in cold weather, this figure gets considerably better, as combustion heat serves two purposes. The (worst case) figure is obtained by dividing the weight of propane containing 1 kWh of energy by the efficiency, then multiplying by the price of propane (using $14 per 20 pound tank fill). The weight of propane that stores 1 kWh of energy is .153 pounds.

[5] If the RVers would probably have stayed at the RV park anyway (because they liked it), and the electricity were included in the rent, then it could be argued that battery charging is free. If however the primary reason for staying at the RV park is to charge batteries, then it could be argued that the entire rent is the cost for energy added to the batteries. If a single 100 Amp hour battery (which contains about 1.2 kWh of energy) were completely charged from a completely discharged state in one day, and the RV park price were $30 per day, then the cost per kWh would be $25. If however the battery had half charge or more upon arrival, then the cost per kWh would be a whopping $50 or more.

Calculating the number of gallons of gasoline containing 1 kWh:

Start with the exchange rate: 130,000,000 Joules = 1 gallon.
Dividing both sides by 130,000,000 yields:

1 Joule = 7.692×10^{-9} gallons.

Multiplying both sides by 3,600,000
(which is the number of Joules in a kiloWatt hour) yields:

3,600,000 Joules = 0.0277 gallons.

Finally:

1 kWh = .0277 gallons.

Calculating the number of pounds of propane containing 1 kWh:

Start with the exchange rate: 98,000,000 Joules = 1 gallon.
Dividing both sides by 98,000,000
and multiplying by the factor converting gallons to pounds yields:

1 Joule = (1.020×10^{-8} gallons) × [(20 pounds) / (4.8 gallons)] =
4.251×10^{-8} pounds.

Multiplying both sides by 3,600,000
(which is the number of Joules in a kiloWatt hour) yields:

3,600,000 Joules = 0.1531 pounds.

Finally:

1 kWh = .1531 pounds.

<u>**Ideal Battery Charging**</u>

Many authors write books telling you the ideal way to charge and use a battery. They go into excruciating detail about optimizing voltages, currents and times, they take into account the battery's varying charge state, the chemistry and properties of various battery types, and they tell you the best way to discharge the battery through appliances. I think this is largely a waste of time. The cost of RV batteries is minuscule compared to other common RV expenses (motor fuel, campsite rent, RV depreciation), and batteries tend to last several years even with moderate abuse. (A new battery only costs the price of 3 or 4 nights at an RV park.) If instead of spending the time needed for gaining the knowledge and habits for "perfect" battery use, you spent that same amount of time flipping burgers at a minimum wage job, you would earn a hefty multiple of the money needed to replace any batteries you trash from less than perfect use.

The main things to remember about batteries are: 1) try to avoid discharging batteries too deeply too often and 2) don't let batteries sit too long in a discharged state---charge them up as soon as possible after discharging. If you practice these two habits reasonably, you should get several years out of your batteries. (I always do.)

<u>A Battery in Like a Pair of Sneakers</u>

I have a habit when I shop for sneakers: To save a trip to the store, I usually buy two pairs at once. I cut the tags off one pair and start wearing them immediately. The other pair I toss into the closet and forget about for awhile. Months later, the pair that I wear is dirty, worn, coming apart at the seams and downright ugly. I then retrieve the other pair from the closet, which still looks brand new, cut the tags off and start wearing them. The moral: Just as: *You can't have your cake and eat it too, You can't have your money and spend it too, You can't have your sneakers and wear them too.* If you wear your sneakers, after awhile, you won't have them anymore. If you want to keep your sneakers forever, you must absolutely never wear them. But what good are sneakers if you never wear them? A battery is like a pair of sneakers. *You can't have your battery and use it too.* The best way to preserve a battery is to put a slow constant charger on it, keep it 100% charged all the time and *never use it.* But what good is a battery if you never use it? Enjoy your electricity once in awhile! After you trash your batteries, buy new ones.

Which Electricity-Making Strategies Work Well Together?

Asking which *one* strategy should be used for making RV electricity may be a bit like asking which *one* tool should you have in your toolbox: a slotted head screwdriver, a Phillips head screwdriver or an adjustable wrench. Just as your toolbox should have several tools, it may be prudent for an RV to have more than one electricity making / battery charging strategy. However, electricity making / battery charging equipment costs quite a bit more than screwdrivers and wrenches, so RVers must ask themselves how much electricity do they really need (or want) and how much money and time are they willing to spend.

Charge-as-you-drive alone can make enough electricity to power small and medium energy conserving RVs without air conditioners. An internal combustion engine generator is a good and necessary compliment to an RV that does have an air conditioner, and if used only to power the air conditioner, can leave an RV quiet most of the time (when temperatures aren't too hot). Anyone having a generator should also invest in multiple batteries and powerful chargers, so batteries can get properly replenished when the generator gets used for some other purpose. For larger RVs, needing more electricity for general (non-air-conditioner) use than charge-as-you-drive can provide, and seeking quiet systems, I recommend a solar panel / thermoelectric generator combination. Solar panels go well with all other charging strategies, but they are unreliable, so alone they can't take up the slack in weak systems at all locations (but they do work really well in deserts). Solar panels and propane fueled thermoelectric generators compliment each other very well, because both make electricity with no (or few) moving parts making no (or very little) noise, and both can provide steady currents for long time spans (building substantial charge in batteries), but differ in that solar panels make electricity that is free but unreliable, whereas thermoelectric generators make electricity that is reliable but not free. Finally, there is one other method of "making" electricity that goes well with and should be used with all other strategies: *conservation.* Alright, technically speaking conservation isn't really a method of *making* anything, but it often has the same effect. As long as some other method exists for making or gathering something (in this case electricity), if less gets used, more of it remains in storage (or it takes longer to run out). Conservation is a negative technique of production, and a very powerful and often overlooked one. It is such a powerful tool in getting RV electrical & appliance systems to work right, that it is a major theme of this book in Chapter 5: "Appliances."

Chapter Conclusion:
The Simplest Might Be the Best

We humans need to work hard to get most of what we have. Tires for cars, gasoline, houses, televisions, cell phone towers… these things don't just happen, we must put meticulous effort into making them happen. Wouldn't it be nice if some of the things we loved just fell from the sky, and all we had to do was pick them up? Like those twigs that I burn in my woodstove to start fires. Wouldn't it be nice if they just fell from the sky?--- Oh that's right--- they do. They fall off the trees onto my lawn and all I need to do is pick them up and put them in my basement, where they dry without me needing to do anything. Or how about that water needed to grow my lawn, plants and that tree in my yard (that keeps giving me all those twigs). Wouldn't it be nice if that water just fell from the sky?--- Oh that's right--- it does. It's called rain. Getting electricity to charge RV batteries in the wilderness is a task that stumps many. Wouldn't it be nice if it just fell from the sky and all we had to do was pick it up? Wouldn't it be nice if everything were that easy?

It never ceases to amaze me how many RVers I meet who don't bother putting that extra battery in their car and charging it as they drive. They drive that car away early in the day, leaving their RV and its battery behind, adding perhaps over a hundred miles to the odometer, then return home to the campsite late--- and what's the first thing they do? They turn on a generator to charge their battery. In doing so, they operate a battery charger that is much weaker and slower than their car's alternator (even if the charger is deemed "heavy duty"), they waste fuel, wear engine parts and make "music" (generator noise) for the neighbors. Their behavior puzzles me. If a person, living in a remote cabin without a well or nearby creek/lake, insisted upon driving dozens of miles to get all his water, his behavior would puzzle me. I'd say: *Hey Einstein, put a barrel underneath the drain of your gutter to catch the rainwater that falls on the roof of your house. It might not gather all the water you need, but you will get some.* So, to all those RVers who insist upon generator-charging all their batteries at campgrounds, or struggle to live with very limited electricity, I say: *Hey Einstein, put an extra battery in your car to "catch" the electricity that your car can and will make as you drive. It might not gather enough to satisfy all your electrical needs, but you will get some.*

Electricity does not fall from the sky like twigs or rain does. But in a way, the next best thing happens. Electricity sort of "falls" from that technological marvel that most of us have and love: the car. Piggybacking an extra battery to a car's system allows one to "catch" it. Sure, you can do what most RVers wanting to go off-grid consider doing: You can try generator-charging or consider solar panels or even a wind turbine. Don't use these as your first line of attack. Don't ignore the obvious. Sometimes the simplest strategy might be the best: Charge batteries as you drive your car.

Chapter 4
Propane

A Natural Fuel Choice

RV appliances typically run form electricity, propane gas or both. While it is possible to run some things from other hydrocarbon fuels or even compressed air, this usually isn't done. Other fuels (heavy oils, alcohol, solid fuel) don't offer any overall advantages over propane. (Heavy oils and solid fuel can be stored in non-pressurized containers, but they make more pollutants when burned, therefore are poorly suited for indoor use. Alcohol is clean burning.) Natural gas and butane have properties so similar to propane, it's difficult to tell them apart. (Natural gas however cannot be liquefied at room temperatures, even under pressure, therefore is difficult to store, therefore not chosen for RVs.) Compressed air requires an air compressor, an additional storage tank and specially designed pneumatic gadgets, which means a lot of extra engineering and expense, with no clear advantages either. Compressed air does however sometimes have use in RV water systems.

Electricity Verses Propane

Electricity and propane gas have very different properties. Electricity is difficult to cheaply store in large enough quantities to do significant tasks, and battery charging can be difficult. But electricity can run TVs, stereos, computers, cell phones and motors; propane can't. Propane on the other hand can be stored in large enough quantities to perform significant tasks, it's cheap, and refilling tanks is easy and doesn't take very long. Therefore, conservation of electricity is extremely important in RVs, whereas conservation of propane is not so important. If an appliance can be made to run from either electricity or propane, it is almost always best to run it from propane.

Properties of Propane

Propane is a colorless odorless flammable poisonous gas. It can kill people and pets if too much of it accumulates inside an occupied enclosed space, and the occupants breath a lot of it in. Accumulated propane in an enclosed space also presents an explosion hazard. Propane is somewhat heavier than air (atomic weight of propane molecule 44, average air molecule about 29), therefore it tends to accumulate in places that are low to the ground. Detecting renegade propane and getting rid of it or evacuating contaminated air spaces is extremely important.

If propane accumulates in an RV, or any portion thereof where it does not belong, turn the main gas valve off, turn all individual appliances off, extinguish all pilot lights and flames, open the windows and roof vents, and remove yourself from the RV until it airs out. Propane can accumulate into an enclosed space by escaping from faulty tubing/ fittings either within appliances or leading up to the appliances, or from some appliance burner that gets turned on without being lit or gets extinguished (but left on) after it is lit.

Detecting propane

By smell: Unfortunately, propane has no smell. So gas companies add a stinky chemical to propane, to aid in its detection. (They also do this with natural gas and butane.) When people say they "smell gas," they don't actually smell the *gas*, they smell this *stinky additive*. This additive almost completely burns up as propane is burned, so exhaust from burned propane tends not to stink, only the unburned propane stinks. While this is a great idea, it is not a perfect tool for detection, because although the stinky additive and the propane can be mixed in proper proportions, they do not stay evenly mixed, because the stinky additive is heavier than the propane. When a tank is full, the unburned propane stinks less, because the stinky chemical tends to settle to the bottom of the tank, whereas the propane drawn for the appliances comes from the top of the tank. When a tank is nearly empty, the propane stinks more, because all that is left in the tank is a small amount of liquid at its bottom, rich in that stinky additive, so that's the time when the last of the tank's contents finally evaporate and make their way out the tubes for appliance use. In addition, remember that RV (and other) propane tanks are filled at filling stations from other much bigger propane tanks, which may also suffer from uneven mixing of propane and the additive. Hence, some tank refills are unusually stinky, whereas other tank refills are unusually clean. (Perhaps stinky tank refills come from nearly empty large gas-supplier tanks, whereas clean tank refills come from nearly full large gas-supplier tanks. Bear in mind too that the stinky additive is put there artificially, so if the proportions aren't mixed right at the gas company, they won't be right in your RV refill tank either.) Hence, both false positive and false negative readings can and sometimes do occur. False positives (it stinks but there's no gas) are more annoying than dangerous, but they can be dangerous, because many false positives might train people to ignore the "smell of gas," which could pose a serious problem when there really is gas. False positives are common when a tank is near empty because even the tiniest amount of escaped gas stinks the place up (this can occur moments after a burner valve opens but before it ignites). Also, not all of the stinky chemical molecules burn up in a flame, so even burned propane, rich in stinky molecules, can stink the place up. False negatives (there's no smell but there is gas) can be very dangerous, because people might not know that there is gas, and might proceed with business as usual, until they are overcome by it. Use the "smell of gas" as a *guide only*, and use other gas detections methods as well.

By electronic detectors: These are small devices similar to smoke detectors. Some have several components, rolled in one, that detect everything bad (smoke, unburned gas, carbon monoxide). All RVs should have complete electronic detectors to keep people safe. Detection is especially important in RVs, because RVs have small amounts of indoor air space (compared to houses), so it takes less gas (therefore is easier) to pollute indoor RV air to critical danger levels than it does to pollute air in regular houses to those levels. A detector can run from a small disposable battery or it may be wired into an RV's central electrical system. I believe that connecting a detector to an RV's electrical system (central battery) is a bad idea, because as stated in Chapter 2 (under "The Octopus Effect"), there are far too many things that can shut an RV electrical system down. If the system shuts down, and outdoor temperatures get cold, that's when the RVers are likely to try to heat their RV with their stove or oven, greatly increasing the risk of poisoning by unburned propane or carbon monoxide. That's no time to be without a working detector. Also, I have read that if an RV's central battery gets substantially run down, its lowered voltage can fool the electronic circuitry of some detectors into falsely detecting a problem, causing them to "scream" nonstop. The only remedy is to cut the wires leading to the detector, rendering it useless. Perhaps a hybrid multi-detector would be best (one that runs from the central RV battery but has its own backup battery), but I don't know if anyone manufactures one. If hybrids are unavailable, detectors are best run by portable batteries, but check them often. A detector with a dead battery is no detector at all.

By your own body: Many experts believe that because propane and other light hydrocarbon gasses have no odor, they cannot be detected by people. I believe this to be false for two reasons: 1) On a couple occasions, I detected propane gas when there was no smell. [It makes me feel tired, dizzy and lightheaded.] 2) Some birds are known to be able to detect hydrocarbon gas. [Is it far fetched to believe that some people could detect gas too?] Coal miners used to take birds deep into the mines. When a bird squawked, fainted and possibly died, the miners knew that significant amounts of poisonous methane gas were in the air, and it was time to get out of that mine ASAP. The birds had these reactions even when the people had no symptoms at all. Some people can detect propane, others can't. Try to determine which category you fall into.

<u>Detecting Propane Without the Smell</u>

Once, while in bed ready to go to sleep, I felt extremely tired and super-relaxed. I believed I was far more relaxed than I should have been. I began to suspect that the air I was breathing was not good, so I opened a window near my head then started breathing the outdoor air. After just several breaths, I started "coming to" and immediately knew that I needed to get out of that trailer fast. As soon as I was quite sure I wouldn't pass out, I went for the door and got outside. There, I sat still and took deep breaths until I believed that I was alright, at which point I shut off the main gas valve, opened all the trailer windows and didn't reoccupy the trailer until it aired out. The next day, I inspected the suspected culprit, which was my furnace. After I turned it on, I sniffed at the tubing leading up to the pilot light. Although there was no smell, I felt dizzy and lightheaded every time I took a whiff. I took the furnace to a gas appliance dealer, who told me that the wrong kind of fittings were on that tubing, and it was no surprise that it leaked. I replaced it with the right kind of tubing and fittings, reinstalled it and "whiff tested" it (no dizziness occurred). That night I used my furnace. No longer did I feel extremely tired and super-relaxed. Problem solved.

By the "soapy water" test: This method doesn't test the *air*, like the above three tests do, it tests the *fittings and joints* in propane tubing. It is a method to be used either after one has already determined there to be a leak (to locate and fix the leak), or as a preventive diagnostic check done on whole systems once in awhile (maybe on RVs taken out of long term storage, before embarking on long trips). One should do this test when one has plenty of time, one is wide awake and preferably during daylight hours. Special chemicals for such tests are available at RV stores, but I think a cupful of heavily soaped up water works just as well (use dishsoap). To do the test, simply drench up a suspect fitting with the soapy water/chemical, using a paintbrush, sponge or a wet paper towel. If gas is leaking, it should start "blowing bubbles" (you will notice some bubbles getting larger, because they fill up with gas). Usually, this situation can be remedied by just tightening the fitting. If this doesn't work, you might have a defective flare, a defective fitting or cracked tubing. You should always keep flare caps on hand. These allow you to shut off the main gas valve, disconnect the tubing to a faulty appliance at its source, cap the hole, then turn the gas back on and use the rest of your good appliances. If you don't have a flare cap, and can't fix a leaky joint by just tightening it, you will need to shut the main gas valve on your propane tank off, which will leave you with no working propane appliances at all. NEVER leave the gas on and try to use appliances, if you know there is a leak somewhere.

Which detection method should be used? I suggest using a combination of all of these. Take the "smell of gas" very seriously. But you may have to ignore it a bit, when a tank is near empty and really stinky. Open windows, just to be safe, if you "smell gas." If you ever feel dizzy, lightheaded, too relaxed for no apparent reason, sick, or you suspect the air is not good, trust your instincts and exit the RV for awhile. Exiting the RV and breathing clean outdoor air for awhile is one of the best ways to test suspect air. If you have symptoms inside the RV, and those symptoms disappear after breathing clean air for a few minutes, you know for sure the indoor air is not good. Always have a working electronic propane detector. And use the soapy water test to locate and fix problems when you've got time.

Combustion Byproducts of Propane

When propane burns, oxygen gets removed from the air, and the byproducts of combustion get added to the air. Some of the byproducts are nasty, some are harmless. One way appliances deal with these byproducts, is to not deal with them at all, by venting all combustion byproducts to the outdoors. (Refrigerators, hot water heaters and many furnaces do this.) Such appliances may draw their intake air from the outside or from the RV interior, but either way, the chemical content of the interior air remains unaffected, because fresh outdoor air enters through seams in the windows and door (even if closed) to replace any air taken up by appliances. Another way appliances deal with combustion byproducts, is to vent them to the indoors, and hope they don't pose problems--- though sometimes they do. (Stoves, ovens, gas lights and some furnaces do this.)

The byproducts of complete combustion of propane are water vapor and carbon dioxide. In addition, there usually are byproducts of incomplete combustion, which are carbon monoxide and unburned hydrocarbons. Water vapor is harmless (though it can be annoying when humidity levels are high; it can also steam windows and walls). Carbon dioxide is annoying but never rises to fatal levels in RVs. Unburned hydrocarbons don't accumulate in enough quantities to be bothersome, if they pass through flame, but as already covered, unburned propane presents a real threat if the flame goes out. The real issue of concern is carbon monoxide, which is deadly and sometimes does kill people.

Carbon Monoxide

Properties: Carbon monoxide (abbreviated: CO) is a colorless odorless deadly gas, made by the incomplete combustion of carbon. It gets released from burning almost anything with carbon in it, including light hydrocarbon gasses (propane, natural gas, butane), gasoline, oil, candle wax, alcohol, coal, charcoal and wood.

Carbon monoxide in RVs: Carbon monoxide is especially dangerous in RVs because RVs are small and thus have small air volumes. Places with small air volumes become polluted to dangerous levels much more easily and rapidly than places with large air volumes (such as regular houses).

How can one prevent carbon monoxide from entering the indoor RV air? There is almost no way to completely prevent carbon monoxide from entering the indoor air, other than by not making any fire inside the RV. However, there are several measures that both the appliances and their users can take to greatly reduce its input: 1) Discharge exhaust gasses to the outdoors whenever possible. This is usually done with primary furnaces, both for RVs and for regular houses. However, this strategy is not perfect, because some furnaces, designed to discharge all their combustion byproducts outdoors, actually leak, so these can still put carbon monoxide indoors, and sometimes such leaky furnaces have been known to kill people. 2) Inspect exhaust vents leading to the outside occasionally. Carbon monoxide deaths have been reported due to blocked vents and chimneys. Make sure vents are not blocked by insects or birds nests or even snow. If they are, get rid of those nests or the snow, but make sure all wasps are dead before removing wasps nests. Also make sure ducts seems don't appear to be loose or have holes in them. If they do, fix or replace them. 3) Burn only clean-burning fuels [if discharging indoors]. In general, light hydrocarbon fuels make less harmful exhaust than heavy hydrocarbon fuels. Natural gas, propane and butane are rather clean burning because they are light hydrocarbon fuels. [Hydrogen is the lightest and the cleanest burning fuel. It contains no carbon, therefore produces no carbon monoxide. But hydrogen is not usually manufactured, stored or commercially sold.] Heavier hydrocarbon fuels [like gasoline, kerosene] make more carbon monoxide when burning, therefore do not use gasoline stoves, gasoline mantle lights, kerosene heaters or generators indoors. Still heavier hydrocarbons [oil and wax] make even more carbon monoxide when burning and may also make put soot or oily compounds into the air and upon surfaces. Use candles and oil lamps sparingly indoors, but some use is permissible, because their flames are not very big. Heavy solid carbon rich materials [wood, charcoal, coal, significant quantities of paper or cardboard] are the worst offenders, and should never be burned indoors, except in outdoor-vented fireplaces and stoves specially designed for such purposes. NEVER take lit charcoal into an RV, a regular house or a garage. 4) Favor cleaner burning appliances. Different appliance flames vary in their burning efficiencies, which means that the ratios of carbon monoxide [incompletely burned carbon] to carbon dioxide [completely burned carbon] in exhausts are not the same for all appliances. Ovens are bad offenders, making lots of carbon monoxide. Stove burners make quite a bit too. "Blue flame" furnaces make less but they do make some. Catalytic burners are the cleanest, making almost no carbon monoxide at all, because catalytic burning is a super-efficient burning process different from conventional burning. 5) Don't burn indoor-discharging appliances longer than necessary. Use the stove and oven to cook only when you need it. Try to use indoor discharging furnaces sparingly [except catalytic burners, which don't usually pose problems]. The exception to this rule is propane gas mantle lights, which may be operated indefinitely, as long as some windows are open.

How can one reduce one's exposure to existing carbon monoxide in the air? 1) If you must cook for prolonged time periods, vacate the RV while your stove/oven is on. Reenter for only brief time periods, to check on your stew or pot roast or that water that you are heating for a shower or dishwashing. Remember that carbon monoxide in an RV poses no problem at all if there are no people or pets in there to breath it. 2) When cooking, try to keep your face out of the exhaust from the stove burners. Standing back helps. So too does holding your breath briefly when placing your head directly above cooking pots with lit burners, while checking on or stirring the contents of those pots.

How can one get rid of carbon monoxide in the indoor RV air? Use some common sense. Make sure that the RV is well ventilated when using indoor-discharging propane appliances (or candles/oil lamps), by keeping some windows and/or roof vents open while burning them. You don't need to be excessive about ventilation, especially if you've got only one thing on, usually one louver window open halfway and one roof vent open 2 inches will do (for one stove burner). If you're operating multiple appliances for prolonged periods, you may want to open all the windows and even the door. Wind is a mitigating factor. In very windy conditions, it may be possible to properly ventilate an RV by opening a window just a crack. Bear in mind that although it is difficult to prevent carbon monoxide from entering the indoor RV air, it is very easy to get rid of it.

The "NEVER LIST" (of most dangerous practices): 1) NEVER go to sleep in an RV with indoor-discharging non-catalytic propane appliances burning. You can't notice symptoms and respond to them if you are sleeping and don't wake up. 2) NEVER close all the windows, roof vents and the door tighter than a drum and burn indoor-discharging propane appliances for extended time periods in an RV that you (or someone else) is occupying. [It is usually permissible to burn *one* thing for a few minutes without opening windows/vents, but don't push this.] 3) NEVER EVER do the above two things simultaneously. The combination of closing windows/vents/doors tighter than a drum, turning indoor-discharging propane appliances on for extended time periods and going to sleep can be lethal. If you do this, be sure your will is in order and you have made proper funeral/burial/cremation arrangements.

How does carbon monoxide poison people? Hemoglobin is the molecule in red blood cells responsible for the transportation of oxygen. It binds with oxygen in the lungs, then releases it in the bloodstream to other cells. Carbon monoxide also binds to hemoglobin in red blood cells, but it does so much more strongly than oxygen, and it is metabolically useless, therefore its presence prevents this "catch and release" of oxygen. A person who has been exposed to lethal levels of carbon monoxide basically suffocates, even though that person may be breathing air containing plenty of oxygen. Also, it can take quite awhile for carbon monoxide to leave the bloodstream, so after a person has been exposed, that person may still be in danger of suffocation, even after that person has gotten to and is breathing fresh air.

What are the symptoms of carbon monoxide poisoning? Light headedness, dizziness, headaches, shortness of breath, perception of "stuffy air," nausea, fatigue, weakness, mental confusion, and possibly (in severe cases) vomiting and chest pain. These symptoms may occur in different combinations in different people. Don't insist upon having *all* these symptoms before concluding that you've been exposed to carbon monoxide and vacating a suspect RV, get out when you have just a couple symptoms. As with propane gas, if you suspect the air you are breathing to be no good, trust your instincts and get to some clean fresh air immediately.

Should every RV have a carbon monoxide detector? Yes, absolutely! But don't rely on readings from a CO detector alone. Their technology is not perfect. They have been known to give both false positive and false negative readings. Also, some people's own human bodies are better at telling them that air is no good than CO detectors. "Abandon ship" if *either* your body or the CO detector sounds an alarm.

What should you do if you suspect you've been exposed to carbon monoxide? Some internet sources recommend dialing 911, requesting medical attention immediately and giving oxygen. This of course is proper advice for serious poisonings, however: 1) Not all poisonings are fatal. For every poisoning that is potentially fatal, there may be a hundred other light poisonings that don't require such drastic measures. [Example: Cigarettes make carbon monoxide when lit. Smokers never die from carbon monoxide released from their cigarettes; they die from other complications, such as lung cancer.] 2) In many wilderness locations, where RVs are likely to be, cell phones don't work. A satellite phone can work, but if you don't have one, calling out for immediate help will be impossible. If you are able to call out for help and choose to do so, be sure you have a G.P.S. and give your exact coordinates so that rescue personnel can locate you. 3) Much internet advice assumes that if you are unconscious, there is some unexposed healthy person to find you (or you are the healthy unexposed person finding someone else unconscious). If you are in the wilderness alone, and you (and everyone else in your group) have become unconscious, most likely you are a goner. And: 4) In many remote areas, it may take too long for help to arrive. The bottom line is: When you are in the wilderness alone, with any serious medical condition, and you can't call out for help, and you're too sick/disoriented to drive yourself to safety, you are likely to die, so *be extra careful and respond to carbon monoxide symptoms immediately before becoming unconscious.*

For light poisonings (not requiring medical attention), I recommend the following: 1) Go outside far enough away from the RV to where you know the air is fresh (even if it's raining or cold). 2) Remain calm and inactive and take deep breaths until your symptoms disappear. Remaining calm and inactive will reduce your body's burning of what little oxygen you have in your bloodstream. Although carbon monoxide bonds to hemoglobin very strongly, it does eventually break loose, so if you have carbon monoxide in your bloodstream, and you start breathing fresh air, and you don't die, you will eventually exhale it. 3) When you're feeling better, return to the RV briefly, open all the windows and roof vents and the door and turn everything off. Then go sit outside for awhile longer.

4) Reoccupy the RV only after it has completely aired out. And: 5) Do not reignite any suspect appliances. The choice of whether or not to seek medical help must be made by you, based upon your individual circumstances. You are responsible for your own safety. Prior experiences will cause you to get really good at knowing how much CO you can tolerate, when you need to open windows and roof vents, when stuff needs to be turned off and when it's time to get out.

The final word on carbon monoxide: Although carbon monoxide can do you in, you should not let it discourage you from using your gas stove, gas lights, oven and furnace in your RV. Just be careful. Many people use these appliances and expose themselves to small doses of CO for decades, without dying from it. I've been using my RV appliances for over three decades now and (as of 2017) I'm still not dead yet.

> You drive a car, don't you? Even though you know you could get into a car accident and die. Why? Because the benefits are huge and the risks are small. You choose to be careful to make those risks even smaller. It would be silly to never drive a car because of the fear of accident. The same is true with propane gas appliances. The benefits of using them are huge and the risks are small--- especially if you are careful. It would be silly to never use propane appliances because of the fear of carbon monoxide.

Carbon dioxide

Properties: Carbon dioxide is a colorless odorless gas. It is one of the main final products of complete combustion of all fossil fuels (including propane), alcohols and wood. In addition to being made by burning stuff, it is also made and released into the air by the metabolic burning of sugars and fats (with oxygen) in humans, animals, bacteria and fungi. Carbon dioxide is naturally present as a trace part of our Earth's atmosphere, and a necessary part of the ecosystem. (Plants use it to produce their own food and structural building blocks. They also change it back into oxygen, which they release into the air.) Carbon dioxide is harmless to people in small atmospheric concentrations. In fact, our bodies need it. We deliberately retain some of our waste carbon dioxide in our blood, to regulate the pH (acid-base) balance of that blood. In moderate atmospheric concentrations, it can make people sick. In very high concentrations, it can actually be deadly.

Carbon dioxide in RVs: In large enclosures or places that are out in the open, carbon dioxide poisoning is rarely a problem, because when it gets released, it dissipates into those large air volumes, so its atmospheric concentrations don't rise very much. In small enclosed spaces however, when it gets released, there is nowhere for it to go, so it can eventually build to toxic levels. Carbon dioxide can be put into the air either as an inevitable final product of combustion or from the breathing of people and animals. The result is the same regardless of its source. In enclosed spaces that have absolutely no ventilation (submarines and spaceships), its concentrations can rise to deadly levels, so special procedures are needed to remove it. Removal is accomplished by reacting it with solid chemicals which capture and retain it, a process called *carbon dioxide scrubbing.* RVs are never air tight, so carbon dioxide always escapes, and probably never gets to deadly levels, however because RVs are small, carbon dioxide can accumulate quickly, especially if the windows are tightly shut and the RV is occupied by several people/pets and/or an indoor-discharging appliance.

Carbon dioxide sickness: Normally, our bodies just get rid of excess carbon dioxide by simply exhaling it. But if there is too much of it in the air, it can become difficult for our bodies to get rid of it (for the same reason that a wet washcloth doesn't dry well in very humid air), so excess quantities can accumulate in our blood, a condition called *hypercapnia.* Some symptoms are: headaches, perception of "stuffy air," panting, nausea, lethargy, confusion and rapid heart rate. Let's consider two major symptoms likely to occur in RVs: 1) Headaches: A long time ago, I noticed that if I stayed in a small room too long, without good ventilation, I got headaches. I told people that these headaches were caused by breathing in my own exhaled air. My mother thought I was nuts. Then one day, I attended a lecture given by a NASA astronaut. This astronaut described a sleeping chamber aboard the space shuttle, emphasizing the need for fans to circulate the air, because without such fans, he would get a "carbon dioxide headache" from inhaling too much of his own exhaled air. Mom, I'm not nuts after all. 2) Panting: When people exercise vigorously, they start breathing very heavily, feeling that they "want more air." The purpose of this biological reflex is not to get more oxygen, it is actually to remove excess carbon dioxide from the blood, made from that exercise. Too much carbon dioxide in the blood inhaled from the air can also trigger that same heavy breathing response, feeling that the person "wants more air." Because RVs are not airtight, even with all windows closed, they naturally ventilate and get rid of carbon dioxide. It is highly unlikely that carbon dioxide levels could ever build enough to kill anyone in an RV, but elevated levels can and do cause people to become uncomfortable in RVs.

Deadly Carbon Dioxide Levels: 1) Some "killer lakes" have suddenly released large amounts of carbon dioxide through natural geologic processes, killing people and animals. Some people, who nearly died but escaped, reported having hallucinations upon exposure. 2) In the disabled Apollo 13 spacecraft, large carbon dioxide amounts added to the air, made by the astronauts own breathing, began poisoning that air and almost killed the astronauts. They finally managed to modify a carbon dioxide removal system to make it work, so they survived. 3) If a child gets trapped inside an abandoned refrigerator, the child suffocates, but death does not occur from the *absence* of oxygen [although this would eventually happen], death occurs due to the *presence* of exhaled carbon dioxide. 4) In 1979, while waiting for admittance to a "Who" rock concert, there were so many people crowded outside in a small area, that they actually poisoned their surrounding air with carbon dioxide from their own breathing. Sadly, several people died.

Trailer Propane Tanks

The propane tanks used for trailer-type RVs are identical to the portable propane tanks used for general household purposes. Tanks may be the standard 20 pound "grill size" or they may be larger (30 pounds typical). A tank's weight designation indicates how much *propane* it stores when full, and does not include the weight of the *tank itself.* Tanks may be made out of steel or they may be made out of aluminum.

Steel tanks are the most common. These are painted, to prevent them from rusting away, but they always get nicked and scratched, and in those places, the exposed bare metal starts to rust, and it looks unsightly. This rusting away can be halted by sanding the rust of these scratches then repainting the bare metal. Steel tanks last much longer if the entire tanks are repainted once in awhile.

Aluminum tanks are less common and more expensive. (Some people who fill propane tanks tell me they have never seen an aluminum tank until I brought them one.) These don't rust at all and are usually not painted. They can last decades.

Weights of tanks: The weight of an empty steel tank is usually about the same as the weight of the propane it stores when full. The weight of an empty aluminum tank is about half that of the steel tank. This makes aluminum tanks a bit easier to lift and carry than steel tanks, both when empty and full.

Tank valve designs have changed several times during my lifetime and could conceivably change again. You will not be able to legally refill tanks with outdated valves anywhere, so if you have them, replace them with new tanks. In some states, a tank is determined to be full by its *weight* (it gets put on a scale while being filled), whereas in other states a tank is determined to be full by noticing that a *bleeder valve* spews out liquid propane into the air (which looks like heavy bluish steam) instead of gaseous propane (which looks like nothing at all). Some people who fill tanks determine a tank to be full when an automatic shut-off valve prevents additional propane from entering the tank. Make sure your propane tanks are fillable in all 50 states. It's infuriating to have an empty tank, and no filling station willing to fill it, because the tank came from another state. (Yes, this happened to me.)

Recertification: Tanks must be periodically tested to be sure they remain safe. It is illegal for stations to refill expired tanks (though many people don't look at the dates and will fill them). New portable tanks are valid for 12 years. After that, they need to be reinspected and pass every 5 years. Expired tanks can be put back into service after being retested and passing.

The gas regulator: Propane in tanks is stored under very high pressure. When a tank valve is open, the propane in the hoses is also at very high pressure--- far too high for typical appliances. The regulator does two things: 1) It lowers the pressure in the downstream tubing heading for the appliances [which usually require only low pressure propane]. And: 2) It keeps the pressure downstream stabilized [until the tank is nearly completely empty--- then it lowers]. Constant pressure makes appliances work better. (Note: There are some high-pressure propane appliances, such as outdoor grills, designed for use without regulators. But these are not typically found in RVs.)

The two tank system: Although some very small trailers still get by with only one propane tank, most use two. *[photo below: two tank system with changeover/regulator assembly (aluminum tanks)]*

A system with two tanks allows one the luxury of *never* running out of propane, something which is a great convenience (running out of propane is the pits) and an absolute necessary for RVs with refrigerators. (RV refrigerators should not be shut down for more than a few minutes. Doing so greatly impedes their performance). A two tank system uses a *changeover valve*. Initially, both tanks are full, both connect to the system and both tank valves are open. The changeover valve draws propane for appliance use from one tank only (called the *service* tank), until that tank completely empties, at which point, it begins drawing propane from the other tank (called the *reserve* tank). *(photo at right: changeover/ regulator assembly hooked up)* This provides uninterrupted propane service to the appliances. When the changeover valve draws from the service tank, a small cylindrical window on the valve assembly near its top indicates green. After the changeover valve begins drawing from the reserve tank, the small cylindrical window indicates red (as in photo), alerting the RV users that the service tank has emptied. The users then flip a switch on the valve assembly, causing the former

reserve tank to become the new service tank and the former service tank to become the new reserve tank. The empty new reserve tank can then be disconnected, removed and transported to a filling station. The valve on the service tank does not need to be shut off while the reserve tank is disconnected, so propane still gets supplied to the RV if the reserve tank is miles away being filled. After the empty tank gets filled, it is brought back and reconnected to the system and its valve is opened. The RV users are then good to go until after the next changeover, when the procedure is repeated. Note that when a tank empties, the users needn't hurry up and fill that tank immediately, because at changeover, the entire system is still half full of propane (one tank is completely empty the other is completely full). RV users should however make it a habit of taking the empty tank into town for filling, the very next time they go to town. Running too low on propane isn't a good policy, especially in very cold weather. The two tank system works beautifully. In all the years I've used it (since 1987), I've only completely run out of propane once, and did so only because all filling stations were closed due to a mad day-after-Thanksgiving holiday.

Changeover tips: 1) When the service tank empties, it is not *absolutely* empty, it still contains a little bit of really stinky propane. Turning the valve on the reserve tank off after changeover, while using appliances, allows every last bit of propane from the service tank to drain and be used. If you don't drain the service tank completely, the leftover stinky chemical gets added to the next fill and makes that next fill and successive fills progressively stinkier. Be sure to relight the refrigerator after you drain the tank and switch to and turn on the other tank, because draining a tank completely causes all appliances to turn off. 2) Because that stinky chemical added to propane tends to get quite strong when the tank gets nearly empty, that smell in your RV is a good way to know that the tank is nearly empty. 3) Consider chaining the propane tanks to the trailer tongue and locking them, to discourage others from stealing them. One tank, not anchored nor locked (while the other tank is away being filled), is especially vulnerable.

A Defect in Many Motorhome Propane Systems

RV electrical and propane systems are quite different. While RV electrical systems tend to be horribly inept, RV propane gas systems tend to be quite good. This does not mean that RV propane systems are perfect, because motorhomes do very commonly have one serious flaw (which is noticeably absent in travel trailers): *the propane storage tanks are permanently attached to the motorhomes and cannot be removed.* Thus, in order to fill a motorhome with propane, *the entire motorhome must be moved.* Why is this a flaw? Because it takes a lot of time, effort and fuel to move a big motorhome. Moving a big RV is something users should do when they want to change locations, not just to fill up with propane. Consider a situation that happened with my mother's RV:

My mom's RV was parked in a National Forest in the mountains of Colorado. When it got low on propane, we had to: 1) pack everything to make it ride-stable, 2) drive that big RV 27½ miles down a winding mountain road to a filling station [which thankfully had propane and was open], 3) have the tank filled [this was the easy part], then 4) drive 27½ miles back up that winding mountain road and finally 5) carefully park and level the RV. While we were driving we burned up an exorbitant amount of gasoline [especially going up]. We probably burned almost as much fuel as we added by buying the propane [gasoline and propane are different kinds of fuels, but this is beside the point], so we had to refill with gasoline at the bottom of that mountain too. We also wondered if the darn RV was going to stall [the engine was quite old and not that reliable] and if we were risking our lives by driving that large vehicle up and down that winding mountain road with steep dropoffs and blind curves. Imagine how much easier this procedure would have been, had we been able to remove a tank and take it for filling *in our car* [which we had with us]!

I believe this design problem has arisen for one main reason: *The people who design RVs don't actually camp in them.* To me, it's obvious that the people who designed such a system (and anyone else who thinks it is acceptable) have never been out to the wilderness. Their only exposure to wilderness is probably from National Parks calendars hanging on their walls or National Parks TV documentaries. Also, those designers have probably never been west of the Mississippi River. Easterners find it is difficult to understand what the western United States is like. The West is still quite unpopulated. Distances between even small towns can be considerable, so driving distances to restock supplies can be considerable and take up lots of time and fuel. Also, many roads in remote western areas are primitive and not well maintained, so driving on them with great big rigs can be difficult and dangerous. If you've got a car along, you certainly don't want to drive your great big motorhome into and out of the wilderness any more often than is necessary (which is usually once to park it when initially arriving, then once out when leaving an area permanently). There is nothing wrong with "city people." But city people should not be designing RVs headed for wilderness.

Proposed Solutions For the Motorhome Defect:

1) "Extend-A-Stay" kit: This kit allows one to hook up an external propane tank to the RV's main tank, allowing longer use of propane appliances and remote filling. However, this system is not as good as the two tank changeover system used for trailers because: A) There is only one portable tank. Either that portable tank must run completely out of propane or it needs to be removed while there still is some propane in it. In order to always have propane, one would need *two* additional portable tanks. This would require a lot of storage space [and propane tanks should never be stored indoors]. B) What is the purpose of having that permanently mounted motorhome tank, if the users need to carry two portable tanks anyway? C) Uninterrupted propane service is not possible, using this system. If only one tank is portable, and the other is permanently attached to the motorhome, one still needs to drive the entire motorhome away, in order to fill the attached tank. D) If there is no changeover, and users are drawing from the portable tank, that tank must run out completely, then they must open the permanent tank valve manually. In all likelihood, the portable tank would run out in the middle of the night, while users were sleeping, and the users would go for several hours without propane before noticing this, which they would do by waking up in a cold RV. They would then need to go outside and open the valve to their permanent tank immediately upon waking, then relight the refrigerator (which won't run well if shut down for several hours), and their furnace, and they probably wouldn't be able to get back to sleep again, because while waiting for their cold uncomfortable RV to warm up, they would likely spend the time bitching about their system. E) External tanks just sitting on the ground not attached or chained to anything are easy to steal.

Despite some drawbacks, this propane delivery strategy is vastly superior to just having one permanently attached tank with no method at all of getting propane to a parked motorhome. In the absence of a fully redesigned system, this is the best choice for motorhomes.

2) The "hanging propane tank" [My own proposal]: In high pressure tanks, at ambient Earth temperatures, propane is a liquid. Liquids settle at the bottom of tanks due to gravity. Suppose that a portable full propane tank were to be hung upside down from an extendable/retractable hook/arm on a motorhome, high above the large permanently mounted tank. If a high pressure hose were to be attached to both tank openings and both valves were opened, gravity should pull the liquid from the portable tank into the permanent tank (filling it). However, a special valve meant to be opened upside down would need to be designed, tested by approved by qualified propane gas engineers. (Ordinary valves might do strange things when upside down. They might not open or they might spew propane out their bleeder valves.) Also, the portable filler tank would need to be stored somewhere, thus taking up valuable storage space. This strategy would probably work, but I like my next idea much better.

3) The "propane drawer" [My own proposal]: Instead of having just one large permanently mounted propane tank underneath the motorhome, I suggest having two half-size detachable tanks there, with a changeover valve assembly, identical to the system used for RV trailers. But the tanks could not be the same type as those used in trailers, because such vertical tanks wouldn't fit underneath motorhomes, and there aren't great other locations in motorhomes to put vertical propane tanks. Mounting tanks outside on the front or back is dangerous, because even a small front end or rear end road collision with those tanks would squash them and cause them to rupture and explode. (This problem does not exist with trailers, because tanks mounted on the front of a trailer in tow are well protected from both front and rear end collisions.) The motorhome tanks could be mounted in some special very well vented side compartment that is completely sealed from the indoor air. However, this might take up quite a bit of precious storage space. Probably the best location for motorhome propane tanks is underneath the floor somewhere, where they already are. Vertical tanks can't sit in this position because they are too tall, however I have read that some portable propane tanks are *designed to sit horizontally*, so such tanks could conceivably fit underneath a motorhome and be removed for filling. I propose that each of the two horizontal tanks be retractable from the RV's underside, in what I call the "propane drawer."

Here's how the "propane drawer" would work: A hinged flap in the RV's siding would hide the tanks most of the time. [It would look and work a lot like the hinged flaps of many outdoor RV storage compartments.] When it's time to remove a tank: 1) the hinged flap would be opened upward. 2) The gas hose to the empty tank would be disconnected. 3) The tank would be unlocked from its position and slid out to where it would be easily accessible [just like opening a drawer]. The tank would sit upon a track with rollers or wheels [similar to a garage door track--- I got this idea one day while in my garage, and the track caught my eye]. 4) The tank would be lifted and removed for filling. [It should have a strap on top, for easy lifting.] Then, after being filled, the tank would be 5) placed back upon the track, 6) the "drawer" would be closed [the tank would be slid underneath the RV] and locked into position, 7) the gas hose would be reattached and the valve would be opened and finally 8) the hinged flap on the RV's siding would be closed and latched. There are two additional design considerations: 1) The propane tanks would need to be surrounded by a metal grate (or a few metal bars), so those tanks couldn't fall off onto the road, if the wheels somehow jumped off of the tracks. 2) The tanks should still be fillable without needing to be removed from the motorhome. This would allow users the choice between removing the tanks for remote filling or filling while the motorhome was in transit.

I believe that such a "propane drawer" is possible to design. I believe it wouldn't add too much cost to the RV. And I believe it would be a great convenience to motorhome users.

Chapter Conclusion: Choose Propane for Heating

Several important RV appliances run from heat sources. These include stoves, furnaces and ammonia absorption refrigerators commonly used in RVs. One could theoretically make the heat to run these things either from burning propane or using electric heating elements. Why choose propane for heating? Because so much more usable heat energy is stored in propane than is stored in batteries. It takes almost 300 pounds of lead-acid batteries to make the same heat as 1 pound of propane stored in a steel tank (including the tank weight) and almost 400 pounds of lead-acid batteries make the same heat as 1 pound of propane stored in an aluminum tank (including the tank weight). (Supporting equation is in the text box on the following page.)

Electricity and propane both compete with and complement each other for running RV appliances. They compete in the sense that several appliances could be run from either electricity or propane. They complement each other much more than they compete, in the sense that they have vastly different properties. Choose electricity for things that beep, flicker, whiz, toot, and compute. Choose propane for things that heat.

Comparing Energy Densities of Propane and Batteries

Step #1: The energy density of RV propane storage needs to be calculated. The energy released from burning one gallon of propane is about 98,000,000 Joules. There are about 4.8 gallons of propane stored in a standard 20 pound (grill size) tank when full. Therefore:

$$\text{(Total Energy Stored in 20 Pound Tank)} =$$
$$\text{(4.8 gallons/tank)} \times \text{(98,000,000 Joules/gallon)} = \text{(470,000,000 Joules/tank)}$$

The empty standard grill size 20 pound tank itself weighs about 18 pounds (if steel) or 9 pounds (if aluminum). Therefore:

(Total Storage Weight) = (Weight of propane) + (Weight of empty tank) =
if steel: (20 pounds) + (18 pounds) = (38 pounds) or
if aluminum: (20 pounds) + (9 pounds) = (29 pounds).

The amount of propane energy stored in one pound, including the tank, is:

(Energy per tank) / (Weight per tank) =
if steel: (470,000,000 Joules/tank) / (38 pounds/tank) = (<u>12,400,000 Joules/pound</u>) or
if aluminum: (470,000,000 Joules/tank) / (29 pounds/tank) = (<u>16,200,000 Joules/pound</u>).

Step #2: The energy density of RV battery storage needs to be calculated. A 75 Amp hour battery weighs about 38 pounds. If discharged 50% (by 37.5 Amp hours), that's about 1 Amp hour per pound. Converting this to energy:

(Energy) = (Voltage) × (Charge),
therefore:
(Energy per pound) = (Voltage) × (Charge per pound) =
(12 Volts) × (1 Amp hour/pound) = (12 Watt hours / pound).

Noting that the energy must be converted to Joules for proper comparison, the energy value per pound is:

(12 Watt hours / pound) × (3600 Joules / Watt hour) = (<u>43,200 Joules/pound</u>).

Step #3: These need to be compared by dividing the propane storage energy densities (from step #1) by the electricity storage energy density (from step #2).

(Energy Storage Ratio) =
(Energy Density of Stored Propane) / (Energy Density of Stored Electricity) =
if steel: (12,400,000 Joules/pound) / (43,200 Joules/pound) = about <u>290</u> or
if aluminum: (16,200,000 Joules/pound) / (43,200 Joules/pound) = about <u>375</u>.

Conclusion: It takes almost <u>300 pounds</u> of lead-acid batteries to make the same heat as 1 pound of propane stored in a steel tank, and almost <u>400 pounds</u> of lead acid batteries to make the same heat as 1 pound of propane stored in an aluminum tank.

Chapter 5
Appliances

Introduction

Do You Speak RV?

When people try to learn to speak a foreign language, the most frequently made mistake is: *they try to speak the new language using the same sounds of their own native language.* This results in improper pronunciations, poor grammar and a clumsy accent that makes it difficult or impossible for others to understand them. The proper way to learn a foreign language is to *clear one's mind of prior learning and babble like a baby until one gets it right.* This approach does not guarantee success, but those who use it almost always do better than those who use the first approach.

In general, we call the acquisition of memories and habits "education," and we almost always view it as an asset. It is something which separates and elevates us from monkeys and stupid criminals. But sometimes old memories and habits limit us rather than help us, because they are inappropriate for new situations, and holding onto them actually blocks our abilities to learn to do things right. In these cases, it might be best to forget some of what we know.

An RV is not a regular house! Learning to live in an RV off-grid is a lot like learning to speak a foreign language. The habits that you've learned for your regular house---*forget them.* The appliances that you know, love and are accustomed to using in your regular house---*forget them too.* I believe the most frequent mistake people new to RVing make is: *they try to live in the RV like it's a regular house.* I believe the most frequent mistake people shopping for their first RV is: *they expect it to look like a regular house.* This is a critical flaw. Some of the appliances that work well in regular houses function poorly or not at all in off-grid RVs. Some of the appliances that work well in off-grid RVs would be strangely out of place in regular houses. If a well designed RV exists, it does not look like a regular house. To be successful at off-grid RVing, one must do the same thing that one does when learning a new language: *clear one's mind, experiment and learn until one gets it right.*

General Criteria for RV Appliances

In order to be successful in an RV, appliances need to be 1) small enough to fit somewhere, 2) lightweight enough to not make the RV too heavy, 3) able to withstand road shaking equivalent to many major earthquakes, 4) reliable, 5) not too specialized to be useful and [most important] 6) conserving of resources which the RV stores in limited supply, which are electricity, propane gas and water (including wastewater tank space). In addition to the general criteria, each individual appliance has its own specific idiosyncrasies.

The Appliances

Lights

Lights are a very important part of any RV. A well lit RV is both a functional and a happy RV. Don't settle for overly dim lighting. Don't sit in the dark. Use the right kinds of lights and you won't have to.

Types of Lights

I.) Incandescent Lights:

What are incandescent lights? How do they work? Incandescent lights are the standard typical common "cheapie" light bulbs that we find and use all over the place in our houses, cars and even flashlights. They work and get their name from making light using a process called *incandescence*, which is a fancy scientific term meaning "glowing in the dark." Any item can incandesce, if hot enough. Consider what happens to an iron bar for example. When an iron bar is cold, it emits no light. Heat it up in a fireplace, then pull it out in the dark, and you'll notice that it glows red. Heat the iron bar up still further (causing it to melt into a puddle, but this is beside the point) and the puddle glows a much brighter yellow or white. Typical light bulbs don't use metals that melt at temperatures too low. They use tiny wire filaments of the metal tungsten to make their light, since tungsten does not melt under the high temperatures needed for white light production. The mechanism that makes a tungsten light bulb work is extremely simple. An electric current passes through the filament, heating the hell out of that filament, causing it glow. As long as the current keeps flowing, the filament remains hot and continues to glow. When the switch is turned off, the current stops flowing, and very quickly, the filament cools back down to room temperature and ceases to emit light.

12 Volt DC incandescent bulb (very typical in RVs)

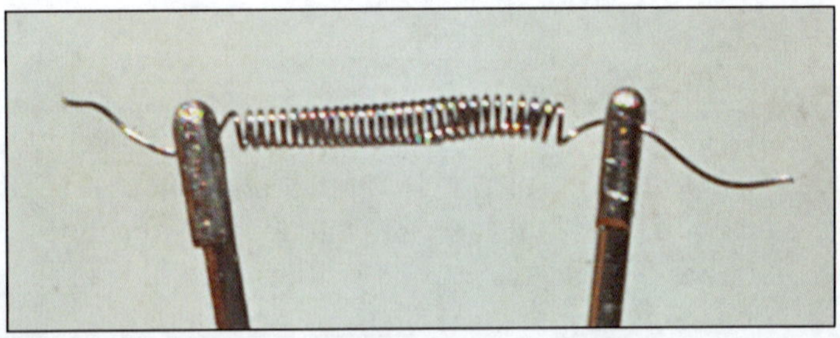

filament of incandescent light bulb

Advantages of incandescent lights: 1) They are cheap (both to make and buy). They are the cheapest of all four lighting types. 2) They are readily available. They can be purchased anyplace where stores and people exist.

Disadvantage of incandescent lights: They are very electricity inefficient. Of all the four lighting types, incandescent lights waste the most electricity to make their light. They are inefficient because, in addition to making light, which people *can* see, they also make heat, which people *can't* see. They don't make enough unwanted heat to really be annoying (still, you usually don't want to touch an incandescent light bulb when it's on), but the problem is: *electricity is needed and used to produce this heat.* Since their heat has no useful purpose, the electricity used to produce this heat is wasted. There really isn't any way, with this type of bulb, to make only the light without also making the heat.

II.) Fluorescent Lights:

What are fluorescent lights? How do they work? Fluorescent lights work and get their name by using a process called *fluorescence*, which is a scientific term meaning changing a higher frequency (energy) light into a lower frequency (energy) light. A fluorescent bulb is filled with a gas (no solid filaments are used). When turned on, an electric current gets sent through this gas, ionizing it. This makes ultraviolet light (which people can't see). The ultraviolet light then strikes and is absorbed by a chemical "frosting" on the inner surface of the bulb, where it changes to and is reemitted as visible light (which people can see).

upper: "cool white" fluorescent light for 12 Volt fixture
lower: "warm white" fluorescent light
(replaces standard 120 Volt AC incandescent bulb)

Properties of fluorescent bulbs: 1) They can be several shapes. They can be straight, circular or coiled over themselves like springs. 2) They can be several colors. This depends upon what kind of bulb is used. Often the color is a cold greenish white. Some have a more full spectrum color, resembling daylight (these are less common). Many made today are made to screw into sockets intended for incandescent lights, and have a warm yellowish color that very closely resembles the color of incandescent lights.

Advantages of fluorescent lights: 1) They are considerably more efficient than incandescent lights. (13 Watts of power is all that is needed to produce the same light as a 60 Watt incandescent bulb. That's over four times as efficient.) They simply don't waste a lot of electricity making unwanted heat, the way incandescent lights do. 2) They are readily available. Although in the past, they were not as common as incandescent light bulbs, they are now becoming the norm [for 120 Volts AC fixtures]. Fluorescent lights are familiar to and widely used by everyone in the technological world.

Disadvantages of fluorescent lights: 1) They are more complicated than incandescent lights. The circuitry of the light fixture, that energizes the bulb, involves a transformer and other discrete electrical components. Hence: 2) They cost more than incandescent lights. 3) In cold temperatures, they don't produce as much light. This isn't much of a disadvantage, because the lights usually warm up within a couple minutes after being turned on, allowing full output. Also, occupied RVs are likely to be heated, so interior lights aren't likely to be operated at the frigid outdoor temperatures that decrease light output. Decreased light output may be a slight issue only for outdoor lights, especially those used in winter. But I've heard that some newer fluorescent lights *do* produce high outputs even at low temperatures.

III.) LEDs:

What are LEDs? The term "LED" is an acronym for "Light Emitting Diode." "LEDs" is the plural form. These are solid crystals which emit light when DC current passes through them.

Typical uses of LEDs: 1) Small indicator lights and glowing numerals and letters on clocks, appliances and other electronic things are formed from LEDs. This is probably the most common and oldest major use of LEDs. Recently, another use for larger LEDs has been catching on: 2) Functional bright lighting. And: 3) Some television screens are now being made from LEDs.

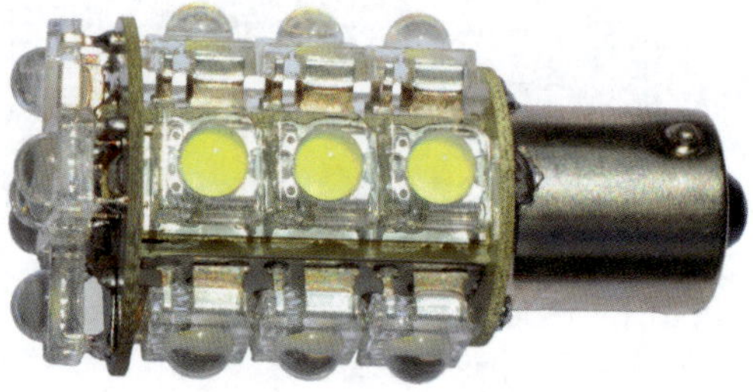

12 Volt DC LED "bulb"
(replaces incandescent bulb)

Properties of LEDs: 1) Their individual light cells are highly directional. And: 2) LEDs are available in several colors. Typically, they can make light in highly saturated intense colors. But lately, people have begun to figure out how to make them white (which is actually a mixture of many different highly saturated colors).

Advantages of LEDs: 1) LEDs are extremely efficient. They can make the same amount of light as incandescent bulbs, consuming as little as 1/10th the electricity. They are even more efficient than fluorescent lights (approximately twice as much). 2) LEDs can be made to fit into sockets intended for incandescent light bulbs. LEDs are not naturally the shape of incandescent light bulbs, usually LEDs are small crystal cylinders. A typical LED "bulb" (not actually a bulb) is a conglomeration of multiple individual light emitting crystals mounted on a single housing, the same general size as a regular incandescent light bulb. 3) Converting an RV from incandescent bulbs to LEDs is easy. If the "bulb" variety is used, converting can be as easy as buying and changing light bulbs. Some LEDs are sold as complete light fixtures containing lenses, on/off switches, and housings that screw into surfaces. Instead of installing "bulbs," one installs the complete fixtures. Complete fixtures are a bit harder to install than "bulbs," but complete fixtures have the advantage that they flicker less than the "bulbs," because properly connected wires and internally soldered connections almost never disconnect, whereas shaking can temporarily interrupt electrical connections where "bulbs" meet their sockets. And: 4) LEDs make very little unwanted heat. You can actually touch them while they are on. They have zero fire clearances and require very little clearances and air flow for heat dissipation.

> Large LEDs have been used to light up the New Year's Eve ball at Times Square in New York City. The 2008 ball, seen by a crowd of up to one million people, used only about as much electricity as ten toasters to make its light.

Disadvantages of LEDs: 1) LEDs are quite expensive. Bright 12 Volt LED "bulbs" (the type that can be used in RVs) can cost $20 each (less brighter ones maybe $7 each), which is quite a bit more money than regular "cheapie" incandescent light bulbs costing less than $1 each. 2) LEDs are not readily available. LEDs, used as bright functional lighting, are still kind of new. Although 12 Volt LEDs can be easily purchased over the internet and shipped anywhere, they might not be in little hardware or camping stores in remote unpopulated areas too soon. RVers, who decide to use them, had best order them before embarking on long highly mobile trips (where getting mail can be a pain). Extra bulbs should also be purchased, however their expense makes purchasing lots of extra LED bulbs unappealing. 3) LEDs can be damaged by heat. For this reason, LEDs should not be used (or stored) too close to furnaces, stove burners, gas mantle lights or (definitely not) in conventional/microwave ovens. Also, since incandescent light bulbs make quite a bit of heat, incandescent and LED bulbs should not be used together in the same light fixtures, since heat from an incandescent bulb can easily transfer to its nearby LED "roommate," ruining it. [Light fixture housings tend to trap heat inside them because they are sealed by lenses. Warranties can be voided if LEDs are used together with incandescent bulbs.] 4) LEDs may flicker. I have noticed that some LEDs rapidly pulsate as they begin to malfunction. This reduces light output and can be quite irritating. Even if most of the individual light cells of a "bulb" (or fixture) are still working, if one or two start flickering, it's probably time to replace the entire "bulb" (or fixture). 5) Light output from LEDs decreases before they "burn out." Like most devices that humans make, LEDs work great when they're new, but not so great when they start to get old. Their great efficiencies taper off toward the ends of their useful lives. 6) 12 Volts DC LEDs may not be all that reliable. LEDs are solid state crystals which must be very pure in order to work right. Good quality control at the factories where they are produced is an absolute must, else they may start flickering, have diminished light output or burn out altogether prematurely. This could frustrate users into opting for fluorescent or incandescent bulbs instead. [I have personally experienced a reliability problem with some green LEDs, on a lamp made for and used in a regular house. The first lamp I returned for an identical replacement. The second lamp developed the same problem. I gave up and threw it away.] 7) "White" LEDs can have unnatural looking colors (especially true of 12 Volt "bulbs"). They are now available in both "cool white" [resembling cloudy day blue sky] and "warm white" [resembling incandescent bulbs or candles]. The "cool whites" may make areas look more like workshops than happy homes. The 12 Volt "warm whites" have more pleasant colors but may still have disproportionate mixes of colors, including hues of magenta or green, and perhaps emit some ultraviolet, which causes some objects to glow due to fluorescence. And: 8) Since LED lights are usually composed of multiple light emitting crystals (called "cells"), each crystal cell has the capability of burning out individually. Burned out cells in an LED conglomeration cannot be individually replaced, so the only remedy is to replace the entire fixture. Burned out cells not only reduce light output of that conglomeration, they can make the whole light structure look funny and imbalanced, especially when multiple LED crystals are arranged in linear patterns. [How irritating is a string of Christmas lights, when just a couple of the bulbs burn out.]

IV.) Gas mantle lights:

What are gas mantle lights? How do they work? Gas mantle lights are lights that run by burning gas fuel, causing ash mantles to glow. When materials are heated, they absorb energy. Certain metallic salts when heated, reemit their absorbed energy as light, a process called *candoluminescence*. The colors vary depending on the metals used. (Fireworks use metallic salts to produce their colors.) Gas mantle lights use salts that make white, yellowish or reddish-yellow white light. (They could be made to glow in other colors but usually aren't.) The fuel for mantle lights is often propane, though natural gas, butane, vapors of gasoline or even heavier oils can be used. (Gasoline isn't used for indoor lights, since burning it releases too much carbon monoxide. Propane is the natural choice for RVs, since it is bought, stored and used for other appliances.)

propane gas mantle light mounted indoors in my RV

What are the components of a gas mantle light? The fixture itself is made from pieces of metal (thin sheet and tubing), a glass "globe," and a ceramic burner nose. The mantle is a thin woven fabric "cup," with many holes in it (having a similar texture to panty hose), that when hung, looks a bit like the end of a very small translucent sock. The mantle is composed of fabric and heavy metal salts that glow when heated. The gas is delivered to and burned at the empty space in the center of the mantle, distributing the heat evenly. Prior to its first use, the fabric of a mantle is burned away, leaving an ash.

upper left: unburned tie-on mantles

above: ceramic burner noses
(left: for tie on mantles)
(right: for preformed mantles)

left: burned mantle on installed burner nose
(globe removed for photo)

Advantages of gas mantle lights: 1) Gas mantle lights use no electricity at all. There is no need to worry about draining or charging (or even having) batteries, when using this type of lights. 2) Gas mantle lights are very bright. According to one of their manufacturers, a single glowing mantle emits as much white light as a 70 Watt incandescent electric bulb. [I use them quite frequently. I don't think they are that bright, but they are very bright. A single glowing mantle is considerably brighter than the single or double 12 Volt incandescent light bulbs, typically found in RVs.] 3) Gas mantle lights consume very slowly. Even a small (20 pound) tank of propane could run the light for *over nine days* straight (24 hour per day operation). Battery operated electric lights cannot even come close to matching that. Thus, one can turn gas mantle lights on for long time periods with little concern for "running out." 4) Gas mantle lights are inexpensive to operate. A single light consumes about 7 cents worth of gas per hour (2016 prices). The cost can be even less, depending upon propane prices.

Disadvantages of gas mantle lights: 1) Gas mantle lights release exhaust into the indoor air. Gas lights work by fire. All fire burns up indoor oxygen and releases exhaust gasses, namely carbon dioxide (which is irritating) and carbon monoxide (which is deadly in significant quantities). If gas lights are used indoors, some outdoor ventilation (at least one open window or roof vent) is necessary, in order for occupants to be safe. However, only a small amount of ventilation is needed. The chance of actually killing yourself with the exhaust is extremely remote. People use gas stoves inside RVs all the time, which also release carbon monoxide, but seldom does anyone die from doing so. 2) Gas mantle lights make heat. All fire makes heat. The heat released by gas mantle lights makes them undesirable or unusable on hot or very warm days/nights. However nighttime, which is when the lights are most likely to be used, also tends to be quite a bit cooler than daytime, so most of the time, temperatures are cool enough to use these lights. 3) Gas mantle lights are somewhat difficult to install (especially when compared to their electric counterparts). The copper tubing that carries the gas to the lights is quite inflexible. It can be bent, but it tends to retain its shape (unlike some electric wire, which just sort of flops around any which way it is shoved). Because the tubing is stiff, it takes more time to install behind walls than electric wiring, if installed before final interior wall panels are put in place, and is almost impossible to run behind walls after those wall panels have been installed. (If installed after an RV is built, wall panel sections might have to be removed, then replaced, in order to get the tubing behind the walls, or the tubing might have to be run in front of the walls, where it would be exposed and might not look too classy, unless covered up with wooden trim). Care must be taken when routing tubing, as the tubing can kink (impeding gas flow), or even crack (rendering it useless). 4) The ash mantles are quite brittle. They can crack from road travel or being touched by any object and can develop holes in them for no apparent reason. However, they don't break very often. And cheap mantles (intended for portable lanterns) can also be used on the indoor lamps. These cost about 50 cents each. 5) Gas mantle lights need to be lit, rather than just flicked on with a switch. This is a bit of an inconvenience, but not that much. They are as easy to light as propane burners on a stove. All RVers light and use propane burners on their stoves. And: 6) The ash mantles gradually darken from carbon accumulation during use. This diminishes their light output. However, this takes quite awhile to happen. When too much black soot forms on a mantle, simply replace it.

Which Light Types are Best Suited For RVs?

I.) Incandescent Lights:

Of all the four lighting types discussed here, incandescent lights are the *least* well suited for RV use. Yet incandescent lights are the *most frequently used* lighting type in RVs today. In fact, most RVs produced use incandescent lights *exclusively.*

Why are incandescent lights inappropriate for RVs? Incandescent lights gobble up massive amounts of electricity compared to the other three lighting types. Batteries don't store enough electricity to make incandescent lighting practical for even moderate off-grid use, and batteries are difficult to charge. So incandescent light users are left with three poor choices: 1) settle for dim lights not left on very long, 2) have bright lights but run the batteries down really quickly or 3) power the lights by running generators, which are noisy, extremely inefficient and make enemies out of neighbors.

Why do RV manufacturers choose incandescent lights? 1) Incandescent lights are the cheapest (of all the light types). This is true of both the bulbs and fixtures. They are far cheaper than any of the other three types of lights. 2) Electric light fixtures (any type) are much easier to install than gas mantle light fixtures. Ease of installation cuts production costs. 3) Incandescent lights are available in lots of different sizes and styles. This is probably because they are so common. Thus, incandescent lights are appealing to those with "designer brains" (individuals who focus their attention on things that *look kind of trendy*, rather than on things that actually *work well in the field*). 4) Incandescent lights are the type that are most familiar to most people. It is always easiest for manufacturers to just grab something familiar, rather than taking the time to become knowledgeable as to what options are available and what properties those options have. Proper research requires reading books, gathering information from the internet, talking to actual RVers, and (heaven forbid!) actually conducting some comparative field tests. Grabbing that which is most familiar eliminates the need for all that nasty studying and testing. 5) Customers tend to seek that which is most familiar. This is especially true if they are uninformed (often the case). This heavily influences what they buy. If they see an incandescent light in an RV, they may think, "that's what a light should look like," and buy the RV, not realizing that the light that works well in a regular house might fail miserably in an RV in wilderness. On the other hand, if they see something unfamiliar, such as a gas mantle light, they may think, "What the hell is that thing?" then walk away without buying the RV, not realizing that they are seeing something that works wonderfully in an RV in wilderness. Familiarity is a lousy barometer for determining suitability, unfortunately that's what many people use. RV Manufacturers legitimately need to make profits, so they may be reluctant to produce things that people are unwilling to buy.

Are incandescent lights really the cheapest? Only if one just considers the installation and replacement of fixtures and bulbs. If the cost of installing and maintaining the electrical systems required to feed the needs of each lighting type are added to the equation, energy-hungry incandescent lights actually are the *most expensive* of all the lighting types.

Who benefits from using incandescent lights? RV manufacturers, certainly not RV users. Manufacturers benefit from using components that are cheapest to buy and install, but only in the short term. If customers realize that manufacturers are cutting corners to boost profits, customers may stop buying from them. This can make a short term gain a long term loss.

Are there any legitimate uses for incandescent lights in off-grid RVs? While diving, yes. Whenever the engine runs, the alternator makes plenty of electricity nonstop. Thus incandescent lights can be effectively used as running lights, brake lights, turn signals and headlights. While parked, perhaps incandescent lights have some value as spare/emergency bulbs, bulbs in out-of-the-way seldom used locations, and portable flashlights. But as a main lighting source, no. The fact that they are used at all (and so widely) indicates widespread illiteracy, ignorance and indifference within the RV designing community.

II.) Fluorescent Lights:

Fluorescent lights represent a *good* lighting choice for RVs, though still not the *best* choice. In my opinion, they should be included in RVs as supplemental lighting, but should not be the *sole* artificial lighting sources. Some RVs have them but the majority don't.

Of the three types of electric lights, fluorescent lights are kind of middle of the road along the fronts of efficiency, availability and cost. *Efficiency:* Fluorescent lights are several (up to four) times as efficient as incandescent lights, but fall slightly short of LEDs (which can be twice as efficient as fluorescents). Fluorescent light fixtures can be made to run from 12 Volts DC, 120 Volts AC or portable batteries. 12 Volt DC lights can be directly connected to an RV's deep cycle battery(s). 120 Volt AC fluorescent lights can be run in RVs by converting battery-stored 12 Volts DC into 120 Volts AC using a power inverter. This is a reasonable option, despite a slight loss of power due to the inefficiency of the inverter, because inverter consumption is small compared to the savings by using fluorescents instead of incandescents. (*Right photo: This 120 Volts AC fluorescent light with the power inverter draws the same current as a single standard 12 Volt incandescent bulb, yet this fluorescent bulb is bright enough to comfortably light half an RV.*) *Availability:* 12 Volt DC fluorescent light fixtures usually need to be purchased at RV stores. Replacement fluorescent bulbs are not as widely available as incandescents, but can be purchased at many stores in out-of-the-way places (unlike 12 Volts LEDs, which usually need to be ordered and shipped by mail or UPS). *Cost:* Fluorescent lights can be quite expensive but not as expensive as LEDs. There are so many shapes and colors of LEDs, and several varieties and colors of

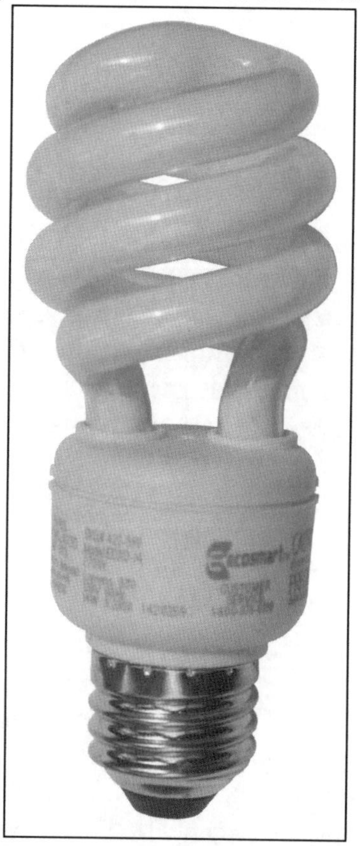

fluorescents, that side by side comparison of the two lighting types may be difficult. LEDs can cost quite a bit more than fluorescents, especially if multiple fixtures (or "bulbs") of light emitting crystals are desired.

Are there reasons for including ***some*** *fluorescents in RVs, rather than using LEDs exclusively (for electric lighting)?* Yes. 1) Fluorescent lights often have pleasing colors that we are used to, whereas LEDs often have colors with strange looking tints. And: 2) Fluorescent lights are not so sensitive to heat, like LEDs are. Thus fluorescents are appropriate to place over stove cooking areas. (This is where I have one in my RV.) 3) Some fluorescent lights are very close to LEDs, in terms of efficiency.

<u>III.) LEDs</u>:

Of all the electric lighting types, LEDs are the *best* choice for RVs. Their inclusion as a main lighting source is a *must* for properly designed wilderness RVs. However, RV designers generally don't like to use them. I went to an RV show in 2010 and saw no LEDs at all.

left photo: The LED (on right) and the incandescent bulb (on left) are about the same size and have the same mount, so...

right photo: the LED can easily replace an incandescent in a fixture.

Why do LEDs work so much better in RVs than incandescents? It's a simple matter of energy efficiency. When hooked up to municipal electricity, LEDs do eventually save money despite their cost. But LEDs really show their might when hooked up to batteries. Using LEDs instead of incandescents can increase run-down times of batteries by as much as tenfold, if such batteries are used for lighting alone. This could make weak lighting systems actually practical, and impossibly clumsy lighting systems possible. Since RVs are powered by batteries, when off grid and generators are not running, LEDs are extremely useful in them.

Can substituting LEDs for incandescents alone save poorly designed RV electrical systems? No, for two reasons: 1) Anything run from batteries eventually runs those batteries down. If there is no plan for charging (or replacing) the batteries, the only thing that electricity conservation can do is prolong the inevitable run-down time. And: 2) If extreme electricity hogs other than lights are hooked up to the same battery(s) as the LEDs, then those other appliances may drain the battery(s) quickly, thus shutting down the entire system of electrical appliances, including the LEDs. This could make the inclusion of LEDs only marginally helpful in improving bad systems as a whole.

Practical considerations for LEDs in RVs: 1) Both bright and dim LEDs have use in RVs. Bright LEDs can be used to replace incandescent light bulbs for functional lighting. Dim LEDs can make very effective closet, cabinet, out-of-the-way-corner or mood lights. Dim LEDs consume so little electricity that they can even be used as night lights and left on for long time periods, something that should never be done with incandescents. 2) Both highly directional and non-directional LEDs have use in RVs. Since the individual crystal cells that make up an LED are highly directional, if all the cells point in the same direction, then the entire light is highly directional, whereas if the individual cells point in multiple directions, then the entire light is non-directional. Highly directional lights can make very effective reading lamps and can be great for illuminating tables or cooking areas. Non-dircctional LEDs are great for evenly lighting general areas. Be aware that one is not condemned to send light only in the direction the crystals point--- adding mirror-like duct tape or even aluminum foil to the fixtures in just the right places can redirect the light in any direction the users want that light to go. 3) Some LEDs benefit from lampshades or reflectors. Highly directional individual LED cells can be rather hard on the eyes if viewed directly, especially since some LEDs are pinpoint light sources. (I find the red and amber dots of some automotive LEDs on other vehicles annoying to look at while driving.) Homemade lampshades made of frosted glass, wax paper or regular white paper, placed in front of the cells can eliminate this problem, although they may cut down light output (not a problem for mood or night lights, which are supposed to be dim). Pointing LEDs at and reflecting them from white ceilings or walls is another option for making soft even light. 4) LEDs can be used in cubbyholes without any realistic possibility of starting fires. This is because they produce very little heat. They can be used in closets or areas where ventilation is poor. 5) The colors of white LEDs can take some getting used to. They are different from incandescent lights. "Cool white" LEDs are functional but annoyingly bluish. They do however blend well when mixed with the yellowish light of gas mantle lights or candles. "Cool white" LEDs are great for outdoor "porch" lights, because the cells in the human eye that respond to low light conditions (called rods) are sensitive primarily to blue light, therefore lights that emit lots of blue effectively illuminate much further into the "yard" than do lights that are short on blue (which are "warm white" LEDs, yellowish incandescents or light made by fire). [A low current (1/6 Amp) cool white LED on the outside of my trailer allows me to see when walking at least 100 feet away from my trailer in the darkest of places.] "Warm white" LEDs are made to resemble the colors of incandescent lights or candles. However, they can have unnatural looking tints of green, magenta or violet. Perhaps with the passage of time, companies that manufacture them will get better at making their colors appear natural. And: 6) Highly color saturated LEDs are great for mood lights or decoration. I have seen them available in red, amber, green and blue. Deeply colored LEDs can also be mixed with "white" LEDs to produce better shades of "white," but lights of several colors can produce multiple shadows of different colors, which also appear unnatural and can be irritating.

IV.) Gas Mantle Lights:

Gas mantle lights are the *best* choice for bright functional RV lighting because they consume no electricity at all. Yet most RV designers, sellers and users seem to be unaware of their existence. This is remarkable, considering the fact that people who camp in tents have been using them for decades, and people who camp in tents often do so right next door to people who camp in RVs. Even more remarkable is the fact that gas mantle lights were commonplace in RVs a half century ago, but in recent times RV designers have done away with using them.

> Usually the only group of people who camp in tents, without gas mantle lights, are those out on their first or second outings. The first time tenter notices neighbors with really bright lights, then walks over and starts asking questions. After some inspection and conversation, that first time tenter decides (s)he absolutely must buy one of them. By the second or third camping outing, that tenter has at least one gas mantle light.

If gas mantle lights work so well, why aren't they commonly used in RVs? My best guess is: RV designers have decided to "go modern," feeling that "fire is for cavemen." My response to this is: "*illiteracy* is also for cavemen." If you (RV designers) read a book once in awhile, you would know that nobody has yet invented a battery that stores enough electricity to allow incandescent electric lights to come close to competing with gas mantle lights off grid. Will the real caveman please stand up?

How much better do gas mantle lights perform off-grid when compared to incandescent lights? This topic will be covered shortly in "Electricity Versus Fire---Which is Better?". Some side by side mathematical comparisons need to be made, in order to understand the great extent to which gas lights outperform incandescent electric lights in the wilderness. If you're short on time or really hate math, skip right to the "Key Points Summary" text box of that section.

How do indoor gas mantle lights differ from portable outdoor gas mantle lights? 1) Unlike portable lights, which aren't attached to anything, indoor gas mantle lights mount permanently to walls, cupboards or specially designed posts for central hanging. 2) Indoor gas mantle lights permanently connect to an RV's central propane system, whereas portable gas mantle lights draw fuel from small portable disposable propane cylinders. Central connecting eliminates the need to purchase and store many small propane gas containers and change those cylinders frequently when they empty. Also, propane purchased in small disposable cylinders typically costs 3-7 times as much as propane purchased as RV tank refills. 3) Indoor gas mantle lights use one mantle, whereas portable gas mantle lights usually use two mantles. 4) Indoor gas mantle lights have a classier look than portable gas mantle lights [in my opinion].

Practical considerations for gas mantle lights in RVs: 1) Gas mantle lights can be used without worrying about conservation. They consume propane so slowly, in comparison to the amounts of propane stored in (even small) tanks, that users can turn them on, walk away, and forget about them, without any realistic fear of running out of propane. Such carefree operation is not possible with bright RV electric lights!
[I sometimes turn a gas light on when going for a walk or swim at night (to serve as a beacon so I can find my camper upon return), when hanging around the campsite and when out attending park ranger programs, to make my trailer look "lived in."] 2) When gas mantle lights are used as the main lighting source, electric lights aren't needed at all. They also are completely functional if the electricity goes out. Gas mantle lights are usually the brightest of all the four lighting types. 3) Gas mantle lights can be used most of the time. *Exception A: When ambient temperatures are very warm.* Because gas lights run by fire, they make quite a bit of heat. Nobody wants to add heat to an already hot RV by making an indoor fire. However, opening all the windows, the door and the roof vents often provides substantial enough air exchange to make them usable, even on moderately warm nights or days. (One needs to have screens to keep insects out.) *Exception B: When ambient temperatures are very cold.* In very cold temperatures, one probably wants to crank up the furnace and shut all windows and roof vents tight to conserve heat. But using gas lights requires at least *some* ventilation, in order to remove the poisonous exhaust gasses that these lights make. One option is to shut all windows and roof vents tight and use only electric lights on very cold nights or days. Another option is to use the gas lights but open a window and roof vent just a crack--- massive ventilation is not needed. When using the second option, one must tolerate a bit of cold drafty air entering the RV. However, that cold draftiness is somewhat mitigated by the heat that gas mantle lights release. *Exception C: In places of high insect infestations.* Some insects are small enough to fit through screens and attracted to light. It can be very annoying to have hundreds of tiny insects all over the place getting into everything. One may want to consider leaving *all* lights off under these conditions. Despite some limitations, in my experience, gas mantle lights overall are usable about 90% of the time. 4) The heat that gas mantle lights make is far more often an advantage than a disadvantage. The typical RVer camps in all kinds of weather, and very often in cold conditions. Though not their intended purpose, the heat from gas lights can help make cold clammy RVs comfortable. 5) Several mantle types are available. A) *Preformed mantles* have their three dimensional shapes formed at the factory. They are somewhat rigid and come attached to ceramic rings which hook onto the burner noses. (Special burner noses, different from those designed for tie-on mantles, must be purchased separately--- they don't come with the lights.) Preformed mantles may be removed and stored upside down in their original boxes for travel, which makes them less likely to break from shaking (shaking can damage them directly or cause them to unhook and fall off their burner noses and crash to the floor). Storing mantles is a royal pain for frequent movers, because one can't use gas lights without mantles, so if one doesn't take the time to put the mantles back in the lights on short stops (rest areas and parking lots), then the gas lights can't be used on short stops. Preformed mantles are rather expensive.
B) *Tie-on mantles from the factory* are flexible and shipped flat, so they take up almost zero storage space prior to being tied to the burner noses then burned. They are less

expensive than the preformed mantles but can still cost quite a bit. C) *Tie-on mantles intended for portable propane lanterns* actually work quite well in indoor gas lights as well. These are readily available and inexpensive (about 2/$1). The light that they make is a bit more yellowish than and probably not as bright as the mantles made for indoor gas lights. (I use these because they are so cheap. I don't get bothered when they break. I always carry plenty of extras in my trailer.) 6) The mantles must be burned to ash prior to use. When new, mantles are cloth. After installation but before turning on the light, users light the cloth and burn it away, leaving behind an ash, which retains the shape of the cloth. The ash is very brittle and must not be touched, else it will break.

Who else uses gas mantle lights? 1) Amish people, who don't believe in using electricity, use them almost exclusively as their indoor lighting sources. 2) People with cottages/cabins in remote areas, where municipal electricity is unavailable, frequently use them. 3) People who camp in tents frequently use outdoor portable gas mantle lights.

Would I (the author) personally venture into the wilderness with an RV not having at least one gas mantle light? Not a chance in hell.

photo: This Photoshopped composite simulation suggests a simple possible solution to the problem of unwanted exhaust from gas lights--- the chimney.

Other uses of gas mantle lights: Though not specifically made for such purposes, gas mantle lights work really well for warming leftover food and drying small amounts of laundry. (Photos included in upcoming sections: "Stoves & Ovens" and "Washing Machines & Dryers").

History of Artificial Lighting

Short Version:

First, humans discovered that fire could make light. Then, humans discovered that electricity could make light. Finally, humans forgot how to use fire to make light.

(What's this? You want a longer version? Alright!)

Long Version:

In 1879, probably the single most significant development in the history of artificial lighting took place. That's when Thomas Edison invented the carbon filament incandescent electric light bulb. Although incandescent light bulbs were invented as early as the 1820s, Edison is usually given credit for the light bulb, because his light bulb was good enough and cheap enough to manufacture and sell commercially. (Prior bulbs burned out quickly--- often after only minutes of use.)

Before the incandescent light bulb: Prior to 1879, for centuries humans used fire to make light, since fire was their only option. For a long time, candles and oil lamps were used. (Whale oil lamps were common in the 1700s.) But fire technology did not stand still. By the latter half of the nineteenth century, it had advanced quite a bit from its early stages.

Gas lighting: By the time the light bulb was invented, gas illumination had become a choice and well established method of making light. The infrastructure for supplying gas for lights was in place and gas lighting was profitable. Old gas lights were different from today's gas lights, in that old gas lights did not use mantles, they made light by directly burning the gas. This was possible because the chemical content of the gas was different from the chemical content of gas commonly used today. Today's light clean burning gasses are natural gas (methane and ethane), propane and butane. Gas used in the old days was dirty burning "coal gas" (manufactured from coal), a mixture of hydrogen, carbon monoxide, methane, some heavy volatile hydrocarbons and some gasses that don't burn. It was the presence of heavy hydrocarbons in coal gas that caused the flame to burn yellow, instead of dark blue, like today's gasses (which contain only light hydrocarbons), hence coal gas was also called "illumination gas." (Candle wax is composed of long chain heavy hydrocarbons. Note that candle flame is yellow, not blue.) The heavy hydrocarbons were a mixed blessing, because those same heavy hydrocarbons, that made the fire burn yellow, unfortunately also produced some black soot, which accumulated on glass fixtures, necessitating frequent cleaning of those glass fixtures.

The early days of the light bulb: When first introduced, Edison's light bulb didn't instantly sweep across the world and dominate. This is probably because back then there were several serious technical problems associated with electric lighting, which have since been solved. *Problem #1:* Getting electricity to people's houses and businesses was a challenge. A) The electric wiring infrastructure did not exist to deliver electricity to houses everywhere back then, and not everybody had a fast flowing river in their back yard, that enabled them to generate their own electricity. Without electricity, people couldn't use electric lights. B) People didn't know how to send electricity long distances back then. Edison tried to send DC to houses. DC can't travel very far and requires really thick wires to travel even short distances. Edison's rival, Nikola Tesla, eventually correctly concluded that it would be better to send AC to houses, and devised a way to send it long distances through comparatively thin wires. He did this by transforming the voltage up (near the generator) then back down again (near the houses). This method is still used today. *Problem #2:* The carbon filament light bulbs were not as good as today's tungsten filament light bulbs. Tungsten bulbs were introduced during the first decade of the 1900s. *Problem #3:* The very notion that all houses should have electricity was not well accepted back then. And: *Problem #4:* Gas lighting remained a fierce competitor for quite awhile after the introduction of the electric light bulb.

Gas Mantle Lighting: In 1891, about a decade after Edison's light bulb, the Welsbach mantle was invented. (Actually, Welsbach had an earlier 1885 version, but it was dim and didn't produce good light color. It produced greenish light. People didn't want it and Welsbach nearly went bankrupt.) Clean burning gas (such as methane, propane, butane) could make pleasing natural looking bright white light by heating these mantles. Mantles could also be made to glow by burning other fuels, such as gasoline or oil. Welsbach mantle lights became popular for several decades following their introduction, but eventually fell into widespread disuse, because electric light bulbs became more popular. But Welsbach mantle lights did not disappear entirely. They are still manufactured and used today--- as outdoor propane lanterns and indoor gas lights. Why, in this "age of electricity," do some people still have such a strong allegiance for gas mantle lights, over one and a quarter centuries since the introduction of Edison's bulb? It has to do with what has *not* been invented.

> Lighthouses, along the shores of lakes and oceans, made bright light by burning whale oil, oil pressed from seeds and finally kerosene. Lighthouses often needed to use fire rather than electricity, because many were located in remote areas far away from electric wires.

*What has **not** been invented?* Even though Thomas Edison's incandescent light bulb was invented in 1879, even today (as of 2017), nobody has yet invented a battery that stores enough electricity to make incandescent lights practical for off-grid use. (Yes, batteries can run incandescent lights, but nobody has yet made a battery that is small enough, lightweight enough, cheap enough and easy enough to recharge, that stores electricity in the amounts that would be needed to effectively run incandescent lights.)

Fluorescent Lights: These were commercially introduced in the 1930s, although earlier versions (which didn't work very well) were made as far back as 1856.

Recent trends: There has been political movement to do away with incandescent lights in regular houses and replace them with fluorescents and LEDs, for environmental reasons, since some electricity is made by burning fossil fuel, which adds carbon to our Earth's atmosphere. Fluorescents and LEDs could save lots of electricity. This would reduce the amounts of fossil fuel needed to make that electricity. Of course, if we humans figure out how to make all our electricity from clean renewable sources, and implement such a program, then conservation will become less of an issue in the future.

Electricity Verses Fire--- Which is Better?

The first two major events in the history of artificial lighting, which were discovering that fire could make light, then electricity could make light, seem natural. The third, forgetting how to use fire to make light altogether, is a bit difficult for someone like me to comprehend. Electricity is wonderful and reliable, when made by municipal power companies, in large quantities by central generators, supplying millions of customers, staffed by full time highly trained professionals who really know what they are doing. In the bush, electricity is still too unreliable and difficult to make and store, in the quantities that we have become accustomed to in our regular houses, without excessive use of noisy smelly extremely inefficient generators. Since humans used fire for lighting for millennia before the electric light bulb was invented, I am bewildered as to why so many people insist upon using only *electric* lights in RVs in wilderness.

Should fire for lighting be given another chance? A side by side comparison between gas mantle and electric lights, taking into account consumption rates, weights, refilling/recharging times, relative efficiencies, and costs, can help answer this question. The contestants are: I.) A gas mantle light, run by a small propane tank. II.) An incandescent electric light bulb, of equal brightness, run an equal time by batteries not recharged. III.) An incandescent electric light bulb, of equal brightness, run by a battery weighing about the same as the full propane tank, charged multiple times. IV.) An incandescent light, of equal brightness, run directly by generator, using fuel of equal weight to the that consumed by the gas light. And: V.) Multiple incandescent lights, each of equal brightness to the gas light, run directly by generator. [Note: II, III, IV and V are all compared to I.]

The Comparison:

(I.) A gas mantle light*, hooked up to a standard full 20 pound "grill size" propane tank.* How bright would it be? According to a leading gas light manufacturer, it would be as bright as a 70 Watt incandescent light bulb. How long would it run? Also according to the manufacturer, the light runs for about 11 hours per pound of propane. Since there are 20 pounds in the tank, the light would run 220 hours from the entire tank, which is a little **over nine days**. [(20 pounds) × (11 hours/pound) = (220 hours) = 9 days and 4 hours.] How much would its fuel system weigh? An empty steel (the common kind) propane tank weighs slightly less than 20 pounds. An empty top of the line aluminum empty tank weight slightly less than 10 pounds. With 20 pounds of propane added, a full tank weighs about 40 pounds (if steel) or 30 pounds (if aluminum). How long would it take to refill the tank? Once at the filling station, the process, including payment, usually takes less than 15 minutes. How much would a new propane tank (including gas) cost? Usually, less than $40 (if steel) or $100 (if aluminum).

(II.) An incandescent light bulb*, of equal brightness to (I), run an equal time, by 12 Volt lead acid batteries, not recharged.* In order to be equal in brightness, the light bulb would need to consume 70 Watts of power. At 12 Volts, this would draw about 5.83 Amps. [(Power) = (Voltage) × (Current), therefore (70 Watts) = (12 Volts) × (5.83 Amps).] The total charge used for 220 hours of operation would be 1283 Amp hours. [(5.83 Amps) × (220 hours) = (1283 Amp hours).] How much would its fuel system weigh? A typical deep cycle battery might store 75 Amp hours. However, batteries should not be discharged to less than 50% of storage capacity, therefore the usable charge is about half that, which is 37.5 Amp hours. Since a typical 75 Amp hour deep cycle RV battery might weigh about 38 pounds, the ratio of weight to charge is about 1 pound/ (Amp hour). [(38 pounds) / (37.5 Amp hours) = (approx.) (1 pound / Amp hour).] Therefore the weight of the batteries to supply the light (equally bright as the gas light) with the needed 1283 Amp hours from a single discharge cycle would be about that same number in pounds, which is **1283 pounds**--- that's right, **over half a ton of batteries**. How many batteries would be required? (1283 Amp hours) / (37.5 Amp hours per battery) = **34 batteries**. How much storage space would 34 batteries require? If such batteries were 10 inches long × 7 inches wide (a typical size that I measured), 34 batteries would take up over **16 square feet** of storage space, if placed next to each other on the floor. To visualize how much floor space this is, look at the following illustration of such a plot superimposed upon the actual floor plan of my 12 foot by 7½ foot camping trailer. Compare it to the tongue space required for the one propane tank needed for equivalent operation of the gas light.

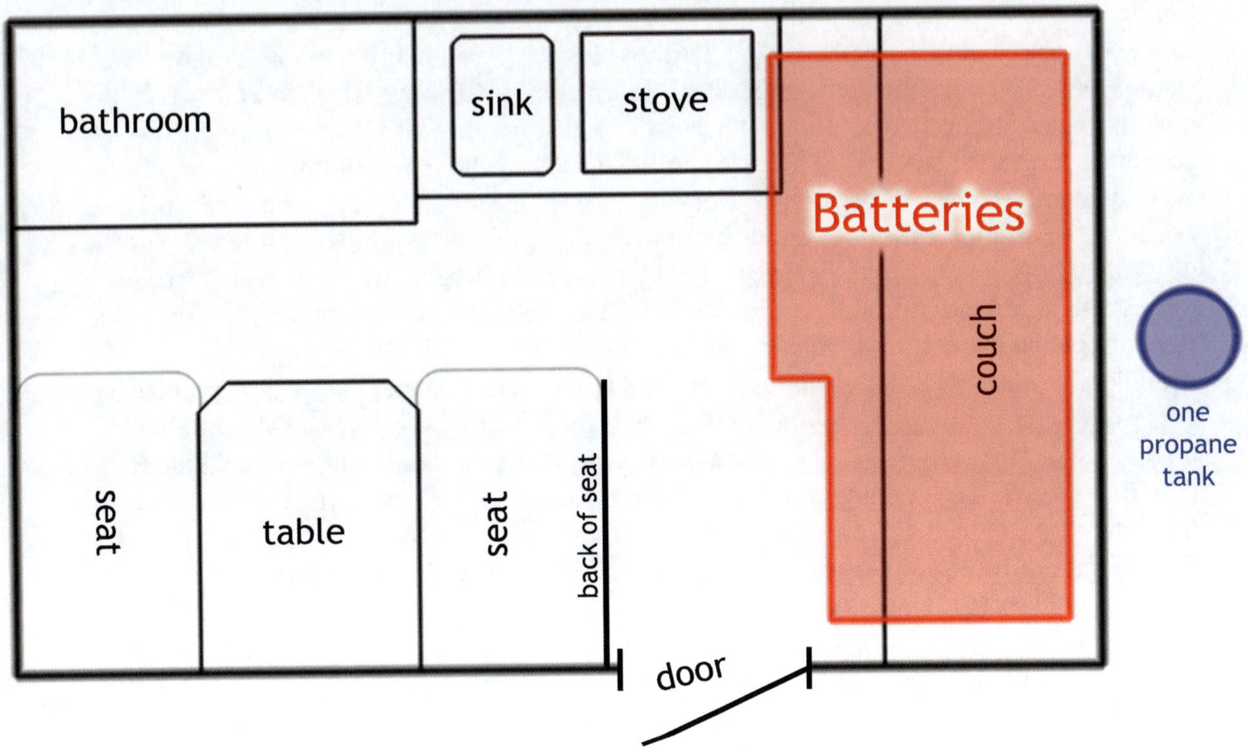

In volumetric measurement, this would take **12 cubic feet** of storage space (assuming a battery height of 8.5 inches). Realistically, that amount of batteries, both in terms of weight and space, could not be stored in an RV. Nor could they be carried or moved by an average human being, even if moved one at a time, in a reasonable amount of time. (The 40 pound filled steel propane tank can be easily lifted by a person not too frail or handicapped.) The only way to cart that amount of batteries around, would be to put them in a separate utility trailer. How long would it take to recharge those batteries? What charging method would be used? I don't even want to get into this! Certainly longer than the 15 minutes that it takes to fill a 20 pound tank of propane! How much would 34 batteries cost? Maybe **$3000**! Final insult: Adding over half a ton of batteries to an RV would significantly increase its weight. This would result in greater consumption of fuel while driving down the road. Cost of this extra fuel could be significant if long trips were taken. Tell you what--- you can do the calculations, if you insist upon using incandescent lights from batteries. I'm done with this one!

(III.) ***An incandescent light bulb, of equal brightness to (I), run by a battery weighing about the same as the full propane tank, charged multiple times***. A single 75 Amp hour battery weighs about the same as a full steel 20 pound propane tank (which is about 38 pounds). How long would the battery run the light, when compared to propane? 37.5 Amp hours of usable charge in the battery could run the light for only about 6.43 hours, which is 6 hours and 26 minutes. [(Charge) = (Current) × (time), therefore: (37.5 Amp hours) = (5.83 Amps) × (6.43 hours).] The propane tank could run the gas light **34 times as long** as the battery could run the electric light. [(220 hours gas light operation) / (6.43 hours electric light operation) = (34.2 gas/electric ratio).] What is the efficiency of charging? The equation comparing the amount of energy in the fuel burned in the generator to the amount of energy transferred to the battery (and ultimately to the electric light) using a 3 Amps charger was already done in Chapter 2 (under Defect #2: Inadequate Chargers), and turns out to be about 1 part in 360. Think of it like this: If a generator runs for 1 hour, the gasoline fed into it for approximately 10 seconds charges the battery. The gasoline fed into it for the other 59 minutes and 50 seconds is used to make noise, make waste heat, wear out engine parts, or the gas just goes out the tailpipe without even burning. How long would it take to charge the battery after the 6 hour 26 minute discharge? Using a 3 Amps charger (whose usable charging rate is only 2.5 Amps after in/out efficiency is taken into account), it would take **15 hours** to recharge the battery for another cycle, which is over twice the time that the light would actually get used. [(37.5 Amp hours) = (2.5 Amps) × (15 hours).] How long would the generator need to run (total) to operate the electric light for 220 hours? This could not be done all at once. The battery would be discharged to run the light, then recharged, then discharged, then recharged, multiple times. The total generator run time, to charge the battery to run the light, would be a whopping 513 hours, which is equivalent to **over 21 days of continuous operation**. [(220 hours of light) × (15 hours charging / 6.43 hours of light) = (513 hours of charging).] That's a bit longer than the maybe 15 minutes needed to fill a propane tank to run the gas mantle light of equal brightness an equal time! What is the relative efficiency between the gas light and the electric light? Calculating this is fundamentally simple. Fuel could be sent directly into the gas mantle light or into the generator to make electricity to charge the battery to run the electric light. Although gas mantle lights and generators often use different fuels (gasoline is usually used for generators), to simplify this comparison, only propane generators will be considered. (Efficiency and cost of gasoline are similar to values for propane.) Consider the 20 pounds of propane, that runs the gas mantle light for 220 hours. Since propane weighs about 4.11 pounds/gallon, 20 pounds is 4.87 gallons. [(20 pounds) / (4.11 pounds per gallon) = (4.87 gallons).] A 3600 Watt Onan propane generator (according to the manufacturer) consumes at least 0.4 gallons/hour. Therefore that propane could run the generator to charge the battery for 12.17 hours. [(4.87 gallons) / (0.4 gallons per hour) = (12.17 hours).] But remember that battery charging takes more time than the actual time the light is run--- for every 15 hours of battery charging, the light only runs for 6.42 hours, therefore this ratio needs to be multiplied. The amount of time the light would run is 5.21 hours, or 5 hours and 12 minutes. [(12.17 hours of generator run time) × (6.42 / 15 light time per generator run time for battery charging) = (5.21 hours).] Recall that 20 pounds of propane sent directly to the gas mantle light will run it for 220 hours. 20 pounds of propane, sent to a generator, to charge a battery, to supply electricity to an incandescent light of equal brightness, will only run the light for 5.21 hours, which is 5 hours and 12 minutes. The relative efficiency of the gas light to the electric light is obtained by dividing the gas run time by the electric run time, which turns out to be about 42. [(220 hours gas light operation) / (5.21 hours electric light operation) = (42.2).] In everyday terms, this means that propane sent to a gas light directly will operate it **42 times as long** as propane sent to a generator to charge a battery to later run the electric light. (And propane burned in a gas light makes neither noise nor enemies from neighbors, like generators do.) **Why the hell would anybody choose a battery operated incandescent light over a gas mantle light for wilderness use?** Doesn't anybody read books or do math but me? Which would you prefer to buy: A) a car that makes 42 miles per gallon or B) a car that makes 1 mile per gallon? I know which one I would buy!

(IV.) An incandescent light, of equal brightness to (I), run directly by generator, using equal weight fuel, *(the step of storing electricity in batteries is skipped here).* Running a generator at the same time a light is run, instead of storing electricity in a battery for later use, can increase efficiency, because: 1) there is a limit to how much current batteries can accept, 2) chargers may be very wimpy and 3) the electricity that comes out of a battery during use is slightly less than the electricity that goes into it during charging. How long would the generator operate the light, burning 20 pounds of propane? 20 pounds of propane has already been established (above III.) to run the generator for 12.17 hours, which is 12 hours and 10 minutes. Therefore, the light would run for that same amount of time. What is the relative efficiency between the gas light and the electric light run by generator? As above, the relative gas/electric efficiency is calculated by dividing the amount of time that 20 pounds of propane fuel runs the gas light by the amount of time that 20 pounds of propane fuel runs the generator for the electric light, which turns out to be about 18. [(220 hours of gas light operation) / (12.17 hours electric light operation) = (18.08 gas/electric ratio).] In everyday terms, this means that propane sent directly to a gas light will operate it **18 times as long** as propane sent to a generator to run a single incandescent electric light of equal brightness. Cost of fuel for operation is that same number, the electric light costs 18 times as much to run as the gas light. Weight and storage space of propane is not significant, therefore won't be considered.

(V.) Multiple lights run directly by a single generator. If more than one light gets used simultaneously, efficiency improves greatly, since most of the fuel burned in a generator is used for overcoming friction. Two lights may be able to run, consuming little more fuel than one. Three, little more than two. Etcetera. But multiple lights cannot run, using exactly the same fuel as a single light because whenever power is drawn from any generator, current passing through coils creates magnetic forces that *slows the generator down.* As more power gets drawn, greater torque gets applied to slow the generator's rotation. Thus, when larger amounts of power are consumed, greater amounts of fuel need to burn, in order to overcome this "evil torque." Suppose the 3600 Watt generator were operated to the max. The number of 70 Watt incandescent light bulbs it could run would be over 51. [(3600 Watts max generator output) / (70 Watts per bulb) = (51.4 bulbs).] At maximum load, according to the manufacturer, the generator would burn 0.7 gallons/hour of propane, instead of 0.4. 20 pounds of propane (which is 4.87 gallons) would operate the generator for about 7 hours. [(4.87 gallons) / (0.7 gallons per hour) = (6.95 hours).] 51.4 light bulbs operating simultaneously is equivalent to one single light bulb operating 51.4 times as long. Therefore, the total number of hours of single light bulb operation is the generator run time multiplied by the number of light bulbs, which turns out to be 357 hours. [(6.95 hours of generator run time) × (51.4 70 Watt light bulbs) = (357 hours of a single 70 Watt light bulb operation).] The electricity/ gas efficiency ratio is calculated by dividing the electric light run time by the gas light run time. The electric lights would be **1.6 times as efficient** as gas lights. [(357 hours of electric light operation) / (220 hours of gas light operation) = (1.6 electric/gas ratio).] In this case, electricity wins! If super efficient LEDs were used instead (which are as much as 10 times as efficient as incandescents), then the electric lights really win, as those figures would also multiply by 10. [514 LED 7 Watt (70 Watt equivalent) "bulbs" could be operated from the generator simultaneously, instead of 51.4 incandescents.] Maximum LEDs, run by maximum generator output would be about **16 times as efficient as the gas light**! But what RVer needs to operate 514 light bulbs at once? What RVer needs to operate 51 incandescent light bulbs at once?

Conclusion to this comparison: *When just a few lights are operated, gas wins* the efficiency contest big time. When really large numbers of lights are simultaneously operated however, electricity wins. Somewhere in between, there is a "magic number" of lights, where the efficiencies equal. But this magic number is far too big for any single RV to realistically need or even use, unless that RV is making electricity for and sharing it with an entire campground. The bottom line is: *If you're going to light up a whole city, **electricity** made by generator is better. If you're going to light up a small campsite or RV, **fire** is better.*

> I have never understood why so many people choose to feed fuel into noisy generators, just to make electricity to run an electric light, when feeding that same fuel directly into a fire light runs the light much longer and doesn't make noise to bother anyone.

Points about the preceding comparison:

1) Consumption rates are very approximate.

2) This comparison assumes a medium generator (3600 Watt) and a weak (3 Amp) battery charger, a combination frequently found in RVs.

3) Most of this comparison [except for some of (V)] assumes *incandescent* lighting. Efficiency of battery powered electric lighting can be greatly increased by using fluorescent or LED lights, but they still lose to gas mantle lights, using the weak one battery and 3 Amps current charger systems.

4) Efficiency of making and storing electricity can be greatly increased by: A) using smaller generators (which burn considerably less fuel and are much better for smaller loads), B) having more batteries (covered in Chapter 2), and C) charging with greater currents (covered in Chapter 2). [Note: Smaller generators can't power big things, like air conditioners. Unless an RV has two generators (I've actually met somebody who has two), at least a medium-duty generator is probably needed, to handle the RV's other non-lighting electrical needs.] Battery powered electric lighting can actually win over gas, if efficient LED lights are powered by multiple battery and high current charger systems, run by generators. But generator run times make noise, gas lights do not.

5) Small numbers of efficient lights don't perform any better than inefficient incandescents, when powered directly by generator, because generators need to consume minimum amounts of fuel when operating, the vast majority of which is just wasted.

6) Calculations are based on figures (brightnesses, fuel consumptions) supplied by manufacturers. If these figures are not completely accurate, or vary due to special field circumstances, the numbers may be slightly off.

7) This analysis assumes that *lights alone* are powered by the generator. Running other appliances simultaneously from the generator dramatically alters the equation, since the generator may be needed anyway to run other things.

8) Many people *don't have gas lights.* If all a person has is a generator and a couple of incandescent light bulbs, that's what the person uses, because the choice is not between electricity and *gas*, the choice is between electricity and *nothing at all.*

Key Points Summary:
(Electricity-Fire Comparison)

1) If propane fuel is fed directly into a gas mantle light, it runs the light 18 times as long as it does if that propane is fed into a generator to power a single electric light of equal brightness.

2) If propane fuel is fed directly into a gas mantle light, it runs the light 42 times as long as it does if that propane is fed into a generator to power a (3 Amps) battery charger, to charge the battery to run an incandescent electric light of equal brightness at a later time.

3) A single grill size propane tank weighs about 38 pounds when full. Lead-acid batteries to run an incandescent electric light of equal brightness for an equal time (from a single discharge cycle) as the gas light (from a single empty cycle) would weigh about 1300 pounds, which is over half a ton.

4) Filling a propane tank to run a gas mantle light takes maybe 15 minutes (once arrived at the filling station). Charging a battery with a (3 Amps) charger to make enough electricity to run an incandescent electric light (for as long as the propane tank runs the gas light) would require over 21 days of continuous generator and charger operation, which is 2016 times as long, and no single battery can even come close to storing that much charge.

5) A brand new full propane tank costs maybe $50. Enough batteries to run an incandescent electric light the same time length (from a single discharge cycle) as the propane tank runs the gas light (from a single empty cycle) would cost maybe $3000, which is 60 times as much.

By the way... Yes, gas mantle lights make some carbon monoxide (open some windows). But generators make more. CO from long term generator use can find its way back into RVs through open windows or even shut windows, even though generator exhausts vent to the outside. Yes, gas mantle lights make heat, which may be unwanted. But generators make more heat. Even though located in insulated compartments, built-in RV generators can add significant unwanted heat to RVs.

Portable Lights

Don't overlook the obvious. RV users also need portable lights not connected to their RV electricity/gas systems. Flashlights, lanterns, and extra batteries are a must for all RVers. Some people, not wanting to deal with RV electrical systems, light their RVs entirely using candles and portable electric lanterns, with much success. Oil lamps also make great dim lighting, both for indoors and outdoors, and are very inexpensive, as well as attractive. (I bought an oil lamp at Wal-Mart not too long ago for only about $5. The oil doesn't cost very much either.) Portable gas mantle lanterns, though usually intended for tenters, are also great for yard or picnic table illumination for RVers, for those times that they desire to light the outdoors as well as the indoors.

Best Overall RV Lighting Plan

Use gas mantle lights as the main lighting source and LED electric lights as the secondary lighting source. With this plan, you can brightly light your RV probably 90% of the time, using no electricity at all, when ambient temperatures aren't too hot. For that 10% of the time when temperatures are too hot, and for cubbyholes, cooking lights, porch lights and supplemental lights (when gas lights are on), you've got the LEDs, which use as little as 10% of the electricity that incandescents do. Fluorescent lights, run from the RV's 12 Volt battery(s) either directly or indirectly (from 120 Volts AC made by a power inverter) use more electricity than the LEDs (but much less than incandescents), but tend to be brighter than 12 Volt LEDs and have more natural looking colors. Fluorescent lights can be used occasionally, in conjunction with LEDs and gas mantle lights, but use them sparingly. Don't use incandescent light bulbs at all. They waste far too much electricity. (It doesn't matter if everybody else uses them, if this is the case, then everybody else is uninformed or just plain stupid.) Finally, be sure you have portable lights for the outdoors, as you might need to make occasional 1:00 AM trips to the outhouses, or you might suddenly remember to "rescue" food items on your picnic table after dark from bears and raccoons (and Park/Forest Rangers who might issue you a summons for attracting bears and raccoons).

The three efficient light types complement one another nicely. You don't need to use just one type of efficient light, you can use them all.

How Much Electricity Can be Saved Using Efficient Lighting?

Estimates on daily electricity requirements for RV lighting vary quite a bit from author to author. Many other authors are too optimistic--- they use unrealistically low figures, assuming that RV users should be content with lights that are dimmer than those in regular houses, and might not be all that much brighter than candles. I like a bright RV. I think that people should be able to see what they're doing, particularly when cooking, eating and washing dishes. I base my calculations accordingly. Because RVs are smaller than regular houses, and lights are likely to be closer to desired illumination areas than they are in regular houses, the power requirements are substantially less. To calculate how much electricity can be saved using efficient lighting, estimates on daily battery drain need to be made, for both standard inefficient lighting systems and efficient lighting systems, then those numbers need to be compared.

Inefficient lighting: I feel a medium RV could be comfortably lit using the equivalent of two 60 Watt incandescent light bulbs. This is about seven typical 12 Volt 18 Watt light bulbs. (The power requirements are exactly the same, regardless of whether 120 Volt or 12 Volt lights are used.) So if 120 Watts were used for one hour in the morning during breakfast, then for another hour during suppertime, this would be 240 Watt hours. Add another 60-120 Watt hours for miscellaneous use (some activities are fine in dimmer light), and this puts daily consumption at 300-360 Watt hours for inefficient incandescent lights, which is 25-30 Amp hours of charge drained from the battery daily.

(Energy) = (Voltage) × (Charge),
therefore:
(300 to 360 Watt hours) = (12 Volts) × (<u>25 to 30 Amp hours</u>).

Efficient lighting: The brightest lighting is probably needed for meals. Entire meals can be prepared, consumed and cleaned using only two propane gas mantle lights, except for the area above the stove, which should have a reasonably bright electric light, so users can see what they're cooking. If one desires brighter light in addition to the gas lights, 60 Watt equivalent 120 Volt LEDs should be chosen. Such bulbs are very bright, have natural looking yellowish color, and use only about 12 Watts of power per bulb (10 Watts for the bulb plus 2 Watts for the power inverter), so this is 1 Amp of current.

(Power) = (Voltage) × (Current),
therefore:
(12 Watts) = (12 Volts) × (1 Amp).

If such a bulb gets used for about 3 hours per day, including over the stove, that's about 3 Amp hours. If another 2 Amp hours per day get used for miscellaneous 12 Volt LED "bulbs," including an average extra consumption for those occasional 10% of days when all electric lights are needed, because it's too hot to use the gas mantle lights, that puts daily lighting requirements at about <u>5 Amp hours</u>. Compared to the 25-30 Amp hours needed for the inefficient lighting, this is a savings of about <u>80-83%</u>.

Calculating more theoretically: Suppose 30 Amp hours are needed per day for inefficient incandescent lighting. What would be needed for efficient lighting? LEDs can make light consuming as little as 10% of everything, so the daily charge needed for efficient LEDs should be about 3 Amp hours. If LED efficiency is slightly overrated, and they use 50% more, the daily consumption number would be 4.5 Amp hours. Suppose that 2/3 of the light gets made by non-electric propane gas mantle lights, instead of electric lights (a very reasonable assumption based on my experience). That would drop the daily consumption number to about 1.5 Amp hours. However, my experience has taught me that, an efficient system works so well, that occupants are likely to both use brighter lights and leave the lights on longer, so let's double that daily consumption number, to compensate for this luxury usage. This raises the daily consumption number back to a final figure of <u>3 Amp hours</u>. This is <u>only 10%</u> of the 30 Amp hours needed for the inefficient incandescent lighting system.

Furnaces

Furnaces are a very important part of any RV. Nobody likes sitting in a cold clammy RV. Furnace failure is probably the number one reason why people give up off-grid RVing, and either head for home, motel rooms or expensive private RV parks, where their electricity-dependent furnaces do work. Don't settle for one of those "furnaces that don't heat." Choose the right equipment and you can stay toasty warm in the wilderness indefinitely.

Furnace Characteristics

Classifying furnaces is a bit tricky for two reasons: 1) Furnaces have a number of characteristics. Like the individual numerals in a phone number, those characteristics can occur in many different combinations, resulting in many (at least hundreds) possible furnace types (although some combinations don't work together). Thus, the necessary task of simplifying matters, by putting all furnaces into a small enough number of categories for the mind to grasp, is somewhat subjective. And: 2) People don't seem to agree on furnace terminology. [Example: The D.O.E. defines furnaces with no blowers, fans nor water pumps as "central warm air furnaces." However, many manufacturers of furnaces misuse this term to include furnaces with blowers or fans, causing confusion.] I've done the best I can to use standard terms that everyone agrees upon, but some of these classifications are a bit my own.

Furnaces may be classified according to the following characteristics:

1) The type of fuel they burn. This might be natural gas, propane, butane, wood, coal, pellets or oil. Some furnaces don't burn any fuel; they make their heat from electricity.

2) Their size. This includes their heat output capabilities.

3) The materials from which they are made. This might be thin sheet metal, heavy cast iron, or even stone/brick.

4) Whether they operate manually or automatically. The simplest of all manually operated furnaces have no thermostatic control, they are just turned on/off by their users. Automatic furnaces (the most common) monitor room temperatures and heat more aggressively when outdoor temperatures are cold. They do this according to thermostatic dials which are adjusted by their users.

5) Whether or not they have motors. Furnaces can move heat by blowing hot air with fans, pumping hot water through pipes and radiators with water pumps or they can allow their heat to flow naturally (by convection) without using any motors at all.

6) Whether or not they have catalytic burners. (Not applicable for electric furnaces and heat pumps.) If a furnace has a catalytic burner, it can be the furnace's only combustion point or a secondary burning point.

7) Whether they discharge their exhaust outdoors or indoors. Some fuels (like natural gas and propane) are clean burning enough such that with adequate ventilation the exhaust from their combustion can be discharged indoors. Other fuels (like coal or wood) produce far too much pollution, so their exhaust must be directed outside.

8) Whether they *make* heat or just *move* heat. Most furnaces make their own heat. One type, called *heat pumps*, move heat by absorbing it from outdoors and dumping it indoors.

Types of Furnaces

I.) Passive Convection Furnaces:

(*D.O.E. Defined as "Central Warm Air Furnaces"*)

These are furnaces that have NO BLOWERS OR FANS. Such a furnace has its burner located in a metal case. When fired up, the metal case gets very hot. The heat transfers to the air, which rises, causing colder air to get sucked up and contact the case, where it becomes warmer, then rises. The process of circulating air by simply heating it is called *convection.* Passive convection furnaces are usually mounted on walls, although some units mount beneath floors, where they allow their warmed air to enter living areas through floor grates. Passive convection furnaces discharge their exhausts to the outside. They may be manually operated or controlled thermostatically.

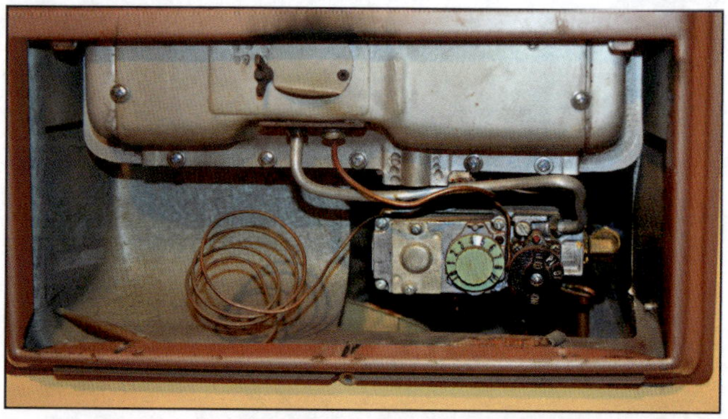

above: burner control and bottom of combustion chamber of the passive convection furnace in my trailer

II.) Central Radiant Furnaces:

These furnaces are a close relative of passive convection furnaces. They also have NO BLOWERS OR FANS. *Difference:* Passive convection furnaces have rather small very hot metal cases located behind protective grates (to prevent people from getting burned through accidental contact), and move their heat primarily through convection of warm air. Central radiant furnaces have larger less hot metal cases located somewhat centrally in a room (not behind any grates), and move much of their heat through infrared radiation.

above: central radiant furnace for a regular house

[They incandesce (glow) in infrared (heat) rays, which move in straight lines, like light.] A cast iron woodstove is a good example of a central radiant furnace. These furnaces may be manually operated (with simple on/off settings) or thermostatically controlled.

III.) Forced-Air Furnaces:

These furnaces move their heated air through ductwork by means of fans (also called blowers). They may also contain second fans which move the air used in combustion into the combustion chambers and outside as exhaust. These are by far the most common types of furnaces used in regular traditional houses almost everywhere (except warm climates, which use heat pumps).

photo: forced-air furnace in RV compartment

IV.) Hot Water/Steam Furnaces:

These furnaces move their heat through a medium of hot water or steam. (The "water" might be just plain water or a solution of antifreeze/antiboil). The water gets pumped to a point near the combustion unit, where it absorbs heat, then moves through pipes into radiators, where it sits awhile, gradually losing that heat, making the radiators warm, transferring that heat into the indoor living space.

V.) Catalytic Furnaces:

These react gaseous fuel with oxygen on the surfaces of catalysts (called catalytic burning), as opposed to having the fuel and oxygen react only with each other and in the air (called non-catalytic burning). [A catalyst is generally defined as "a substance which participates in a chemical reaction, without getting used up itself." After a catalyst participates in a reaction, it is exactly as before, so it can participate in an unlimited number of identical repeat reactions.] Although sometimes catalytic burners are secondary burning points that exhaust to the outdoors (as in some woodstove chimneys and always in catalytic converters of automobiles), the catalytic furnaces that are likely to be used in RVs and tents almost always use their catalysts as the sole burning points and almost always discharge their exhausts indoors. Catalytic burning has two properties: 1) it takes place at lower temperatures than non-catalytic burning and 2) burning is more complete. Often, catalytic furnaces are portable (not built into the RVs). (*photo: portable catalytic heater*) Often, platinum is used as the catalyst.

VI.) Direct Discharge (non-catalytic) Furnaces:

These are close relatives of passive combustion furnaces and central radiant furnaces, except (as their name suggests) direct discharge furnaces release their exhausts indoors instead of outdoors.

VII.) Electric Space Heaters:

These make heat by passing electric current through wires of moderate resistance (called "heating elements"). The current causes the wires to get hot. That heat then radiates into the living areas. Electric space heaters can be made in any size. They can be portable or they can be built right into houses or RVs. Often, when permanently installed in houses, they travel the entire lengths of rooms, near floors, where they are referred to as "baseboard heaters."

VIII.) Heat Pumps:

Heat pumps work using the same basic principle as household refrigerators. When a gas is compressed (force is needed), heat gets released. When a gas is allowed to decompress (this occurs naturally, no force is needed), heat gets absorbed. Thus if a sealed gas (called a refrigerant) alternately compresses and decompresses in different locations, heat is absorbed at one location then deposited in another location. Except for some heat energy, associated with running the compressors, these systems don't *make* heat (nor cold), they simply *move* heat from one place to another. Heat pumps differ drastically from every other furnace type in one respect: *they can be designed to move heat in either direction.* On hot days, they serve as air conditioners, moving heat from the insides to the outsides of houses or RVs. On cold days, they can serve as heaters, moving heat from the outsides to the insides of houses or RVs. These systems aren't all that great at heating, but in warm climates, they are ideal, because not all that much heat is needed, even on cold nights. (In Florida, heat pumps are almost exclusively used. Such units function both as furnaces and air conditioners.)

Which Furnace Types are Best Suited For RVs?

Before discussing the merits of each furnace type on my somewhat simplified list, certain characteristics should be immediately eliminated for RV consideration.

Almost anything made for a regular house would be too big, too heavy and put out too much heat for a much smaller RV. However specially designed RV versions, scaled down in terms of size, weight, fuel consumption and heat output, might work well.

Fuels: Solid fuels such as wood or coal are not appropriate because they take up too much storage space. (Also, in some places it is illegal to gather or transport wood for environmental reasons.) Heating oil has been used in the past by some vintage older RVs, and probably works well. However it may be difficult to buy in some out of the way places. Natural gas has almost identical burning properties to propane. It is not used in RVs, because it cannot be liquefied at normal outdoor temperatures (because it is composed of very small molecules). However compressed natural gas has been proposed and successfully used as an alternative motor fuel in specially modified vehicles, and thus could potentially also be used as RV fuel both for the engine and for household use. Propane or butane are the remaining RV furnace fuels of choice, since both are clean burning and easy to store in pressurized tanks. Propane is the obvious choice, since almost all RVs these days buy store and use propane, and it is readily available near almost all places that people camp.

Automatic control is needed for anything except the most primitive weekend camping huts. A furnace with only on/off settings might be adequate when occupants are awake to alternate those settings, but when they sleep, they are likely to be either freezing or frying.

I.) Passive Convection Furnaces:

Of all the furnace types, passive convection furnaces are the *best* choice for RVs, because they are simple, relatively inexpensive and extremely reliable. Although they were commonly used in RVs decades ago, today's RV designers have all but done away with them, for reasons I cannot comprehend.

How do temperature controlled passive convection furnaces work? They work exactly the same way that most gas hot water heaters for regular houses do. A standing pilot light remains lit all the time the furnace is ready for use (a safety feature shuts off the gas if the pilot light goes out). The furnace regulates its heat output by turning its main burner on and off multiple times. The more time the main burner spends "on," the more heat gets delivered. When the furnace is on standby, a thermostatic bulb monitors temperature. When the temperature at the bulb gets below a certain critical low, the gas valve to the main heating burner opens, firing that burner up, causing the furnace and its surrounding vicinity to heat up. When the temperature at the bulb exceeds a certain critical high, this gas valve closes, shutting the main heating burner down, causing the furnace and its surrounding vicinity to cool. Temperatures alternate between the critical low and critical high (and are always between these two values). The critical low and critical high are adjusted up or down together by means of a single thermostatic dial.

photo: burner control of my passive convection furnace

Advantages: 1) These furnaces do not use electricity. RV users with this type of furnace don't need to worry about making/storing electricity for their furnaces, nor do they need to worry about being left in the cold if their electricity systems fail for any reason, including temporary battery run-downs. 2) These furnaces are extremely reliable. They can work off-grid for months, problem free, even in bitter cold winter weather, as long as the propane tanks keep getting refilled.

> Gas hot water heaters for regular houses use burner systems fundamentally identical to passive convection furnaces for RVs. (They have a pilot light with a thermocouple controlled shut off valve and a temperature sensitive main burner that turns itself on and off.) They work without electricity and can function for years without breakdowns.

Major disadvantage: They are becoming increasingly unavailable. For quite some time, RV furnace manufacturers have been making less of them, and RV manufacturers have been ceasing to include them in their designs. The trend is to favor and use only forced air furnaces. No matter how good something may be, it can be difficult to obtain if manufacturers unite and refuse to produce it.

Practical considerations: 1) Location: In order to heat an RV as evenly as possible, a single furnace should be located about halfway between the front and back of the RV and as close to the floor as possible. If located too close to the front or back, one end of the RV is likely to get too hot, while the other end is likely to stay too cold. Since warm air rises, a location low to the floor facilitates natural convection air currents within the dwelling. If located too high, air near the ceiling is likely to become too hot while air near the floor is likely to remain too cold. 2) Larger RVs might need more than one unit for even heating. Because these furnaces do not use fans, they are poor at moving heat horizontally. Having more than one unit results in more even heating. Having more than one unit also allows an RV to still be heated (though not uniformly) if one of the furnaces somehow fails.

II.) Central Radiant Furnaces:

These heaters, larger and more out in the open than passive convection furnaces, might be a good choice for larger RVs, but because of their size, are probably not appropriate for smaller RVs, because they would take up too much space.

Advantages: 1) They use no electricity. Thus: 2) They are extremely reliable.

Disadvantage: I don't know if anybody makes such units small enough for RVs with automatic controls, although they are commonly made for regular houses.

Practical Consideration: They should be centrally located, low by the floor. Some of them have windows, allowing users to see their flames, making them resemble indoor fireplaces, giving them a much classier look than passive convection furnaces.

III.) Forced-Air Furnaces:

(*"furnaces that don't heat"*)

Like incandescent lights, forced-air furnaces are *very poorly suited* for RV use, yet they are *almost universally chosen* for installation by RV designers/manufacturers. I can almost guarantee that any non-custom RV you look at has a forced-air furnace in it.

Major disadvantage of forced-air furnaces: They don't heat. Really! I'm not kidding! If you have any doubts about this, read the "Furnace Stories" (actual accounts) soon to follow.

Why don't force-air furnaces heat? These furnaces use fans to move air through them. Those fans absolutely require electricity--- the furnaces don't work at all without that electricity. Nobody who designs or chooses these furnaces ever bothers to make sure that the RVs in which they are installed have enough batteries to run them long enough to be practical, nor any discernable plan for charging those batteries. Just as "a chain is only as strong as its weakest link," any device that requires electricity is only as reliable as the electricity source that runs it. Since these furnaces run down RV batteries very quickly, and chargers (on average) typically can't replace the electricity as fast as these furnaces use it, these furnaces are useless in wilderness, except for very brief stays of only one or two nights, between RV parks with municipal electrical outlets.

> A typical 3 Amps charger might require five to ten hours of generator run time per day to replace the electricity drained from the battery each cold night by even an "energy conserving" furnace (which draws about 3 Amps).

Other disadvantages: 1) A forced-air furnace can shut everything else in an RV off. After a forced-air furnace runs down the battery(s), almost everything else in the RV shuts off, leaving the occupants with no lights, no water coming out of the faucets, no refrigeration (amazing, considering that the refrigerator runs on gas) and an inability to get the generator started. 2) These furnaces are actually dangerous. Because they work so poorly, RVers often can't or choose not to use them. Instead they choose to use their stoves or ovens as heaters, where they risk death by putting carbon monoxide or unburned propane [if the burner somehow goes out] into their RV air. 3) The fans in forced-air RV furnaces are noisy. Not only is the noise annoying, it can wake people up, especially if the furnace is located under a bed where someone sleeps (often the case).

Advantages of forced air furnaces: 1) Forced-air furnaces can distribute heat more evenly than passive convection furnaces. A fan can move heated air horizontally through ductwork, allowing that heated air to enter a dwelling at several locations away from the furnace, providing more even heating. In some elite installations, hot air can even be blown downward through super-insulated ductwork in or under floors, emerging near the floors. (This is called "basement heating.") Since hot air rises, introducing it as close to floors as possible provides the most even heating possible. 2) Forced-air furnaces can be made smaller and have less fire clearances than passive convection furnaces. If air moves fast through a furnace, by being blown with a fan, it carries away heat with it. This means the furnace doesn't become as hot, so it can be smaller and have less fire clearances. Smaller furnaces are easier to find locations for and install than larger furnaces. Since a lot of air carries away the heat, that air too doesn't become as hot, reducing the risk of that hot air igniting something.

Could a better quieter RV electrical system be designed, to adequately run a forced-air furnace? It could, but such an overall system would cost thousands of dollars more to install and maintain than a simpler non-electric passive convection furnace. Such a system would also consume quite a bit more fuel, be noisier and considerably less reliable than a passive convection furnace.

Are there any overall advantages to forced-air furnaces over passive convection furnaces in RVs? Nope--- at least not for the RV users and not for those who camp off grid. Any possible advantages are outweighed by their unreliability probably a hundred times over.

If forced-air furnaces offer no advantages for RV users, why do manufacturers choose them? They may offer advantages to someone other than RV users.
1) Forced-air furnaces allow RV designers and manufacturers to cut corners in order to boost profits. If forced-air furnaces are small, easy to find places for and easy to install, they may be very attractive to RV designers/manufacturers who care not one iota about what happens to the RVs or the people who buy them after the time of sale. Sure, RV users would probably prefer furnaces that actually heat, instead of furnaces that install easily, but what RV users want usually doesn't even enter the RV manufacturing equation. 2) Designing RVs to "die in the field" boosts business for private RV park owners. If RVs "die" shortly after being disconnected from electricity, this almost forces their owners to stay in expensive RV parks, which can then charge whatever they want to for sites. No single appliance type more effectively causes RVs to "die in the field," than forced-air furnaces. Have RV parks united and somehow influenced RV manufacturers? I don't know, but it seems possible.

Has the situation always been this bad? No. From the early days of RVs until sometime around the 1970s, nearly all RVs had the simpler more reliable passive convection furnaces. Then, some designers decided that passive convection furnaces had to go (and be replaced with forced-air furnaces exclusively).

Would I (the author) personally venture into the wilderness with an RV having only a forced-air furnace? Not a chance in hell. If a forced-air furnace were my only option, I would give up RVing altogether.

IV.) Hot Water/Steam Furnaces:

These furnaces are *probably not appropriate* for RVs. They are complicated and even simple systems would need to have lots of parts (pipes, radiators and water pumps). But they offer some possible advantages. This is definitely a do-it-yourself job for any RV user who wants such a system.

System Variations: Water can circulate through the system either as a hot liquid or as steam. The liquid can be either water or antifreeze. If antifreeze is used, it probably needs to remain a liquid, because antifreeze is also anti-boil. If water is used, then the system can be connected to the same RV hot water heater that sends hot water to the faucets at the sinks, or it can be completely separate from the drinking water system.

Possible Advantages: 1) A water system might use significantly less electricity than a forced-air furnace. Water stores a lot of heat. Once hot water gets moved into a radiator, that water can stay put awhile, slowly losing its heat to the indoor environment. Thus, an electric water pump might only need to run for short time periods, followed by long periods of "off" time. This would save electricity. 2) Rather even heating could be achieved, if radiators of various sizes were properly placed. Small radiators could be placed in corners or cubbyholes. Valves in the tubing could allow users or computers to decide where the heat should go and in what proportions. 3) It may be possible for a water furnace to *make its own electricity* to move the water. Water cooled *thermopiles* could be designed into the system to do this. [See the "Thermoelectric Generators" section in Chapter 3 for greater detail.] And: 4) The water heater could be mounted underneath the RV rather than in it. This would save space and allow greater flexibility in finding a location for the water heating unit.

Disadvantages: 1) The RV user would probably need to design and install such a system him(her)self. I don't know if any RV (or RV parts) companies would be willing to make or install such systems, as there would be many components (pipes and radiators) that would need to be routed and hidden. It could be an installation nightmare. 2) If water were used, A) all the water would need to be removed from all pipes before cold winters (a process called "winterization"), to prevent that water from freezing in and bursting those pipes. [Winterization usually needs to be done once a year while an RV is in storage. But it would need to be done more often in an RV used during the winter.] This would require an awful lot of work. If antifreeze were used, winterizing would be unnecessary, however B) two separate water heaters would be needed. One would be used to heat the antifreeze going to the radiators and the other would be used to heat the drinking water going to the faucets. 3) The hot water furnace would still need electricity to pump its water. If it needed too much, then this system would suffer from the same problem that forced-air furnaces do, making it yet another "furnace that doesn't heat." 4) Heating an RV with a hot water tank may be rather propane inefficient. Too much of the heat may end up outside of the RV. And: 5) The water pump might be rather noisy. This might not only irritate the occupants, it might wake them up at night as well.

Some people believe that water heating systems for RVs are a lot on nonsense. Others steadfastly believe that water systems are a viable option. (I found quite some discussion about them on the internet.) I tend to believe that choosing water systems (over passive convection furnaces) adds complication and unreliability without real clear benefits.

V.) Catalytic Furnaces:

Catalytic (direct discharge) heaters are appropriate for *some* of the heating needs of RVs, but they should not be used as the sole furnaces. This is because, although they are very clean burning, they do remove oxygen from the indoor air and replace it with carbon dioxide and water vapor.

Advantages: 1) Catalytic burning is very clean burning (little pollution), therefore 2) their exhausts can be released indoors. And: 3) Catalytic heaters are great for cubbyholes and supplemental heating. This is because they can be made small. This is also because they burn at lower temperatures than conventional burners, so they pose less fire hazards.

Disadvantages: 1) Releasing exhaust indoors pollutes the indoor air, no matter how clean that exhaust is. Therefore: 2) some ventilation is need. At least one window and/or roof vent should be opened during use to let the exhaust-enriched indoor air out and replace it with fresh air. Fresh air from the outside [on nights when furnaces are needed] is cold, so letting too much of it in partially defeats the purpose of heating. 3) Catalytic heaters may not work well at high altitudes. As a safety feature, many catalytic heaters have oxygen sensors that shut them down, if the indoor oxygen levels get too low. At high altitudes, there is considerably less oxygen in the air. Consequently, catalytic heaters may shut down prematurely, because they use what little oxygen exists in the high altitude air up more quickly.

How is clean burning achieved? How clean something burns is a function of two variables: 1) fuel type and 2) how completely that fuel gets burned. Let's explore each. 1) *Fuel type:* Many fuel types (such as wood, coal or oil) release quite a bit of soot, because they contain large long chain molecules that often don't break up and burn completely. Therefore the exhausts from burning these fuels must be directed outside. Cleaner burning fuels, such as propane, butane and natural gas, produce very little soot when burned (even non-catalytically), because these molecules are small and lightweight, so even if they pass through flames without burning, they float away as gas. Therefore, sometimes the exhausts from lightweight fuels may be released indoors. However, non-catalytic burning of even these lightweights does produce some carbon monoxide, which is a deadly gas. 2) *How completely that fuel gets burned:* Burning that takes place on the surface of a catalyst, rather than in the air, is much more complete. Since deadly carbon monoxide and unburned hydrocarbons get produced from incomplete combustion, very little of these poisonous gasses gets released from catalytic burning. Instead, carbon dioxide and water vapor get released into the indoor air as end products of combustion. Carbon dioxide is only toxic in rather large quantities, whereas carbon monoxide and unburned hydrocarbons are toxic in very small quantities. Water vapor is quite harmless, although too much of it for too long released indoors can cause mildew.

Possible uses for catalytic heaters: 1) Supplemental heating: Whether connected to RV gas systems or as independent units with their own small portable propane tanks, catalytic heaters are great for putting small or moderate heat amounts in cold corners. 2) Emergency RV heating when regular furnaces fail. 3) "Last resort" heaters for RVers who don't want to bother trying to use their forced air furnaces that don't heat. 4) Heating of enclosed places outside the RV, such as tent interiors, or parked cars.

Conclusion: If the exhaust absolutely must end up indoors, catalytic heaters are clearly the best choice. Heating an RV with a catalytic heater is always better than trying to heat an RV with a gas cooking stove or a direct discharge non-catalytic furnace. However, dumping large amounts of carbon dioxide into any indoor air without adequate ventilation is never a good idea.

VI.) Direct Discharge (non-catalytic) Furnaces:

Direct discharge (non-catalytic) furnaces are probably not appropriate for RVs, except for very occasional use with ventilation, and are definitely not appropriate as the only or main furnaces, because they poison the indoor air.

In technical write-ups and ads, often heaters are described as being "ventless" or "direct discharge" (synonymous terms), as if this were some kind of great feature. My definition for these terms is: "somebody was too lazy or too stupid to install a chimney, so that person decided to dump all the toxic exhaust from the fire indoors instead of outdoors where it belongs." There is nothing great about dumping the toxic exhaust from a fire indoors. Heating an RV with one of these furnaces is only slightly better than heating an RV with a propane cooking stove. These furnaces do however shut down if the fire somehow goes out. This is one definite advantage over propane cooking stoves, which don't do this. If you use a direct discharge non-catalytic furnace in your RV occasionally, please do not go to sleep in there with the furnace running. You might never wake up.

The First Engineers To Use Convection

Like fire, convection currents occur in nature (in the atmosphere and in many caves), so for humans, controlled convection was more of a *discovery* than an actual *invention*. Who discovered convection? When? Under what circumstances? (Like fire, controlled convection was used long before historical records were kept, so much of this is speculation.) A documentary I saw on TV may shed some light.

This documentary was about a large colony of leaf cutting ants. Leaf cutting ants are a hard working species indeed. They march out of their colony, climb trees, remove leaves, cut those leaves into pieces small enough to be carried by single ants, then march those leaf fragments deep into their underground chambers. There they chew them up and spit them out. Upon the spit out pulp grows a fungus, which is the only source of food for these ants. Researchers studying this colony were puzzled by one thing. The amount of fungus needed and actually grown for the colony would release huge amounts of carbon dioxide while growing. Given the known amount of fungus, its growth rate and the depths of the tunnels, it was absolutely impossible for the carbon dioxide to escape fast enough through diffusion alone. That carbon dioxide should accumulate to toxic levels and kill the ants. But the ants weren't dying; both they and their fungus farms were doing well. Researchers realized that there had to be some mechanism for pumping the carbon dioxide out to the surface. Researchers studied the air flow at the many entrance holes of the colony, and quickly discerned that the entire tunnel system was actually "breathing," and quite quickly--- it was as if the ants had electric fans down there circulating the air. Some surface holes were exit vents while others were intake vents, and there seemed to be no rhyme nor reason as to which was which. Then, the researchers' curiosity got the best of them. They decided that human knowledge was more important than the continued existence of a single ant colony. So one day they showed up with a large vat of wet cement. It was a good day for the advancement of science. It was a bad day for the ants. (A really bad day!) The researchers poured this wet cement into all the holes of the ant colony. After it hardened, the researchers began excavating the dirt from around the concrete pillars, being very careful to physically support them, so the structure would not shift or collapse. This enabled them to "see" the structure of the (now extinct) anthill in the "negative:" what used to be empty space (the tunnels) had become solid concrete, what used to be solid (the ground) had become empty space. It took the researchers awhile, but by the time the excavation was complete, they found the answer they were looking for. These ants placed their organic garbage as certain strategic locations within their underground system of caverns. The organic garbage naturally decomposed, releasing heat. The heat warmed the nearby air, causing that air to rise though the vertical tunnels and exit to the outside, just like the chimney of a fireplace does. As the warm air rose and exited, it created a vacuum, which sucked cooler fresh air deep into the underground chambers, through colder intake tunnels. As the deep interior air moved out to the surface, it took with it the massive amounts of poisonous carbon dioxide produced by the fungus, thus preventing its accumulation and keeping the air breathable. Yes, these ants used *convection* to aerate their network of tunnels and chambers. Thus, convection almost certainly was first put to some artificial use eons ago not by humans but by *some non-human species*.

This story raises two questions: 1) If a bunch of stupid ants can figure out that carbon dioxide doesn't belong in their house, why can't we humans? And: 2) Isn't it a bit embarrassing that a bunch of stupid ants can came up with a reliable way of moving air, but human RV designers can't do this with their furnaces? (Don't expect the fans in your RV forced-air furnace to work as well as an anthill, because it won't.)

VII.) Electric Space Heaters:

Electric space heaters are *completely inappropriate* for RVs, because they use up massive amounts of electricity. Nobody has yet invented a battery that stores more than a tiny fraction of the electricity that would be needed to successfully run them. They could be used by running powerful generators, however: 1) such generators would consume far more fuel to produce the same interior heat as passive convection or central radiant furnaces, 2) the generators would make lots of carbon monoxide [which could leak into RV interiors, especially if those generators were built into the RVs (usually the case)] and 3) the generators would make lots of noise, instead of being completely quiet like passive convection or radiant furnaces. Also, furnaces/heaters are generally needed most at night, since nighttime is colder than daytime, and generators should never be operated at night wherever camping neighbors are present.

Could electric space heaters be run from municipal electricity in RV parks? Some could (especially smaller ones) but this still presents problems. Substantial heating requires so much electricity, that the outdoor wires leading to the RV sites would have to be big and thick. If wires too thin were used, excessive heat would build. This could lead to wire damage (melting of wires or their plastic casings), fires and circuit breakers shutting off. (One private seasonal campground, that my parents patronized, forbade even possessing electric space heaters, because their wiring couldn't handle the electrical demands of such heaters. Even if not used, anyone caught with an electric heater would be asked to leave with no refund.)

What should a potential RV buyer do upon realizing that an RV has only an electric heater? If any off-grid camping is even a remote possibility, <u>don't buy that RV</u>.

<u>VIII.) Heat Pumps</u>:

These are *completely inappropriate* for RVs as furnaces, because they use way too much electricity for batteries to store practically. A generator would need to run simultaneously the entire time that a heat pump furnacc would run. Anything requiring constant generator use is not a design of choice for RVs, if the same task could be done with no generator use. However, most RV air conditioners are heat pumps. They could be designed to be reversed as heaters, but they would only work in RV parks with electricity and they would be quite noisy (passive convection furnaces could be run anywhere and make no noise). Air conditioning is generally needed during the day, a time when some generator noise is usually tolerated. Heating is generally needed at night, a time when all generators everywhere should remain off.

<u>Insulation and Ventilation</u>

Regardless of furnace type, furnace efficiency is greatly affected by two external variables: 1) insulation [in the walls, ceiling and floor] and 2) ventilation [unintentional indoor/outdoor air exchange occurring at "closed" leaky windows]. Window glass has a poor insulation (R value), so having many windows can cause an RV to lose lots of heat, both due to air leakage and the low R value of the glass itself. For people who are diehard cold weather campers, both these problems could be greatly reduced by installing temporary storm windows or even plastic garbage bags or bubble wrap over the regular windows (as viewed from the inside). Care must be taken though not to make an RV too airtight, as some indoor/outdoor air exchange is necessary to prevent indoor air from becoming stale and to decrease the risk of poisoning and death due to accumulation of toxic gasses. The topics of insulation and ventilation are covered in greater detail later in this chapter ["Air Conditioners" (passive)].

above (both photos): plastic garbage bag "storm windows"

Cutting plastic garbage bags and putting them over window screens is a great way to reduce indoor/outdoor air exchange and keep warm air inside. I do this often in bitter cold. Garbage bags may be either taped (left photo) or snapped (right photo) in place. Bubble wrap can work well too.

Best Overall RV Heating Plan

Primary heaters (types and placement): For small RVs, one propane fueled passive convection furnace is enough. It should be located centrally and low near the floor (or underneath it). For larger RVs, two passive convection furnaces may be better (located at the 1/3 and 2/3 distance points between front and back). For really large RVs, three passive convection furnaces may be best (located at the 1/4, 1/2 and 3/4 distance points between front and back). In large RVs, a central radiant furnace may be substituted for one of the passive convection furnaces, for a classier look, if space permits.

Secondary heaters (type and placement): Larger RVs could benefit by also having one or more small indoor-discharging catalytic heaters placed in corners that are likely to get cold or drafty. These may be especially useful for RVs that have slideouts, as these areas could be particularly prone to coldness and draftiness. Catalytic heaters should be supplemental heaters only (the main furnaces should be completely capable of reasonably heating the RV on their own). Catalytic heaters should be equipped with low oxygen sensor shut off valves, thermocouple controlled safety shut off valves (to prevent unburned propane from entering the RV, if the burners somehow get extinguished or fail to light) and programmable timers that limit burning times then shut down. It is alright for small catalytic furnaces to have only two settings (on and off).

Heating capability: The total furnace package should be able to warm the RV to at least 70° F when outdoor temperatures are 15° F.

Automatic control: All primary furnaces should have thermostatically controlled temperature dials that turn their main burners on and off multiple times automatically. Only two main burner settings (on and off) is not appropriate.

Forced-air furnaces: These should not be put into RVs. I nicknamed them "furnaces that don't heat" for a reason. If you've bought a used or new non-custom RV, almost certainly you've got a forced-air furnace already installed in it. Don't rely on it and don't even use it. It's airtight to keep it in there, if you don't need the space for some other purpose. But you may want to consider just removing it, to avoid the temptation of thinking that you have a working furnace, when in fact you do not.

Detectors: All RVs should have carbon monoxide and unburned propane gas detectors, operated by independent small batteries, just in case.

How Much Electricity Could be Saved Using Efficient Heating?

Since heating can be done with no electricity at all, and this task is technologically easy, the percentage savings by using non-electric furnaces instead of forced-air furnaces is a phenomenal 100%. It may however be more meaningful to consider the daily number of Amp hours that a forced-air might use and consider what else that electricity might be used for.

Inefficient electricity-dependent furnaces: Furnace use is heavily dependent upon outdoor temperatures. Since outdoor temperatures vary drastically in real world use, only very rough estimates of daily forced-air furnace Amp hour consumptions are possible. Note that the fan goes on whenever the burner goes on. Based on the calculations in the text box on the next page, I'm going to use 15-30 Amp hours as the final estimate for daily battery drain for a forced-air furnace on nights when it is actually needed.

What else could 15-30 Amp hours be used for? 15-30 Amp hours could run an inefficient lighting system for one day, or an efficient lighting system for 3-10 days. 15-30 Amp hours could run a (4 Amps) TV for 4-7 hours. 15-30 Amp hours could run a water pump for one to two weeks (at 15 minutes of water pumping per day). 15-30 Amp hours could run a 1200 Watt microwave oven (an extreme energy gobbler) for 8-15 minutes. Note that 15-30 Amp hours represents a battery drain of 30-60% of the usable charge stored in a standard 100 Amp hour battery, assuming that the battery is only drained to 50%.

Forced-Air Furnace Electricity Consumption Calculations

Semi-cold nights: On these nights, furnace use might be occasional, just to take the chill out. If the fan draws 3 Amps, and is on for slightly over 3 hours (max), 10 Amp hours or less might be a good daily estimate.

Moderately cold nights: Suppose a furnace needs to be on standby for 12 hours, with the fan going on 1/3 of the time (4 hours). If the fan draws 3 Amps, daily consumption would be 12 Amp hours.

Very cold nights: Suppose a furnace needs to be on standby for 24 hours per day, with the 3 Amp fan going on for 1/3 of the time during the warmer 12 hours of daytime (4 hours) and half the time during the colder 12 hours of nighttime (6 hours). The daily consumption would be 12 Amp hours for daytime and 18 Amp hours for nighttime, totaling 30 Amp hours.

Bitter cold nights: Suppose the fan needs to be on constantly for 12 hours during the nighttime, and half the time during the 12 hours of daytime (6 hours), totaling 18 hours. The daily consumption for the 3 Amp fan would be a whopping 54 Amp hours.

Note #1: These estimates are based on the premise that the furnace doesn't draw significant current when on standby. A high phantom standby current would cause greater consumption.

Note #2: These estimates are based on the 3 Amp consumption figure provided by the *manufacturers*. If the fans actually use more current, daily consumption figures will be greater.

Note #3: Some fans in bigger RVs use as much as 8 Amps of current (not 3 Amps). Daily electricity consumption for these furnaces could be--- "the sky is the limit." Check the furnace consumption currents in the technical manuals for specific models under question. Or better yet, measure current with an ammeter. Adjust equations accordingly.

Calculating average use: Avoid the temptation to consider the *average* daily use to be the *actual* daily use. Because cold nights tend to bunch together, several consecutive days of heavy furnace usage are commonplace. A system that is only functional under average temperature conditions is going to be inept in places and seasons where and when prolonged cold spells occur.

Furnace Stories
(actual accounts)

#1: "My Stove is Also My Furnace"

Before I knew much of anything about RV furnace types, I had a conversation with another camper about the cold night that lie immediately ahead. This person told me that he did not like using the furnace in his trailer. When I asked why, he said, "The fan in my furnace runs down my battery and causes that battery to be low the next morning." He further stated that he warmed his trailer by lighting one of the burners on his gas stove and leaving it on all night. I told him that this was an extremely dangerous practice and strongly urged him not to do this anymore. He then stated, "I leave a window open." Folks, please heed my warning: **Do not use your cooking stove as a heater!** (Especially with the windows closed or while sleeping.) Prolonged combustion of propane gas in a poorly ventilated RV causes poisonous carbon monoxide gas to accumulate in the indoor living space, which can cause **DEATH**. In addition, stoves (unlike furnaces) have no thermocouple controlled shut-off valves (a safety feature) which shut the gas off if the flames from burners somehow get extinguished. Unburned propane added to the air is very poisonous and explosive as well. It can be ignited from other pilot lights, spark starters from other appliances or even from static sparks on sheets and cloths. If you *must* heat with a fire indoors, use a catalytic furnace and leave at least one window open.

Post Script: There may be a reasonably safe way of heating an RV with a cooking stove after all. Remember that fire basically makes two things: 1) heat, which we *do* want indoors [on cold nights] and 2) poisonous exhaust gasses, which we *don't* want indoors [anytime]. So the engineering question is: *How can we separate the heat from the exhaust, and get the heat to stay indoors, while getting those nasty poisonous exhaust gasses to go outdoors?* Obviously the best solution for the RV is a furnace with a thin metal chimney or vent, but in the absence of such, try this trick: 1) Fill several large metal pots with water and place one pot on each stove burner. The water doesn't need to be drinkable, water from lakes or creeks will do. 2) Turn all the stove burners on high and light them. 3) Open several windows and roof vents. 4) Vacate the RV of all people and pets [sit outside]. Return for only brief periods to check to see if the water is boiling. 5) When the water boils, shut the burners, cover the pots [to prevent them from steaming up the RV], vacate the RV, and wait for five to ten minutes for the carbon monoxide to escape to the outdoors. 6) Return to the RV, close all the windows and roof vents and the door. 7) Enjoy the heat! Have you figured out what's happening here? The exhaust gasses end up *outside* the RV. Much of the heat from the stove burners ends up *inside* the RV. This is because the heat goes into the water, gets stored there for awhile, then after the windows get shut, the heat slowly diffuses into the RV as the water gradually cools over the next several hours. Also, some of the warmth from the exhaust transfers into the walls and cupboards before that exhaust heads out the windows. The heat stored in those walls and cupboards stays inside for awhile after the windows get shut. I call this trick the "poor man's hot water furnace."

#2: "Some Like It Cold!"

I met a soon-to-be retired couple taking a short two week camping trip in a large trailer in the desert. They spoke of plans to take a much longer trip soon. They were new to camping, so they were just starting to get their act together. Adventurers they were. People with a zest for life they were. Electricians they were not. But they figured out that by not using their furnace, they could conserve on what limited amounts of electricity their battery stored, enabling that electricity to last the better part of a week. This would keep their lights, their water pump and their propane refrigerator working. When I brought the subject of heating up, they told me that they *never* use their furnace. The man said: "We like to sleep in the cold. We bought some really warm sleeping bags and blankets. Even if we could figure out how to get our furnace to work, we wouldn't use it." I am not so convinced that they endured the cold because they *liked* it. Perhaps they just learned to like the cold because they realized they *had no other choice!*

#3: "I Just Can't Figure Out…"

One October, in the mountains of Nevada, nighttime temperatures started getting kind of cold. Light snow fell to the ground and stayed there a couple of times for a couple of days. One of my camping neighbors told me how he used his furnace: "I turn my thermostat to a setting no higher than 50° F. That way, the fan doesn't kick on too often. This conserves the electricity in my battery. It's cold and uncomfortable in there but I don't know what else to do." This camper lived less than a hundred miles away. His plan was for a short (less than two weeks) camping voyage. One rather cold day, he took off for home, unexpectedly soon. I didn't get a chance to ask him why he left, but I suspect that he had run his battery dead, because the evening before leaving, he said to me: "I get it. The furnace works for a day or two. But then it runs my battery down and I just can't figure out how to get that battery charged up again." He is not alone. Every RV user who goes off-grid, faces this same dilemma when they try to use one of those forced-air "furnaces that don't heat."

Story #4: "My Old Reliable Furnace"

In 1989, I decided to replace the original furnace in my trailer, because that furnace was too small, non-adjustable (it was either "on" or "off") and malfunctioned frequently. I visited an off-the-beaten-path RV parts store and asked the dealer if he had a convection unit, with no electric anything. He soon took me into the back room and showed me a furnace that was exactly what I had asked for. It was missing a rear grate but it did come with an owner's manual. The dealer told me that the furnace was in "new" condition (never "used" nor installed in any RV--- only test fired up once at the factory), but was actually quite old, as it had been sitting around somewhere in his warehouse for maybe 20 years. He offered to sell it to me for $200. I accepted without even haggling. I believe the dealer was being truthful when he said that the furnace had never been used, because the first time I fired it up, clouds of toxic white smoke filled my trailer (because the original factory paint burned off--- after several subsequent firings, no more toxic smoke got produced). I also believe the dealer was being truthful when he said that the furnace was old, because none of the phone numbers in the owner's manual worked anymore, even though that manufacturer was still in business. When I finally got a hold of a valid phone number and contacted the manufacturer, attempting to buy the missing grate, the woman on the other end sounded bewildered that anyone would

homemade outdoor furnace grate

made in 1989
still in service at publishing date
(2016 photo)

even request a part for a furnace that old, stating that they no longer made (nor had) parts for it, because: "That model of furnace hasn't been manufactured in years." Fortunately, the owner's manual had a clear picture of the missing grate. I *made* a grate out of a piece of sheet aluminum, by copying that picture. I sliced, scored and bent the metal. It's not pretty but it works. During the installation process, I had to do some sawing, drilling and rebuilding of the wood and panel structure around the furnace, to get it in there, but I got it in there. Once installed, it fit so well that it looked as though it were specifically made for that trailer.

Once installed, this furnace fit so well that it looks like it was made for this trailer.

How well has this furnace worked? It worked reliably and put lots of heat in my trailer when I needed it for about a decade, with no maintenance, other than occasionally taking it apart to clean out cobwebs and rust (not a big job). It kept me warm as toast every night in Yellowstone one October, when temperatures routinely dipped to about 25° F. In 2000, after about ten years of operation, the flame of the main burner kept getting smaller and smaller. So I took the valve assembly to a gas appliance repair dealer. After examining it, the technician told me that the valves were old and shot, and would need to be replaced. He said: "The valves get like that after awhile. They only last so long." The new valve assembly would cost $200. Naturally, I told him to fix it, since $200 of maintenance every ten years isn't a lot of money. After reinstalling it, the furnace once again worked perfectly and continued to do so to for about ten years, needing only occasional cleaning. In October 2009, I put it to a real test in Idaho, when my trailer (and everything else in the vicinity) got buried under nearly a foot of snow. Temperatures dipped into the teens at night. During the days, temperatures went above freezing for only a few hours. I had to refill my propane tanks often, but that furnace kept both me and my dog nice and warm in that trailer for the whole week that we were there. In 2010, ten years after the new valve assembly had been installed, once again the flame of the main burner got kind of small. So once again, I took the valve assembly to the (same) gas appliance repair dealer, expecting to replace the whole valve assembly. Without even giving me an estimate before fixing it, they left a message on my telephone answering machine: "Mr. Childs, your furnace is done." What was going on? No estimate? When I went to pick the furnace up, I was told, "It only needed cleaning." The bill was $38. I put it back in my trailer, where it still works perfectly to this day (2016).

When I park somewhere in wilderness and a brand new RV arrives and parks nearby, one thought always goes through my mind: *I would not trade places with those RV owners, if given the chance.* I know that my antique discontinued half-century-old furnace will keep me warm much longer than the hi-tech silliness they've got in that brand new RV. I will probably be relaxing in my warm trailer, days after unhitching, while they are shivering in the cold, packing to head home or for some expensive private RV park with electricity. In fact, if I live long enough and stay healthy, my old reliable furnace will probably be keeping me warm when their brand new RV (and everything in it) is sitting and decaying in some junkyard. Oh, by the way--- don't bother contacting me with any attempts to buy my old reliable furnace. It is NOT FOR SALE.

The 99 Night Field Test

Every furnace must be able to heat properly and reliably. If a furnace quits every week, or worse yet, every few days, or is nonfunctional more often than functional, it is not much of a furnace at all. It is either a piece of junk or something in need of serious redesigning. I believe that all RV furnaces (actually any furnaces) should be able to heat for 99 consecutive cold nights without failure, else they should not be installed. I base this opinion on three observations that make me absolutely certain that such performance is both possible and reasonable: 1) The forced-air furnace in my regular house has heated reliably for well over 99 nights. 2) The furnaces in the regular houses of most people I know perform with similar reliability. 3) The passive convection furnace in my small trailer has heated that trailer well over 99 nights (actually it worked a substantial multiple of that time span), needing only occasional cleaning and once-in-a-blue-moon service (and of course frequent propane tank refilling).

If an RV forced-air furnace can pass a rigorous field test that reasonably represents actual RV live-in conditions, I will be a good sport and reconsider my view on forced-air furnaces. Until then, I will continue to call them "furnaces that don't heat," and recommend that they be crushed and sold or given away to junk dealers as scrap metal.

The 99 Night Field Test

To Pass:

During a live-in RV field test, an installed furnace must perform without failure for 99 consecutive cold nights without municipal electricity.

Definition of terms:

Night: A 24 hour period beginning at 12:00 Noon and ending at Noon the next day.

Cold Night: Any night in which the main burner of a furnace needs to fire at least 3 hours total, in order to keep indoor RV temperatures (measured halfway between the floor and ceiling) at least 70° F. Furnace should be on standby (ready to automatically fire) for considerably longer than the three hours, preferably for the entire time that the occupants sleep.

Warm Night: Any night in which the furnace is not needed at all, or the main burner of the furnace needs to fire less for than three hours total to keep the RV warm (70° F).

Consecutive Cold Night: A night occurring as soon as possible after a previous cold night. Warm nights do not break the consecutiveness; they simply do not count. (Example: Suppose during an unbroken 15 night period, 4 of the nights are too warm to warrant furnace usage, but the other 11 nights are cold enough for the furnace to be needed. This period of 15 actual nights is considered to be 11 consecutive cold nights.)

Without Municipal Electricity: Since this is an off-grid test, the occupant/testers are encouraged to never plug into municipal electricity. If they feel they must plug into electricity and do so: A) Any nights that the furnace runs with municipal electricity do not count (like warm nights). B) Municipal electricity must not be used to charge batteries for furnace use. Since plugging into electricity charges batteries, after unplugging, all coach batteries must be drained to the charge levels they had before arriving and first plugging in, before the test may be resumed. C) Five nights must be taken off the count of successful nights as a penalty for each time the occupant/testers plug into municipal electricity. D) Batteries may not be removed from vehicles and charged elsewhere using municipal electricity. E) Occupants may not plug into municipal electricity more than 3 times during the 99 night test period.

The Test

There is a saying: "A chain is only as strong as its weakest link." In engineering, any device, that is part of a system of components, is only as reliable as the weakest component of the system. It is not acceptable to test only the furnace, in some parking lot of a factory, using a constant municipal electricity source, then pass and sell it with two wires hanging from its end and a tag saying: "Connect to 12 Volts DC source," and expect that leprechauns will use magic to provide the electricity to run it.

(text box continued next page)

<u>Definition of Failure</u>:

The furnace is deemed to have failed if there is a failure of either the *furnace itself* or *any external component of the system that supports it*. Specifically:

Furnace itself: If for any reason the furnace itself malfunctions, causing it to not properly heat the RV on a cold night, the furnace fails.

Battery: If for any reason the battery, that supplies electricity to the furnace, fails to do so, causing that furnace to fail to properly heat the RV on a cold night, the furnace fails. (This is true even if the furnace itself would probably be functional if adequate electricity were present.)

Charger: If for any reason the charger, that charges the battery, fails to do so, resulting in a battery being in a discharged state, causing that battery to fail to provide electricity to the furnace, causing that furnace to fail to properly heat the RV on a cold night, the furnace fails. (This is true even if the battery itself is deemed fully capable of accepting and storing charge, were an adequate charger provided and used.)

Energy Source (that runs the charger): If for any reason the energy source, that runs the charger, fails to power that charger, causing the charger to not charge, resulting in a battery being in a discharged state, causing that battery to fail to provide electricity to the furnace, causing that furnace to fail to properly heat the RV on a cold night, the furnace fails. (This is true even if the charger itself is deemed fully capable of charging, were an adequate energy source provided.)

Propane (or other heating fuel): Tanks must be large enough to store sufficient propane to operate the furnace for reasonable time periods without running out. One propane run-out is allowed during the test period due to carelessness of the occupants (must take 5 days off the count as a penalty), and one additional propane run-out due to temporary vehicle immobility (occupants can't travel to get propane because of vehicle mechanical problems or vehicle gets pinned/stuck in place by mud/snow or blocked exit road). If propane runs out more often, the tanks are deemed too small and the furnace fails.

Municipal Electricity: Municipal electricity must not be used to either power any component of the furnace or charge batteries to run the furnace. If this is done, the furnace fails. (Any deviation beyond the exceptions on previous page constitutes failure.)

Generators: Generators must not be run excessively just to charge batteries. Only two hours of generator use solely for battery charging are allowed per week. Generators may however supply electricity for battery charging while being used to power other things (for up to one hour per day for general use, up to six hours per hot day for air conditioning). No generator may be used between 10:00 PM and 7:00 AM anywhere. If a generator gets used beyond these limits, the furnace fails. (Thermoelectric generators are a different type of device. They are exempt from all these rules.)

Alternators: Extra automotive engine use (either from idling in place or from taking extra unnecessary drives), solely for the purpose of battery charging, is prohibited. If any extra engine idling is done (more than 10 accidental minutes per week), the furnace fails. However, there is no limit to how much battery charging may take place, during drives that a vehicle would take for other purposes.

(text box continued next page)

Exceptions:

Vacation or Storage time: If the continuity of the test period needs to be broken, because the testers want or need to be somewhere else, all unoccupied nights are like warm nights--- they do not count. If the batteries were charged during storage, they must be drained to their charge levels prior to storage, before the test can be resumed. Up to two such periods of non-occupation lasting up to two weeks each are permitted. More non-occupation constitutes an aborted test, in which case, a new test must be restarted (the next cold night the furnace successfully heats is counted on night #1).

Unoccupied time: Testers may routinely vacate the RV for periods less than 18 hours, to do things that RVers normally do, such as shopping, sightseeing, visiting. The furnace does not fail if the RV gets cold while the people are not present. However: 1) The furnace must keep all components of the water system from freezing. And: 2) Upon returning to the RV, the furnace must be able to heat the RV to comfortable temperatures in reasonable times [one hour on moderately cold nights, three hours on bitter cold nights]. If the furnace does not do (1) or (2), the furnace fails.

Other:

Documentation: A detailed test report must be made. Inadequately documented tests are considered invalid. 1) The report must list and describe both the furnace and all supporting equipment that helps run the furnace. 2) Important technical specifications of all supporting equipment must be included [examples: how many Amp hours get stored in the batteries, how much current the battery charger charges with]. 3) Detailed information must be included on the procedures of use for all supporting equipment. 4) If other things are run from the same batteries, reasonable consumption estimates/measurements of these things must be made and included in the report. 5) Daily approximate temperature measurements and estimates must be made and included in the report. 6) The report must be made available as a public document. A copy should accompany all subsequent new furnaces of that type sold. 7) The test report must be available printed on paper. Electronic documents may also be available for those who prefer e-versions, but paper is generally better, as computers tend to crash and most hardware/software becomes obsolete after a few years, causing files to disappear.

After the test:

If the furnace fails: The test may be redone an indeterminate number of times after failures. Either the same supporting equipment or different (hopefully improved) equipment may be used. The counting starts over again. (The next successfully heated cold night is considered night #1.)

If the furnace passes: The furnace only passes and is recommended when used with the same (or reasonably similar) supporting equipment. (Example: Suppose a furnace passes when used with a 3 square foot solar panel and a battery bank with a 400 Amp hour storage capacity. The furnace is considered neither passed nor recommended for an RV with no solar panels and only a 100 Amp hour storage capacity battery.) The RV make and model should be noted, but is not considered part of the test--- a passing furnace may be used in any RV. Adjustments for recommended heat output should however be made for different sized RVs.

Progress--- But In What Direction?

The RV forced-air furnace is a classic example of technology gone mad. It is a classic case of reverse progress. Manufacturing trends may have even made the predecessor to the forced-air furnace (the passive convection furnace) artificially obsolete.

As technology progresses, it is natural for new inventions to replace older ones and make the older ones obsolete. But common sense dictates that newer inventions are supposed to be *better* than the ones they replace, not *worse*! When simple inexpensive reliable devices get replaced by ones that are more complicated, expensive and prone to failure, reverse progress takes place. Taken to an extreme level, the even more bizarre process of artificial obsolescence can occur. In this process, items of sound design become increasingly unavailable, not because of technological inferiority (the driving force behind natural obsolescence), but only because designers decide to stop making them.

Reverse progress starts when designers and manufacturers somehow depart from reality. They ignore all advice and pleas from customers and even consciences people within their own ranks, thinking they are somehow far superior to those mere mortals. Their designs are then "shoved down people's throats" on a like-it-or-not basis. They figure, if they are stubborn enough in steadfastly refusing production and/or inclusion of items of certain designs, they can artificially force those items to become obsolete, and make customers just give up and buy their trash anyway, because those customers have no other choice. Unfortunately, sometimes they win. (Have they won with RV furnaces? Must they win? Could their trend be reversed?) Customers too unwittingly play a role in reverse progress, by being uninformed and too tolerant of poor designs. Keep in mind: *As long as customers keep buying crap, somebody will keep making that crap.* Keep the opposite also in mind: *If customers stop buying crap, manufacturers will stop making that crap.* Customers need to be informed, demanding of products of reasonable design, and stubborn enough to not allow themselves to be bullied into allowing manufacturers decide what they take home. Customers, not manufacturers and salespeople, must ultimately set standards and design trends, if the tide of degenerative invention is ever to be shifted, once again sending "progress" in the proper direction.

Final Thought on Furnaces

In the early days of RVs, furnaces were simple. They didn't need to be fancy, they just needed to work. Then, somebody had to ruin everything by doing away with those simple furnaces that worked. It is so ironic that the simplest technical task of staying warm indoors, now poses the biggest challenge for today's RVs, considering that the problem of indoor heating was solved centuries before humans began playing with electric lights, refrigerators, air conditioners, motors, computers and TVs. Fortunately, indoor RV heating could be a very easy problem to fix, if somebody within the furnace or RV manufacturing trades could be persuaded to listen to reason.

Designing an RV furnace that works is not a complicated matter. Model it after the simplest and oldest of all heater types: stone fireplaces/chimneys and cast iron woodstoves. Replace the heavy stone or cast iron with thin sheet metal. Replace the solid wood fuel with propane gas. Add a temperature controlled on/off switch. Finally, downsize it to a scale that is appropriate for RVs. The result is the daughter of these centuries-old designs: a small, lightweight, durable structure, that burns fuel that is also lightweight. Perfect for RVs!

Nothing fancy is needed to heat an RV, just a fire in a metal can that vents to the outside. The old reliable RV convection furnaces need to make a comeback. Simplicity and reliability need to make a comeback.

Stoves & Ovens

Stoves/ovens are a very important part of any RV. Cooking is an essential part of RVing because: 1) occupants get to choose the foods they like [which they can't do with restaurant food], 2) eating foods that occupants are accustomed to boosts their moral and happiness levels and even makes them feel "at home," 3) there is often not much to do in the wilderness, so cooking good meals is a fun engaging activity and 4) being able to prepare and eat food at the RV allows people to stay in the wilderness for awhile, without needing to spend lots of time driving out long distances to find eateries.

Types of Stoves/Ovens

Indoor Propane Stoves:

Stoves are one of the few appliances in RVs that are simple and actually do what they are supposed to do. There isn't a heck of a lot to say about RV propane gas stoves other than you light them and they cook food. I can think of only four points: 1) A stove should have burner grates that fasten in place during travel (so they don't fly around) yet are easily removable so the stove can be cleaned. 2) The burners should make good sized flames, for fast cooking. 3) Burner valves should have adjustable flame settings. Valves that can only be "on" or "off" are difficult to use. They either burn food or they don't cook it at all. 4) The easiest way to light stove burners (on stoves without pilot lights or spark igniters) is with long spark or long butane hand held igniters.

above: indoor RV stove

Outdoor Fold-Down Stoves:

Some RVs have a second stove located on the outside, for outdoor cooking. These fold up vertically for travel (or when not in use), latch and merge with the RV siding. These are a very good idea, because some days are too hot to cook indoors. Cooking indoors might necessitate using air conditioners on days that air conditioners otherwise might not be needed. Users may not want to run air conditioners if they don't need to, either because of generator noise or because users may not even have air conditioners. Outdoor stoves also are convenient for times when people choose to eat outdoors.

Outdoor Portable Propane Stoves:

Do not despair if your RV doesn't have a built-in outdoor stove. Portable gas stoves, that run from small disposable tanks of propane, are available at reasonable prices. (One burner models are less than $20, two burner models are less than $40). People who camp in tents use these stoves all the time. They are small and set up quickly. They are an absolute necessity for outdoor RV cooking on hot days, and for high odor foods (like fish). In addition, they are great for picnics, and for emergency cooking during power outages (in regular houses having only electric stoves). I have encountered only one problem: sometimes the attachment between the small propane tank and the stove leaks gas at the stove junction. This gas might (and sometimes does) catch on fire. The remedy is to reposition the attachment and propane tank, until the right spot gets found where the leaking stops (and sometimes tightening the fitting helps too). Because these stoves and fittings are highly prone to leaking, the primary valve at the portable propane tank should always be kept off when the stove is not in immediate use.

my outdoor portable propane stove

For years, I struggled to eat at my trailer home on hot days. I had to either eat cold food (usually leftover) or make outdoor fires of charcoal or wood for cooking. Charcoal takes up lots of storage space, it takes a long time to light and I didn't always have it. Firewood is not available everywhere (in some places it doesn't exist, in other places, it's illegal to collect and burn), it takes up lots of storage space and sometimes won't light if too wet. Then, I found out about portable outdoor propane stoves. I wish somebody else had written this book and given me a copy years ago, so I wouldn't have had to learn everything the hard way.

Conventional Ovens:

Bigger RV gas stoves often also contain conventional gas ovens. These ovens may contain pilot lights and main burners that turn themselves on and off multiple times, controlled by adjustable temperature dials. There is not much to say about ovens, other than they are good to have. Don't try to use them as furnaces for comfort heat though because they make quite a bit of carbon monoxide.

Microwave Ovens:

The microwave oven is the only type of electric cooking appliance to which there is no gas burning counterpart. Food cooks much faster when bombarded by microwaves than it does by just applying heat made by burning gas, because microwaves penetrate objects deeply and quite evenly, whereas heat from burning gas gets absorbed at the surfaces of objects then takes quite a bit of time work its way through those objects. Microwave ovens are great to have for those quick warm-ups, both in regular houses and in RVs. Unfortunately, they use *lots* of electricity, so off-grid RV users who want to use them have only two options: 1) use them sparingly or 2) operate generators while in use.

Using them sparingly: I suggest using microwave ovens for no more than a minute or two at a time from batteries (connected to power inverters) and limit the daily run time to just a few minutes. This way, a battery charging system may be able to keep pace with consumption. But beware: A single minute of operation of a 1200 Watt microwave oven drains a battery by about 2 Amp hours, which is the same as a whole hour of operation of a large bright double fluorescent bulb light fixture. Using a microwave oven for just eight minutes per day from the battery might *double* the overall electricity consumption of an entire (otherwise) electricity conserving system.

How much battery drain results from
one minute of 1200 Watt microwave oven use?

Calculation:

(Energy) = (Power) × (time),
therefore:
(1200 Watt minutes) = (1200 Watts) × (1 minute).

The 1200 Watt minutes needs to be converted to Watt hours by dividing by 60, noting that 60 minutes = 1 hour.

(20 Watt hours) = (1200 Watt minutes) × (1 hour / 60 minutes).

The power inverter usually consumes as additional 20%, therefore the energy consumption needs to be multiplied by 1.2.

(24 Watt hours) = (20 Watt hours) × (1.2 efficiency factor)

Since:
(Energy) = (Voltage) × (Charge),
and:
(Charge) = (Energy) / (Voltage),
thus:
(2 Amp hours) = (24 Watt hours) / (12 Volts).

Answer: It uses 2 Amp hours.

Operating generators while in use: Turning a generator on just before using a microwave oven, then off immediately after use, causes zero battery drain, so this may be a better option, especially for extensive microwave oven use. Microwave ovens generally operate for only brief time periods (a few minutes), so such short generator runs are unlikely to bother anybody and won't consume much fuel. However, engines (including generators) wear out to much greater extents from being *started* than they do from prolonged *continuous operating periods.* [I read that starting up a cold automobile engine does as much wear to it as 500 miles of highway driving.] This is because when an engine is sitting, the parts actually touch. When started, microscopic "mountains" on the metal get sheared off, as the motion starts. Once running, a thin coat of oil separates the moving parts, so the moving parts hardly touch each other at all, hence very little wear takes place. If you start a generator few times per day, it probably will last awhile. But if you make a habit of starting it many times per day, its lifespan may shorten.

Once, when traveling with my mother in her RV, we stopped for burgers and fries. We ate in the RV. My French fries were rather cold. So I turned on the generator, put the fries in the microwave oven, "nuked" them, then shut the generator off, all in just a couple minutes of time. I then enjoyed warm French fries. As much of a generaphobe as I am, I must admit, generators really come in handy for microwave ovens.

Charcoal Grills:

Although this chapter is primarily about appliances that are inside RVs and attached to them, one should remember that a big part of RV living includes gadgets that are used outside and/or are portable. Small charcoal grills are great portable appliances for RVers to have. Having only a small one saves storage space and is more than adequate to cook for limited numbers of people. (RVs are often occupied by two people. If big gatherings of people occur, have *someone else* bring a really big grill, then have that other person take it home when the gathering is over.) Charcoal grills take awhile to get lit, and it takes time to set up a fire, but once the fire gets going, they can cook food fast and hot (for those who like crispy chicken skins and well-done meat), or slowly with less heat (for those who like rare meat). Heat is controlled by placing the lid on and limiting air flow. By severely restricting air flow, the inside of the grill can also be used either as a smoker or a make-shift oven. To use as an oven, protect the food from smoke by wrapping it very well in several layers of aluminum foil (in case you don't like smoke flavor in your baked apples). To use as a smoker, just place the food on the grill before putting the lid on.

Some people succeed at smoking by wetting pieces of fuel wood, so the wood smolders instead of catching fire and burning quickly. There are entire books written on charcoal grilling, so I don't want to go into too much detail here.

If you don't have an oven…

Make a small temporary oven out of your cooking stove: Place a large cooking pot over a burner upside down (*shown right*), then turn the burner on very low. Make sure there is air space under the bottom, so the flame doesn't get smothered. The hot air from the flame enters the pot at the opening in the bottom and gets trapped within it. Temperature control needs to be done manually, by lighting and extinguishing the burner multiple times, or using a very low flame, and can only be estimated. Inside the pot put some kind of a wire rack to suspend the food high above the flame. (Metal coat hangers work well for this. They bend and cut easily.)

Wrap food to be cooked in aluminum foil, to avoid spattering and messy clean ups. This "poor man's oven" is usually good only for warming things. Slow roasting or prolonged cooking is too tricky and messy.

Lamp-warm food: The intended purpose of propane gas mantle lights is to make light. They also can be used to warm food, and work well for this, though warming food is obviously not their intended purpose.

left: fruit in foil warming on top of a gas mantle light
above: Ahh! Warm fruit for breakfast.

Gas lights have metal lids that deflect heat and exhaust from their burning gas horizontally. These lids and the space immediately above them get pretty hot but not exceptionally hot, so if food is placed there, that food can get rather warm but isn't likely to burn too quickly. Wrap the food well in aluminum foil and place upon the lid. (Poorly wrapped food starts oozing out and dripping on the lights. Cleaning burned food off these lights is a bitch.) Flip several times with a potholder and occasionally remove from the lid, to prevent overcooking. Lamp-warming food has a couple of advantages: 1) It frees up burners on the stove, which might be needed when cooking multiple course meals or for several people. 2) Usually at least one lamp is on at night anyway (such is my habit), so since the heat is there, why not make use of it? I have lamp-warmed leftover rice, potatoes, vegetables, pork chops, chicken, hamburgers, meatballs, spaghetti and pizza. But my all-time favorite use is warming fresh fruit for pancakes. I think that putting cold fruit on a warm pancake is some sort of culinary sacrilege. I just love a pancake, served with real maple syrup and lamp-warmed fresh fruit, oozing in its own juices, that comes steaming out its foil package, eaten somewhere in the woods. I've served them to guests, who often rave about them.

Use an outdoor portable gas grill: Small outdoor grills, fueled by small disposable propane tanks, can also make effective ovens. The problem is: they take up valuable storage space. Big RVs have enough storage space to carry them, however these grills probably won't be needed, because big RVs are likely to have ovens included with their stoves. Small RVs could benefit from having portable outdoor grills to double as ovens, because small RVs are unlikely to have conventional ovens, but storing anything in a small RV is a big issue. This one is your judgment call.

Alternatives to Cooking with Electricity:

I have one single word of advice for anyone considering using electricity to cook in an RV off-grid: Don't. With the exception of microwaving, any electric cooking can be done just as effectively with gas, therefore there are no advantages to cooking with electricity instead of gas, and there are plenty of disadvantages. Recall from the equation at the end of Chapter 4, that it takes between 300 and 400 pounds of batteries to store (and eventually make) the same amount of heat energy as 1 pound of propane (including the weight of the tank that stores the propane). Recall that battery charging is difficult, time consuming and often unreliable, whereas propane filling is easy, fast and usually problem free. Also, electric heating elements consume massive amounts of electricity compared to other electric gadgets (such as lights, water pumps, TVs, small motors), so cooking with electricity multiplies an RV's daily electrical needs many times over, severely stressing a system, usually beyond its breaking point. One could effectively bypass batteries and cook with electricity made directly from a generator, but there is no advantage to doing this either. Far more fuel gets consumed by a generator to do electric cooking, than gets consumed if fuel gets sent directly into a stove burner, to do gas cooking. And generators make noise which bothers people, whereas gas stove burners are quiet.

Suggestions for using gas to cook things often cooked with electricity:

Coffee: Cook it in a coffeepot that sits upon the burner of your gas stove. The only possible advantage an electric coffeepot offers over a stove top coffeepot is: it shuts off automatically when the coffee is done. This feature can easily be duplicated with a stove top coffeepot and a cheap battery-operated electronic kitchen timer. Here's how to learn such a method: 1) Cook a pot of coffee in your favorite electric coffeepot at your regular home or in your RV when plugged into the grid. Use the timer to measure the time it takes to cook between plug-in and shut-off. 2) Cook a pot of coffee on your RV stove burner, set on maximum, using the same amounts of coffee grounds and water in a similar sized stove top pot. Set the timer for the previously measured time. Start the timer when you turn on the burner, then shut the burner when the timer beeps. 3) Sample the coffee. If it is too weak or too strong, make appropriate adjustments to the time. The strength of your first pot should however be pretty close. 4) Continue making adjustments to the time. By your third or fourth pot, you'll be cooking coffee like a pro. You are likely to forget about cooking with an electric coffeepot altogether. Be aware that you may also need to make additional adjustments in cooking time to compensate for altitude changes.

Toast: Method #1: There actually are inexpensive camp toasters that direct the heat of a propane gas burner to sliced bread or rolls to make toast. They are made primarily for people who camp in tents, but they work equally well for RV users. You need a good sized burner flame to properly use these, otherwise they just dry out the bread without actually toasting it. *Method #2:* Make toast on the surfaces of frying pans heated by gas flames. You can butter or oil the pans and get very good results. (This is the method I most prefer.) I recommend against using Teflon-coated pans to make toast. Teflon pans work just fine when there is something in them with water, because the water starts boiling away as the pan gets heated, limiting the temperature to about 212° F. Heating a dry Teflon pan without water can cause the temperature of that pan to get several hundred degrees higher than the boiling point of water. At those temperatures, Teflon begins to decompose into a poisonous gas. If ventilation is poor or indoor air space is limited (always the case in RVs), people can inhale this gas and get flu-like symptoms (which could also be mistaken for carbon monoxide poisoning). Decomposing Teflon with high heat isn't good for your frying pans either. *Method #3:* Hold the item to be toasted over a gas stove burner with tongues or a spatula, keeping it in constant motion, for even heating. (This is my preferred method for waffles.) *Method #4:* Place the bread or roll on a charcoal grill (assuming you've got charcoal burning to cook something else). Check it frequently, it cooks fast. As with method #1, the fire needs to be hot enough, else the bread/roll just dries out without actually toasting (good for making homemade croutons though). If the fire is too hot or there are lots of flames, the bread/roll can blacken very easily. (Sometimes the surface turns black, whereas the subsurface doesn't toast at all.)

Popcorn: Cook it the old fashioned way. Use a pan with some oil on top of a gas stove. People cooked popcorn for a long time before we had hot-air-poppers or those handy microwave oven bags. Remember when Jiffy-Pop was popular? I hadn't seen it for a long time, so I figured the microwave popcorn bags caused Jiffy-Pop to disappear like Blockbuster or Atari, but no, as of 2017, Jiffy-Pop is still available.

Refrigerators

Refrigerators are an important part of RVs as well as houses in general. The ability to keep food cold allows RVers to eat for awhile, thus permitting them the luxury of staying in the wilderness for long periods of time. A freezer expands this ability considerably, so I highly recommend having a freezer as well (which is usually as part of the refrigerator). A freezer allows RVers to stay in the wilderness for long times because it allows them to prepare cuisine in large amounts, portion it into small amounts, freeze those small amounts, store them indefinitely, then reheat them easily on days when the RVers want to spend time doing things other than cooking.

> Refrigerators are a relatively new invention in the grand scheme of things. For centuries, nobody had them. It's hard to imagine how people stored food before they had the ability to keep it cold--- they must have done a lot of canning, drying and salting. Early success at keeping things cold came in the 19th century, when people learned enough about insulation to store natural ice for long time periods and ship it great distances. Ice was cut from frozen lakes in the winter and saved for ice box use in the summer. Artificial ice eventually displaced the need for natural ice. It first became popular during the Civil War. Ice boxes, stocked by centrally made and delivered artificial ice, continued to be a common method of keeping household food cold throughout the first half of the 20th century. Finally, household compression refrigerators came along. Today, we can't live without them.

Types of refrigerators

I.) Ice Boxes & Coolers:

Come on, we all know what ice boxes and coolers are. Either water or carbon dioxide is frozen artificially (or found frozen in nature) then placed in a well insulated box to keep food (or other) items cold.

II.) Compression Refrigerators:

These work by compressing then decompressing gasses indifferent places. Often the gas becomes a liquid, then boils to become a gas again in the interior of the fridge. When a gas is compressed (force is needed), heat gets released. When a gas is allowed to decompress (occurs naturally, no force is needed), heat gets absorbed. Thus if a sealed gas (called a refrigerant) gets compressed, is allowed to cool to room temperature, then decompresses in another place, heat is absorbed at the location of decompression and released at the location of compression. These refrigerators don't *make cold*, they *move heat* from the insides to the outsides of refrigerators. Compression refrigerators are the most common type of refrigerator today. They are found in almost all regular houses.

III.) Ammonia Absorption Refrigerators:

These use the circulation of a fluid/gas mixture to absorb heat inside the refrigerator and dump that heat outside the refrigerator. They do not use motors or pumps to circulate the fluid/gas mixture, instead they use a heat source located near the refrigerator's bottom. That heat source can be from an electric heating element, but more often, it is from a small flame made by burning flammable gas (usually propane) or kerosene. Because ammonia absorption refrigerators can be designed to use no electricity at all, they are used where electricity is unavailable. Besides being commonly used in RVs, they are used in houses too remote to be connected to the grid, and by Amish people, who for religious/lifestyle reasons, choose to live without electricity.

ammonia absorption RV refrigerator interior

IV.) Thermoelectric Coolers:

These move heat using no moving parts nor liquids/gasses. A thermoelectric cooler moves heat by passing DC electric current through *thermopiles.* When a DC current is passed through a thermopile, it causes one side to absorb heat (becoming cold) and the other side to deposit heat (becoming warm).

Which Refrigerator Types are Best Suited For RVs?

I.) Ice Boxes & Coolers:

Ice boxes can be appropriate, as the only way to keep food cold, for very simple RVs that are only used occasionally and for rather short periods of time. However, ice boxes are difficult to use and don't keep food very cold, and thus are not appropriate as the sole form of refrigeration for more luxurious RVs or for extended trips. However, ice boxes can come in handy as supplemental coolers for RVs with fridges.

How do ice boxes work? Cold ice brought from somewhere else is placed in a container along with objects desired to remain or become cold. The ice absorbs heat, both from those objects, and from the exterior. Because the exterior walls of the container are insulated (meaning that heat has difficulty getting through them), it takes awhile for the ice and ice box contents to warm up. However, usually this warm up does not occur, because the ice box tends to get restocked with fresh ice before it gets a chance to warm up too much. Because water ice melts at 32° F, objects in close continuous contact with solid ice can never get much warmer than 32° F.

Advantages of ice boxes/coolers: 1) Ice boxes & coolers are extremely cheap. They cost just a few dollars, whereas refrigerators cost hundreds (or thousands) of dollars. 2) Ice is cheap. If the ice box is used only occasionally and for short time periods, its owners probably won't need to take out a second mortgage on their house in order to stock it. 3) Ice coolers are great as supplemental coolers. Occasionally, RV occupants may want to entertain large gatherings of people, thus they may run out of refrigerator space. Also, RV users in remote locations (quite far from supermarkets) may not be able to shop for groceries very often, so they may elect to stock up on groceries. They too may need cooler space additional to what their refrigerators provide, especially the first few days after shopping sprees. Vegetables and fruits don't need to be super cold, so they survive well in ice coolers. 4) Ice coolers are great at keeping food cold while being transported from the supermarket to the RV. In remote areas, a return trip from the supermarket to the RV may take a couple hours. Some foods could start to spoil and frozen foods may thaw during this time period, if not kept cool during transit.

Disadvantages of ice boxes/coolers: 1) Temperatures of food items are difficult to control. Items in continuous contact with the ice (or its water) may remain quite cold, but some items are likely to be away from the ice. These items may become unacceptably warm. 2) Ice boxes/coolers don't keep food very cold. They certainly don't work as well as refrigerators, especially in hot weather. 3) Food may become contaminated by water from melted ice. If water is allowed to accumulate, and non-sealed items are submerged in that water (like cream cheese, lunch meat, cut vegetables), food particles accumulate in that water and can start to rot. Then, bacteria contaminate all food submerged in that water. Also, many items (especially cream cheese) "drowned" in ice water are disgusting. 4) Ice boxes need to be restocked often. Driving to a store every day to buy ice is very time consuming. If users don't get to the store or can't (because of a vehicle failure), and the ice melts, food starts spoiling quickly. 5) Keeping an ice box continuously stocked can be costly. One or two bags of ice bought occasionally don't cost very much. But many bags of ice for long term operation can cost plenty. Also, if special trips to stores are needed to get ice, then the cost of gasoline for travel becomes part of the ice box's operational expense. Travel distances to get ice may be considerable in some wilderness areas. Then there is the lost time due to traveling to get ice--- time that RVers might want to spend doing something else. Lost time is also lost money, since RV trips can be costly. Often, the cost of keeping an ice box stocked with ice exceeds the value of the food items within it.

> The greatest difficulty, with using the ice box is the need to continually restock it with fresh ice almost daily.

In this 1986 photo, taken during my first long RV voyage, I "mined" ice for my ice box. I did not yet have a refrigerator back then. Ice was so hard to get and needed to be obtained so often, that I mined ice from this mountain snow field shortly after a swim in a mountain pond, while still in my swim suit. Believe it or not, the water in the pond was comfortably warm, despite being close to the snow bank. Restocking the ice box was such a bother, that I bought and installed a refrigerator shortly after this trip ended.

Types of ice boxes/coolers:

A) Ice Only: In this type, water from melting ice is allowed to drain out the bottom of the container.

B) Ice & Water: In this type, water from melting ice is retained in the container.

C) Saltwater: Pure water ice melts at 32° F. Adding salt to ice causes it to melt at temperatures significantly below 32° F. (This is why roads are salted in winter.) Adding salt to ice in a box causes most or all of the ice to melt, but the coldness from the ice is retained within that liquid (which is not allowed to drain from the container). Sealed items (in leakproof containers) submerged in this super chilled water can remain frozen. (The guy who gave me this tip told me that he drove hundreds of miles with a cooler containing some frozen venison, which lasted days without thawing, using this method.)

D) Dry Ice: Dry ice can be used alone to keep things frozen rather than just cold, or in conjunction with water ice, to extend the life of that water ice. Dry ice is frozen carbon dioxide. It does not melt into any liquid, instead it sublimates (evaporates) into a gas, which is harmless in the small quantities likely to be released from small-scale household cooler usage. Dry ice is very cold (–110° F), therefore it should not be handled without gloves (else it can cause frostbite injury). It is becoming increasingly available. I've seen it at many supermarkets near National Parks and other remote areas.

Ice Only box verses Ice & Water box: Should one drain and discard the meltwater or should one retain that water in the box? There are advantages and disadvantages to each. Draining the water can be an advantage because it keeps food cleaner. Retaining the water can be an advantage because it causes the ice box to remain cool much longer with less ice. This is because water stores quite a bit of heat in both solid and liquid form, so cold liquid water absorbs quite a bit of heat from other items before warming up. Cold water is great for keeping items not spoiled by "drowning" cool (like cans and bottled beverages--- also many uncut raw fruits and vegetables survive fine submerged in water). Unlike ice cubes, cold liquid water surrounds items completely. If both ice and water are present, the mixture is 32° F, therefore everything in that liquid/ice mixture will be at about a chilly 32° F. Retaining the water can be a disadvantage if: A) that water becomes contaminated by non-sealed food items, B) that water takes up too much space and that space is needed for restocking the cooler with fresh ice or C) that water gets too warm after the ice completely melts. Discard the water if any of these three occur.

Final tip: If all ice melts (or sublimates), avoid opening the cooler too often. Food items retain coldness for quite awhile even without ice.

II.) Compression Refrigerators:

Compression refrigerators are completely inappropriate for RVs because they require too much electricity. They do however work great in regular houses where electricity is always available.

Compression refrigerator don't work in RVs because: 1) They use more electricity than can be realistically provided by batteries. And: 2) That electricity needs to be available continuously, something that generators can't provide.

Advantages of compression refrigerators (in regular houses): 1) They work really quickly. After turning one on, it takes just minutes for the freezer compartment to start frosting. This type of refrigerator cools faster than any of the other three types. 2) They are inexpensive. Used working compression refrigerators can be purchased for just a couple hundred dollars. Sometimes people just give or throw old units away even while they still work. 3) They can have long lifespans. My family had a compression refrigerator for over half a century. It still worked when my brother threw it away.

III.) Ammonia Absorption Refrigerators:

Ammonia absorption refrigerators (also called "gas absorption refrigerators") are *extremely appropriate* for RVs, because they can be run from propane gas, instead of electricity. Fortunately, somebody within the RV designing community was able to see this, and RV manufacturers do choose to put them in RVs. However a design flaw was implemented in some of these refrigerators, which seriously compromises their performance.

Interior cooling fins of RV ammonia absorption fridge

How does the cooling unit in an ammonia absorption refrigerator work? 1) Ammonia gas is boiled out at the heat source from a dissolved state in a liquid mixture containing water, ammonia and a rust inhibitor. 2) The ammonia gas condenses into liquid ammonia at high pressure, then cools. 3) The liquid ammonia enters the pipes inside the refrigerator, where it evaporates into a hydrogen atmosphere, absorbing heat. 4) The evaporated ammonia exits the refrigerator, then gets reabsorbed by the water/ rust inhibitor, and also loses its heat. The cycle then repeats. Got all that? It's a bit much to understand--- I've been reading about them quite awhile and still don't completely get how they work. So here's a simplified explanation of how they work: The heat source, combined with gravity, drives the liquid/gas mixture in the cooling unit in a manner very similar to the way a coffeepot moves liquid and boiled water to make coffee. In a coffeepot, a heat source at the bottom boils and percolates the water upward, then gravity pulls the water back down again, but only after it passes through the coffee grounds, picking up their flavor and gradually becoming coffee, as the cycle repeats. In the ammonia absorption refrigerator, the heat source near the bottom boils ammonia out of the water (technically not "percolating" because the bubbles are tiny), driving the circulation of the liquids and gasses, which absorb heat (rather than coffee flavor) from inside the refrigerator, then deposit that heat outside. The process is very similar to convection, except that "convection" is a term usually applied to gasses alone. Now here's a simplified explanation of the principle that makes them work: When a liquid evaporates, heat gets absorbed. (When you put on a wet shirt on a hot summer day, what happens?) When a liquid condenses, heat gets released. So the goal is to get the liquid (ammonia) to evaporate *inside* the refrigerator, then condense *outside* the refrigerator.

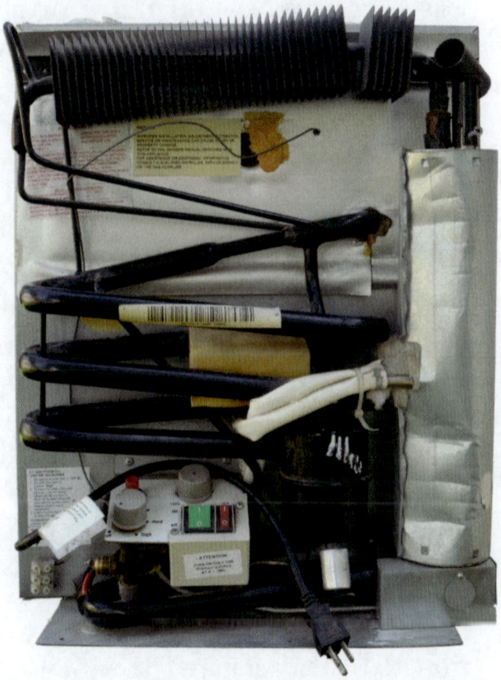

RV ammonia absorption refrigerator (back view of functional apparatus)

The heat source (that drives the refrigerator): Just like a coffeepot, the refrigerator can be run from the heat of either a flame or an electric heating element. Usually a small propane gas flame is used, which is barely larger than the flame of a butane cigarette lighter. While on the road or parked hooked up to municipal electricity, an electric heating element is often used instead. The electric heating element can run from either 12 Volts DC (while driving) or 120 Volts AC (while parked). A refrigerator capable of running from either gas or one type of electricity is called a "2-way." A refrigerator capable of being run from gas or either of the two types of electricity (120 Volts AC or 12 Volts DC) is called a "3-way."

Advantages of ammonia absorption refrigerators: 1) They have no moving parts [since no motors or pumps are needed]. 2) [main advantage] No electricity is needed. Therefore, they are the only realistic option for RVs that move and get used, as all RVs at least occasionally go off grid. 3) They are very quiet. There is absolutely no noise resulting from operation of these refrigerators. The only way to know they are on, is by the cold they produce. [Although on absolutely silent nights it is possible to sometimes hear a barely audible fizzing sound from boiling ammonia or the flickering of the flame.] 4) They consume propane very slowly. A small refrigerator, such as the one I have in my RV, can easily be run for at least two weeks from a single fill of a 20 pound propane tank. 5) They are very reliable. I had a unit that worked for 20 years.

Disadvantages of ammonia absorption refrigerators: 1) They are slow to begin cooling. This is in stark contrast to the compression refrigerators, found in regular houses, which frost up only minutes after being turned on from room temperature. It is best to start an ammonia absorption RV refrigerator up at least a day before stocking it up with cold food and actually using it. If the refrigerator shuts off accidentally while camping [perhaps from running out of propane or an electrical problem], it should be turned back on as soon as possible, and opened as infrequently as possible [for the next half day] until it begins cooling really well again. [In my personal experience, I've found that my refrigerator works alright the first few days, then starts cooling really well after about a week, when substantial ice builds on the cooling unit.] 2) They cool slowly but steadily. Slow cooling is only a disadvantage when users want to cool a lot of stuff from room temperature fast. [Sometimes people continuously put warm cans of pop/beer into their RV refrigerators and remove cold ones. This tends to rob the refrigerators of their coldness and overwork them. Use an ice chest or submerge sealed pop/beer cans in cold creek/well water for this purpose.] Slow steady cooling is in stark contrast to the way compression refrigerators cool, which is in spurts, on an as needed basis, controlled by a thermostat. 3) They need to be level to operate. Leveling, therefore, becomes a necessary chore for the RVer, not just a matter of comfort. Leveling is necessary because gravity causes fluids in the cooling unit to gradually move down. "Down" becomes a different direction, if the refrigerator is too far off level. If attempted to be operated off level, the fluids might not move right, which can cause the refrigerator not to work. [If the RV *must* be parked off level for awhile, the refrigerator should be turned off until moved again.] 4) They can actually be destroyed by operating off level. Heat in the non-circulating fluids, built up near the cooling unit's heating source, can cause the rust inhibitor to decompose into tiny crystals, which can at a later time clog the flow of fluids and stop the cooling action. This can permanently destroy the cooling unit. Cooling units can cost lots of money to replace. Despite the fact that such refrigerators can be destroyed by off level operation, this doesn't happen all that often, and is not such an easy thing to do. [This has never happened to me.] 5) They are expensive. However, they are not usually more expensive than new compression refrigerators designed for regular houses.

Repair tip: If the refrigerator isn't working, even though on, and some exterior parts of the cooling unit get too warm (indicating disproportional heat distribution), it may be due to internal clogging. Shut it down and try the following: 1) Disconnect the refrigerator and remove it from the RV. 2) Beat up the cooling unit with a rubber mallet (not enough to break it but enough to shake it up). 3) Turn the whole refrigerator upside down (leave it that way for at least ten minutes) then rightside up again. Repeat several times. 4) Then put the refrigerator back and reconnect it to the gas and electric lines.

This may remove a mild blockage. I tried it. It really works!

Practical Considerations: These refrigerators are operated differently under different conditions. 1) When the RV is parked away from electricity (and no generator is running), the refrigerator always runs on propane gas. 2) When the RV is plugged into municipal electricity (or the generator is running), the gas flame is shut off and the refrigerator runs from a 120 Volts AC heating element (in models designed for 120 Volt operation). Some refrigerators make the gas/electric transition back and forth automatically, others need to be switched manually by their users. 3) When the RV is driving down the road, the refrigerator is often run from 12 Volts DC [although some models still run from propane while driving]. This is the only circumstance under which 12 Volt operation is feasible, since as long as the vehicle drives, its alternator keeps making electricity. There are disadvantages to propane gas operation while driving: A) High winds, due to the RV moving through the air, can blow away much of the heat from the gas flame, causing the refrigerator to not work so well. High winds can also blow out the gas flame entirely. [If either of these conditions occur, try using duct tape to temporarily block some of the ventilation slots while driving. The wind from high speed travel will usually provide enough ventilation for proper operation. But be sure to remove the duct tape vent blockages as soon as parked.] B) Leaving the propane gas on during travel could cause a fire in the event of a collision. Some people always shut the propane gas valve off during travel, others don't. [I don't.] 4) These refrigerators need to be defrosted once in awhile [maybe once per month]. They should be emptied and shut off for a few hours until all the ice dislodges from their internal cooling components. Then, the ice should be removed and the meltwater wiped away. One should have a cooler to temporarily store refrigerator items while defrosting.

Could the refrigerator be run (while parked) from batteries instead of propane? In theory, yes. In practical reality, no. Running even a small (95 Watt) heating element would draw about 8 Amps of current. [(96 Watts) = (12 Volts) × (8 Amps).] That would use 192 Amp hours per day [(192 Amp hours) = (8 Amps) × (24 hours)], or 1344 Amp hours per week [(192 Amp hours/day) × (7 days/week) = (1344 Amp hours/week)]. At about (1 pound) / (Amp hour) [equation done in the "Lights" section, sub-section "Electricity Verses Fire"], that's 1344 pounds of batteries (for one week) or 2688 pounds (for two weeks). At 37.5 Amp hours per battery (drained halfway), that would require 36 car batteries for one week's operation or 72 batteries for two week's operation. By the way, 72 batteries would cost over $5000. Running the refrigerator for two weeks from a 40 pound (when full) tank of propane, costing $10-$18 to fill is starting to look not so bad. Nobody in their right mind ever attempts to run these refrigerators from batteries.

The design flaw: Electronic "brains" (circuit boards) and igniters were added to some models to make operational decisions and ignite their gas flames. Of course, these refrigerators were then made absolutely *dependent* upon electricity, and thus designed to shut down if the electricity goes off. Not a good idea, considering that RV electricity is unreliable. (Recall the list of "20 Things That Can Shut an RV Electrical System Down," in Chapter 2, section "The Octopus Effect.") Not a good idea for another reason: Unlike other appliances (like lights, furnaces or TVs), refrigerators need to work continuously. If a refrigerator goes off for too long, food starts spoiling and must be discarded. That food must then be replaced. It takes money, time and gasoline (for trips to stores) to replace food. Also, major inconvenience can occur, should the owners fail to replace some important items, and not realize it until they are home at the RV, miles away from civilization and ready for supper. I cannot comprehend how somebody decided that it would be an "improvement" to redesign a refrigerator that *runs on gas* to shut down when the *electricity* goes off!

> Personally, I don't see why circuit boards were "needed" in these refrigerators, in the first place. I've had a total of three small simple model refrigerators without circuit boards over the years. They all worked just fine. They didn't even connect to the RV electrical system while operating on gas.

Are there advantages to having circuit boards in refrigerators? Like the fans added to furnaces, there may be some *slight* advantages to adding circuit boards to refrigerators: 1) These refrigerators can light themselves [both initially or in the event of their flames going out accidentally]. Electronic igniting/re-igniting however can also be achieved by adding an electronic re-igniter to the refrigerator. This device, hooked up to the RV battery, electronically senses the flame, and if it goes out, immediately makes many sparks to re-ignite. It works before the thermocouple controlled safety shutoff has a chance to shut the gas valve off. And: 2) These "brains" allow the refrigerators to respond to different temperature and humidity conditions and adjust their gas flames accordingly. However, like adding fans to furnaces, any possible advantages achieved by adding circuit boards to refrigerators are overshadowed a hundred times by the disadvantage of having these refrigerators shut down every time the electricity goes off.

How could this design flaw be corrected? 1) During electrical outages, refrigerators with circuit boards need to be designed to revert to several user adjustable "default" flame settings, thus keeping them on. 2) Refrigerators should have backup procedures for igniting their flames to start their operations under zero electricity conditions.

Final comment on the design flaw: Reasonable people always have a "Plan B." Reasonable people always plan a little bit for the unexpected. That's why people wear seat belts. That's why cars carry spare tires. That's why skydivers usually take two parachutes when stepping out of airplanes. Reasonable people don't design things to work only in a perfect world. Reasonable people didn't redesign gas refrigerators to shut down when the electricity shuts off, idiots did.

IV.) Thermoelectric Coolers:

Thermoelectric coolers are not appropriate for RVs, because their operation requires quite a bit of electricity nonstop--- more than RV batteries can effectively store. While a vehicle is driving however, its alternator makes electricity that never runs out. Therefore, small thermoelectric coolers may have limited RV use, as they can plug into cigarette lighter receptacles of RVs or automobiles while driving, to keep stuff cool for short trips or picnics. Many portable automotive/RV thermoelectric coolers have switches that also enable them to be reversed and used as heaters instead of coolers.

Best Overall Refrigeration Plan

Have a refrigerator that doesn't contain a circuit board and in not dependent upon electricity. It should not be too small and if you take extended trips, it should have a freezer too. Since these refrigerators need to be level to operate properly, be sure you have jacks to level your RV (and you are capable of using them). In addition, if space permits, have a cooler. A cooler and a bag of ice allows you to transport food long distances from civilized areas back to the RV without spoilage and without thawing frozen items. A cooler also allows extra space to store lots of cold fruits and vegetables, which you might need if you go into wilderness areas for extended time periods.

Air Conditioners

Air conditioning is something that is nice to have in RVs but is usually not an absolute necessity. It may however be a necessity if the users have a habit of camping in very hot weather and places--- especially when parking in the sun. It may also become a necessity for RVs that are poorly insulated from the heat of the sun.

Active Air Conditioning Verses Passive Air Conditioning

Active air conditioning uses external power sources (electricity or flammable fuel), whereas *passive* air conditioning uses no external power sources.

Active air conditioning is done by discrete noticeable machines designed specifically for this purpose, whereas *passive* air conditioning is usually done subtly by structures that exist primarily for other purposes.

Active air conditioning is about *removing* heat, whereas passive air conditioning is usually about *preventing* unwanted heat from entering dwellings in the first place. (There are exceptions--- opening windows to let out hot air is a passive way of removing heat.)

When considering air conditioning, most people think only of the active components--- they think of those hard working noisy energy-gobbling machines. Few give much thought to the passive heat blocking components of the walls, windows and even the interior objects. This is a critical mistake, because both are needed to make RV living areas comfortable and both work hand in hand. I don't believe that it is the job of an air conditioner to remove an extra 20° F of heat that never should have been admitted into a dwelling to begin with.

Active Air Conditioning
(uses power sources)

Types of Air Conditioners

I.) Compression Air Conditioners:

These work by the exact same principle that compression refrigerators do: by compressing then decompressing gasses. (details in "Compression Refrigerators.") Like refrigerators, air conditioners don't *make cold*, they *move heat* from the insides to the outsides of dwellings. These air conditioners differ from compression refrigerators only in three non-fundamental ways: 1) Air conditioners cool much larger volumes than refrigerators usually do. 2) Air conditioners don't need to lower the temperatures as much as refrigerators do--- refrigerators usually need to make things very cold or freeze them, whereas air conditioners only need to cool to comfortable levels [about 70° F]. And: 3) Unlike refrigerators, which must work constantly, air conditioners only need to go on during hot temperatures [often only at midday]. *(photo: RV compression air conditioner)*

II.) Evaporative Air Conditioners:

(also called "swamp coolers")

These work by absorbing heat through the process of evaporation of water. In chemistry, whenever a substance changes from a liquid to a gaseous phase, heat gets absorbed. Water is used since it is abundant, extremely cheap, non-toxic and often evaporates with little prompting. Evaporative cooling is used by biological organisms to prevent overheating. [This is why people sweat on hot days or after exercise (which produces internal body heat through metabolism). It is also why dogs pant under the same conditions (fast moving air on their wet tongues evaporates their saliva, cooling their circulating blood).] In an RV evaporative air conditioner, an absorbent mat resembling a large thin sponge gets drenched with water. An electric fan slowly intakes outdoor air and blows it through the mat. The air gradually absorbs the water through evaporation. The air that exits the unit and enters the dwelling is cooler and more humid than the intake air. The mat must be kept continuously wet, in order for this process to keep working.

III.) Ammonia Absorption Air Conditioners:

These are basically scaled up versions of the much smaller ammonia absorption refrigerators that almost all RVers have come to know and love.

IV.) Solar Powered Fans:

These are simply small fans, powered by solar panels on the RV roofs, which suck up interior warm air near RV ceilings and discharge that air to the outside.

Which Air Conditioner Types are Best Suited For RVs?

I.) Compression Air Conditioners:

Compression air conditioners are rather well suited for upscale RVs. This is the air conditioner type most frequently used in RVs. But they do have some problems.

Advantages: 1) They remove lots of heat. They can therefore work in very hot weather. 2) They start working quickly. You can feel the cool air blowing on you only minutes after turning one on. 3) They are fairly reliable. And: 4) They are readily available for RVs.

Disadvantages: 1) They use enormous amounts of electrical energy. Running such a unit from a battery is unthinkable. A typical RV air conditioner might use 1300 to 1800 Watts of power. If run from a power inverter, and a 12 Volt fully charged 100 Amp hour battery, it would drain 90% of a that battery in only 30 to 40 minutes. Therefore: 2) They require generators while being used. Both users and their neighbors must tolerate generator noise in order to use them. Also, be aware that some generators are not powerful enough to operate these air conditioners. [Compare power ratings before attempting to use.] 3) Generators installed in RVs add additional heat to those RVs from combustion while operating. This extra heat must also be removed by the air conditioners. 4) These air conditioners are rather noisy, even when plugged into municipal electricity sources [where there is no generator noise]. 5) These air conditioners work so well, they may encourage poor RV designs. The skin and window design of RVs should prevent excess heat from entering them on sunny days. (This is covered shortly in the "Passive Air Conditioning" section.) This type of air conditioner can remove that excess heat so effectively that designers may not properly design RVs to block that heat. This can necessitate air conditioner use on days when air conditioners shouldn't be needed, therefore causing generator noise. Also, since generator use is prohibited at certain times and places, if laws are obeyed, poorly designed RVs may become unbearably hot. 6) These air conditioners may encourage rain water to enter RVs. An RV air conditioner typically sits upon the roof of the RV. But an air conditioner is heavy, so if the roof is flat, it may actually bend that flat roof into a concave roof, making the point where the air conditioner sits the lowest point. This downward indentation, combined with the fact that the air conditioner is installed over a large square hole in the roof, causes the roof to act as a funnel and dump water into the RV. [This happened to my mother's RV. There was absolutely no way to stop the rain from coming in, other than placing a tarp over the entire RV, something we didn't want to do. The best we could do was put buckets inside underneath the leakage points to catch the water.]

Practical considerations: 1) Small compression air conditioners may be mounted on RV roofs or on their sides at the windows. Use only roof mounted units for RVs. Side mounted units are not practical, unless those RVs are permanently parked and do not move. [One would need to remove a side mounted unit every time one takes a ride, or design a special slideout so the unit could be retracted for travel.] 2) These RV units don't cool completely evenly. However, this is rarely a problem unless somebody sits directly under one in the path of the air conditioner's cold blowing air. Some units do blow air sideways at the ceiling. This helps to cool an RV more evenly. 3) Some units don't turn on/off automatically. It is annoying to need to leave an air conditioner on continuously [especially if taking a nap] or to not have it working at all. 4) Larger more luxurious RVs may have two air conditioners. Two air conditioners may gobble up so much power that they may not be able to operate simultaneously from a generator of normal power. One solution of course is to install a super-powerful generator, capable of running both units at once. This however costs some really big bucks, and may not be worth doing, because the RV might not encounter that many really hot occupied days. Furthermore, really powerful generators waste more fuel, especially when used to power gadgets of more modest electrical consumptions. Another solution, that some RVs use, is to automatically cycle the two air conditioners when powered by a generator. The first unit operates for a specified number of minutes, while the second is off, then the second unit operates for a specified number of minutes, while the first is off. The cycle then repeats itself an indefinite number of times.

Conclusion: Compression air conditioners work well and are commonly used in RVs. But off grid, they require constant generator use, so both owners and neighbors must be willing to tolerate long periods of generator noise in order to use them.

II.) Evaporative Air Conditioners:

Evaporative air conditioners can be rather well suited for RVs, when used in hot dry climates (the desert southwest), but are not suited for use in hot humid climates (like Florida and Georgia). They don't cool to the extent that compression air conditioners do.

Advantages: 1) They have simple designs. Therefore: 2) They require less maintenance [than compression air conditioners do]. 3) They are available for RVs. RV owners don't need to purchase units for regular houses and modify them to fit their RVs [though some people have done this]. 4) They are energy efficient. The fan and extreme intermittent use of the water pump consumes only a tiny fraction of the electricity that is needed to run a compression air conditioner of equal cooling capability. Therefore: 5) They can be run without simultaneous generator use. But beware: users still need decent equipment and strategies for charging batteries in the bush. If the only way to charge a battery is with a generator powering a 120 Volt AC battery charger, and the charger is weak and slow, more generator run time may be needed for battery charging than the actual run time of the air conditioner (though generators and air conditioners run at different times). It may thus be misleading to for manufacturers to state that these air conditioners can be run without generator use. These air conditioners *can* be used in the bush without generator use, but *only if quiet well designed battery charging equipment and strategies are in place*. 6) They can be run with no stored external electricity on sunny days. If a solar panel is used [I highly recommend having one--- about 80 Watts], that panel can more than satisfy the needs of the fan, making the air conditioner's drain on the battery zero [or less than zero]. 7) They are relatively quiet. The only moving component of an evaporative air conditioner is its fan. Fans don't make much noise. Compression air conditioners, combined with the generators required to operate simultaneously to power them, are extremely noisy. 8) They move fresh clean air through the RV. This is because they intake their air from the outdoors. They also filter out some dust and pollen. Compression air conditioners recirculate interior air, so if they are used, inside air can get stale.

Disadvantages: 1) They can't cool as well as compression air conditioners can. 2) They don't cool areas as quickly as compression air conditioners do. Evaporative air conditioners are meant to stay running for long time periods. They are not great for suddenly being tuned on and cooling RVs at midday that have sitting in the sun for hours with the windows closed. 3) They don't work when humidity in the air is high. The relative humidity of the air must be about 50% or lower for them to function. Therefore: 4) They are unsuitable for hot humid climates. Consider your probable frequent camping destinations carefully before choosing an evaporative air conditioner as your only air conditioning unit. 5) They make the air humid. This can make RV interiors feel "sticky," "muggy" or "swampy" [this is probably why they are called "swamp coolers"--- they don't work well and aren't used in actual swamps]. Prolonged excess humidity in the indoor air from extended use can cause mildew or even mold to grow. This can stink the place up or even make indoor environments toxic to humans and pets. 6) Units mounted on flat roofs can encourage rain water to enter RVs. As already mentioned in the previous section on compression air conditioners, any heavy object on a flat roof mounted over a hole can bend the roof into a funnel, causing rain to enter. It makes no difference which type of air conditioner is used. Whether or not rain enters, [and if so, how much] is dependent only upon the roof shape, the weight of the object and how well the seams are sealed. And: 7) They are incompatible for simultaneous use with compression air conditioners.

Practical considerations: 1) Compression and evaporative air conditioners are incompatible for simultaneous use because they are very different. A) Compression A.C.s are closed systems. This means they intake indoor air and recirculate it after cooling it. Hence all the windows should be closed. Evaporative A.C.s are open systems. This means they intake outdoor air and release it indoors after cooling it. They must take their air from the outdoors because they need dry air for the evaporation process to work, as the indoor air has too much moisture. Hence some windows should be open to allow old air to escape, in order to make room for the new cool air that gets added. Also, ventilation allows cool air to migrate from the A.C. unit to and through other parts of the RV. B) Compression A.C.s *remove* humidity from the air [from condensation on the cooling unit] whereas evaporative A.C.s *add* humidity to the air. Adding humidity to the air can be an advantage because dry air aggravates allergies and asthma. Also, many people find humid air to be more comfortable to occupy than dry air [provided that the air isn't completely saturated with water]. 2) Evaporative air conditioners shouldn't be powered by generators. Because: A) An operating genset [generator installed in an RV] adds heat to an RV. This heat added is probably too much to be overcome by an evaporative air conditioner. B) The fans of evaporative A.C.s don't need that much current [4-6 Amps is typical]. Solar panels or batteries can handle such current loads reasonably well, at least for awhile. C) Evaporative A.C.s probably should be left running all day [and possibly during the night as well]. RV Generators shouldn't be left running for such long time durations [and never at night]. 3) Their mats must be kept continuously wet in order for the evaporative cooling process to work. There are several ways that this can be accomplished: A) A unit can tap into the RV's existing water system, opening a valve when water is needed, causing the demand pump to pump water from the freshwater tank to the A.C. unit, just as it would to an open faucet. [Be careful not to run your freshwater tank dry.] B) A pressurized water tank can do the same. An air compressor is needed to build pressure in such a tank. The air compressor does not need to [and probably won't] run at the same time that the A.C. unit needs and draws its water. C) Water may be fed into the unit manually. This is probably tedious because it must be repeated frequently. 4) Evaporative air conditioners can cool incoming outdoor air by 20°-30° F before releasing it indoors. There is however a limit as to how much they can cool. [The temperature of the incoming dry air is called the *dry bulb temperature*. The lowest possible temperature achieved by evaporation is called the *wet bulb temperature*.] One person reported on the internet that his evaporative air conditioner kept his RV comfortable when outdoor temperatures were around 100° F and still relatively comfortable when outdoor temperatures were 120° F. [This person had an Airstream, which reflects the heat of the sun, due to its metallic surface.] 5) Heavy mineral deposits can build up on their mats quickly. Remember that when water evaporates, it leaves its dissolved minerals behind. Built up minerals can make a mat begin to look like a miniature coral reef and can eventually clog it. The mat should be removed and cleaned about once per month. I don't know if soaking it in acid is recommended, but this would probably work--- check with the manufacturer. I get calcium build up in my cooking pans when I boil tap water frequently for tea without cleaning them. Soaking a pan in a weak acid solution such as vinegar for a couple hours dissolves and removes that lime build-up very well. The A.C. mat might get clean from being soaked in a dilute solution of hydrochloric acid, also known as muriatic acid. This acid is readily available from swimming pool suppliers and evaporates completely, leaving no residue. Another possibility for addressing mineral build up is to avoid it by using distilled water [which has no minerals] for the air conditioner mat. Bottled distilled water is readily available from supermarkets. However, this remedy could get to be expensive with extended use. Jugs of water take up lots of space. Also, the RV would need a separate water tank [and pump] just to hold the distilled water for the air conditioner. 6) Big RVs could have both types of air conditioners. [But both shouldn't be used simultaneously.] This would give a big RV the best of both worlds. The evaporative A.C. could be used when off-grid when moderate heat removal is needed in relatively dry climates. The compression A.C. could be used when more aggressive or rapid heat removal is needed and/or in more humid climates. The compression air conditioner could be powered by a generator, or used when hooked up to municipal electricity, where generator use isn't needed and noise isn't an issue.

Evaporative cooling is the oldest cooling technique known to man. In ancient Egypt, people hung wet blankets in their houses during hot summers, to cool those houses. Kings also used slaves to fan the blankets, to speed up the evaporation process, to keep the buildings still cooler. I don't know if this fanning technique by humans actually worked though. True, fanning a wet blanket would cause evaporation and cooling to occur more quickly, however the human slave would need to burn energy to do the fanning, hence the slave's body might add more heat to the environment than got removed by the fanning process.

Salt-water evaporative air conditioners: This is a variation of the standard simpler evaporative air conditioner design. In this setup: A) Water is evaporated into air incoming from the outside, cooling it. B) A fine mist of salt water containing a desiccant [usually calcium chloride and lithium chloride] is sprayed through the air. This salt water absorbs most of the moisture from the air but also unfortunately heats it. C) Water is once again evaporated into this dry air, thus cooling it again, before this air gets released indoors. Salt water evaporative air conditioners cool more efficiently than standard evaporative air conditioners (else they wouldn't be made) but they are more complicated. With added complication comes added cost. More important to RVs, additional chambers for evaporation and drying translate into larger spaces and weights for such units. Although these units exist for regular houses, I don't know if anyone makes or recommends them for RVs. Also, as the recirculating salt solution absorbs more and more water, its saltiness decreases, and so too does its effectiveness at removing further moisture. This salt solution would need to be "cooked" occasionally to remove that extra water. Since "cooking" adds heat, and the solution must cool before use, "cooking" should only take place at night and outside the RV.

Homemade air conditioners: You can use evaporation to cool an RV or other enclosed spaces with little technology. 1) You can use a plant sprayer to mist general areas. As the mist dries, it absorbs heat. This technique probably won't absorb substantial amounts of heat though. 2) You can put on a wet shirt and wear it until it dries. This technique works well to keep a person's body cool but is a bit uncomfortable. Furthermore, the water from the shirt can transfer to other objects that you would rather have remain dry, such as couches and beds. 3) You can remove the RV's drapes, soak them in cold water, then hang them back up while wet. As the water dries over the following couple hours, heat gets absorbed. Once dry, the procedure can be repeated. This technique work rather well, if there is a cold water source nearby, such as a stream or lake. [In the desert southwest, water evaporates quickly, hence streams and lakes tend to be rather cold, even if the surrounding air is hot.]

Conclusion: An evaporative air conditioner can be a good comfort cooling choice for RV users who want to camp off grid frequently but don't want to annoy themselves and others by running generators frequently and for long time periods. But users must be willing to accept some drawbacks.

Important Note For Both Compression and Evaporative Air Conditioners: All rooftop air conditioners block roof vents, regardless of their type. Unobstructed roof vents are very important for ridding RVs of excess heat and exhaust from combustion from stoves, ovens, gas mantle lights and interior-discharging heaters. It is not a good idea to block all roof vents. All small RVs should have at least one unobstructed roof vent. Larger RVs should have more than one. Fortunately, additional roof vents aren't difficult to add to existing RVs. One only needs to cut a square hole in both the roof and ceiling between the roof rafters, add two short pieces of lumber perpendicular to the rafters, install the vent kit, then apply sealant. Vent kits are readily available and inexpensive.

III.) Ammonia Absorption Air Conditioners:

If you've got a reasonably good mechanical and logical mind, and have read and understood the recent section on RV refrigerators, by now you may be asking: "If an ammonia absorption refrigerator can work well in an RV without needing any electricity at all, is it possible to design a larger ammonia absorption air conditioner to cool an entire RV without needing any electricity at all?" In theory, yes. In practice, this would be extremely difficult and probably impossible. Ammonia absorption air conditioners are therefore not appropriate for RVs. Let's cover the reasons.

Problems: 1) The cooling unit piping would need to be very large. It would take up too much space in the RV. It would be so large that the entire RV layout [living space, furniture, appliances] would need to be designed around such a unit. 2) The large cooling unit would add too much weight to the RV. RVs should be kept as light as possible. 3) Ammonia absorption cooling units are extremely slow to get started. Recall that ammonia absorption refrigerators can take at least half a day after being tuned on to really get cooling. Compare this to compression refrigerators, which start frosting up only minutes after being turned on. Since an air conditioner would be much larger than a refrigerator, an RV ammonia absorption air conditioner would probably take even longer to really get cooling--- maybe several days. Users would need to anticipate the times when air conditioning would be needed and turn their units on several days in advance. It is impossible to accurately anticipate temperatures that far in advance. And: 4) Ammonia absorption cooling units work slowly but steadily. People want air conditioning quickly, heavily and in spurts. An ammonia absorption air conditioner would need to be left running continuously, something which people don't want. People want air conditioning during the hot times of the day, not during cold nights. People usually decide on a whim when to turn their air conditioners on. People want and expect cooling to start only minutes after turning their units on [which is the way the compression air conditioners work]. People have grown accustomed to the way compression air conditioners work and would reject ammonia absorption air conditioners in RVs.

Suitability in regular houses: Although ammonia absorption air conditioners are not appropriate for RVs, they may work well in regular houses/cabins that are too remote to be connected to municipal electricity. The first two above problems of large sizes and weights are not issues for permanent structures that don't move and are located on properties with lots of acreages. The third problem, of being extremely slow to get started, and the fourth problem, of cooling slowly but steadily, actually have a simple common solution. In such units, the air conditioner is left operating continuously. However, it does not cool the entire house. Instead, it cools only one specially designed very well insulated "cold storage room." Inside that room is stored a super-chilled liquid sealed in pipes. (Water is not used, since it would freeze. The anti-freeze liquid is kept at temperatures *below* the freezing point of water.) When the house gets too warm and air conditioning is needed, the liquid gets pumped outside the room through pipes and "inverse radiators" located in the rest of the house. This design is basically a hot water furnace working in reverse (removing heat from living areas rather than adding it). One difference is the preferred location of the pipes. In hot water furnaces, the pipes and radiators should be located low near the floor (or below it under metal grates) because hot air tends to rise upward. Since cold air tends to descend downward, the pipes and "inverse radiators" of ammonia absorption air conditioners should be located high near the ceiling (or above it, possibly above metal grates, for aesthetic purposes). The pipes should travel through long thin devices that look like baseboard heaters (except located on the walls high near the ceiling rather than low near the floor). The "inverse radiators" work exactly the way regular radiator do (in hot water furnaces), except that they *absorb* heat rather than *releasing* it. Some electricity is needed to operate the pump, but not that much. That electricity can be made from solar panels and effectively stored in batteries. When the house is cold or of moderate temperature and air conditioning is not needed nor wanted, the water pump is simply turned off and the coldness made by the air conditioner gets confined to the "cold storage room," where that coldness builds up and gets stored for later quick release during hot days. As long as the *average* cooling of the unit is less than the needs of the house, and the water pump works (and has electricity to do so), and there are no leaks in the system, the air conditioner works. Ammonia absorption air conditioners for regular stationary houses/buildings do exist. In fact, they have been proposed for use in third world countries where electrical grids don't exist. They can even use sunshine, focused by special parabolic and moving mirrors, to act as their heat sources, on sunny days. (On cloudy days and at night, propane or flammable liquid gets burned for their heat sources.)

Conclusion: Ammonia absorption air conditioners do exist. They could work well for off grid stationary houses that are fairly large. But they are probably not for RVs.

IV.) Solar Powered Fans:

Fans, powered by solar panels on roofs, could be well suited for RVs, but don't expect too much from them, because technically they are not air conditioners. They do not and cannot cool the inside RV air to temperatures *below* that of the outside air, all they can theoretically do is lower the indoor temperatures to those that *equal* that of the outside air (and even indoor/outdoor temperature equality isn't possible). Solar fans accomplish cooling by blowing indoor air that is unusually hot (from the sun shining directly on RV roofs and sides) to the outside. Removing air creates lower air pressure, which sucks cooler ambient temperature outdoor air through open windows to the inside of the RV to replace it.

Advantages: They are: 1) simple, [therefore] 2) inexpensive, 3) small, 4) able to work for rather long timespans [a small fan, that operates for most of the day, does the same cooling as a large fan that works for a short timespan], 5) reliable and 6) long lasting [I've read that such a setup can last a couple of decades]. 7) They can cool RVs parked in outdoor storage where no electricity is available. This is a real plus, and is probably their biggest advantage, because RVs in storage are often neglected. Excessive heat in RVs, unoccupied for long time periods, can cause plastics and rubbers inside those RVs to decompose prematurely, making them become brittle and start to crack. Removing some of this heat, at least by opening windows, is a good idea. 8) They can prevent RVs without working air conditioners from becoming excessively hot. And: 9) They tend to work best when needed, and fail when not needed. Heat is most likely to build inside an RV when the sun shines directly upon it. This is the time when solar panels will work to make electricity to run the fan. On cloudy, rainy or snowy days, or at night, there won't be sunshine to make electricity, so the fan won't work. But this doesn't matter, because when the sun isn't shining, the indoor RV air isn't likely to become hotter than the outdoor air, so the fan and interior cooling won't be needed.

Disadvantage: They are not air conditioners. Don't expect them to remove lots of heat. They shouldn't be used as substitutes for air conditioners.

V.) Thermoelectric Air Conditioners:

It is theoretically possible to cool a large indoor environment by sending DC electric current through thermopiles. On a practical lever however, doing so would require so much more electricity than batteries can store, so this notion isn't worth contemplating for RVs.

Passive Air Conditioning

> "An ounce of prevention is worth a pound of cure."
>
> ---old piece of wisdom

If an active air conditioner were a part of speech, it would be a *noun*, because it is an actual physical object. If passive air conditioning were a part of speech, it would most likely be an *adjective*, because it describes the properties of some other existing object, specifically that object's ability to block or remove heat.

Passive Air Conditioning Variables

I.) Reflection:

When light hits any object, two things can happen: 1) it can be reflected away from the object or 2) it can be absorbed. Light is a vehicle that contains and carries energy. *Reflected* light carries away all of its energy, therefore does not heat the object it strikes at all. *Absorbed* light transfers all of its energy into the object and heats it. *Incident* light refers to the total amount of light striking an object. In most instances, both reflection and absorption take place. *Reflectivity* refers to the ratio of reflected/incident light. A reflectivity of 1 describes an object which reflects all incident light; 0 describes a totally black object which absorbs all incident light. This ratio determines the extent to which incident light heats an object.

What does sunshine consist of? In addition to the light that we humans can see (referred to as "visible light"), there are invisible forms of radiation that we can't see, called infrared radiation (heat rays) and ultraviolet radiation. These also carry energy, which can be reflected away or absorbed. Sunlight is a mixture of invisible infrared radiation (heat rays), visible light and invisible ultraviolet radiation.

How does an RV's reflectivity effect it? The skin of an RV (roof and sides) has a substantial surface area. When direct sunshine hits the skin of an RV, if too of that sunshine gets absorbed, it can turn that RV into a "hot box." If the goal is to keep the RV interior as cool as possible, it is best that the RV's skin be as reflective as possible. Indirect light and heat radiation, coming from nearby objects, also has a warming effect when it hits an RV, but there is much less indirect light around than direct sunshine, so the warming effect of indirect light is negligible. However all light and invisible radiation reflected away from an RV carries away its energy, so it is always best for an RV to be as reflective as possible.

What is the mechanism by which sunlight heats an RV? Once the skin is heated by absorbed sunlight, some of that heat energy gets reradiated to the outside of the RV (fortunately), but much of it transfers from atom to atom and eventually works its way through the skin, the insulation and the interior ceiling and walls. There it either gets reradiated inside the RV as infrared radiation or it gets transferred to the interior air molecules. Either way, it can make the place HOT. The ceiling and walls can feel like a radiator. (You can actually feel the heat rays emanating from the walls and ceiling of a poorly designed RV parked in the sun.) When air molecules hit a warm ceiling and walls, their motion increases. Increased molecule motion is defined as an increase in temperature.

> Q: What is the largest organ of the human body?
> A: The skin.
>
> Q: What is the largest "organ" of an RV?
> A: The skin.
>
> An RV's skin is of utmost importance. It provides a barrier of temperature, keeps out rain, wind, insects, animal and even human intruders, and provides privacy.

Which sides of an RV receive the most sunlight? There are six sides to every RV: four vertical sides (including the front and the back), a roof and a bottom. The roof gets the most direct sunshine. Of the four vertical sides, the one that gets the most sunshine is the side which faces South (in the Northern Hemisphere). The two sides which face East and West each receive about half that amount. The side that faces North receives almost no direct sunshine at all. From the viewpoint of RV design, all vertical sides should be equally reflective, because the directions they face while parked are likely to be random and equally divided (although users can sometimes choose how to orient their RV when they park). The bottom of the RV receives no sunshine at all. Normally the bottom transmits no heat into the RV because the ground beneath it is cooler than everything else around during the day, due to the fact that it is in shadow from the sun by the RV itself. The bottom can however receive and transmit infrared radiation into the RV from sun heated asphalt shortly after being parked upon it.

What materials are best at reflecting light? Shiny metals with mirror-like surfaces are the best at reflecting light. Aluminum is excellent. So too is stainless steel. Rubber, plastic and fiberglass can absorb quite a bit of light. Color is important for non-metallic materials. White material reflects light well. Moderately colored material does not. The worst offender for absorbing heat is black material. Note that many RV surfaces are painted. Since light does not penetrate surfaces very deeply, the paint may determine the reflectivity. (Metal coated with paint may not reflect light/heat away as well as plastic/rubber coated with metallic paint.). Note also that not all "light" is visible to the human eye. Two materials or paints, that may look identical to our eyes, may absorb heat from sunshine differently, because one may be "white" in the infrared or ultraviolet, whereas the other may be "black" in the infrared or ultraviolet.

What materials are used to make the roof and sides of RVs? Very often, RV roofs are made from white rubber. This is unfortunate because these rubber roofs become notoriously hot when exposed to prolonged direct sunlight, even though they are white. RV sides are often made from plastic or fiberglass. While these materials have some nice properties (fiberglass can often absorb impacts without permanently bending or breaking), they do absorb quite a bit of sunshine and can really heat RVs up. Older RVs almost always had painted sheet aluminum for their siding. This works very well for reflecting away the heat from sunshine. The roofs of some older RVs were made from galvanized steel. Top-of-the-line RVs are often made from highly reflective almost mirror-like sheet metal. This is the best for reflecting away sunshine.

What materials should be used to make the roofs and sides of RVs? I believe that aluminum is one of the best choices. Since the roofs get the most sunshine, I believe that all RV roofs should be either unpainted aluminum or shiny stainless steel. Rubber roofs (which are almost always used) probably should not be used, unless they are painted and it can be demonstrated that the paint adequately reflects light to prevent overheating, or shiny sheet metal is glued onto those rubber roofs so the direct sunshine doesn't hit the rubber. (Bear in mind that paint can peel or abrade away. This would require repainting, else the roof's reflectivity would diminish over time.) Painted aluminum siding is good. (This is what I have on my trailer's sides. My trailer withstands direct sunshine without overheating quite well.) The colors of the paints should be white or light colors. Dark colors shouldn't be used for large surface areas, since dark colors absorb light, however some small dark patches and stripes are alright as stylish accents, as long as they aren't overdone. Also, colored paint absorbs some frequencies of *light*. Since we humans cannot see infrared and ultraviolet radiation, all paints and exposed non-metallic siding/roof should be "white" in the infrared and ultraviolet (as reflective of it as possible). While bare aluminum for roofs and painted aluminum for sides are really good choices, unpainted mirror-like sheet metal is the best for reflecting away sunshine. If an RV skin and sides were all made from perfect mirrors (and the windows too were shaded by perfect mirrors), then that RV would never get any hotter than the air around it, no matter how long it sat in the direct sunshine.

photo: Taping aluminum foil over windows reflects away all of the light and heat. Louver windows such as these can still be opened when "foiled."

Aluminum Foil:
The Perfect Material --- The Perfect Mirror

While still a child, years before my RVing days, I remember reading a warning on a box containing a roll of aluminum foil: "Caution! Do not put aluminum foil too close to the heating element of your oven. Aluminum foil reflects heat so well that it could cause overheating of the oven's heating element. This could damage your oven." I don't know why my brain chose to remember this tidbit of information, but it did.

Years later…

I discovered just how well aluminum foil reflects light and heat one hot sunny day in the desert. To keep my trailer as cool as possible, I decided to temporarily tape aluminum foil over the glass on my open louver windows, to prevent the sunshine from entering my trailer and heating it. It worked--- no sunshine entered my trailer. I then noticed something amazing about the foil and the windows: Even though the direct unfiltered desert sun was hitting the foil constantly, neither the foil nor the glass beneath it warmed even slightly! It must have reflected away nearly 100% of the light and heat. If you "foiled" your entire plastic/fiberglass/rubber RV, it wouldn't warm in the sun either! Unfortunately, your RV would look like garbage.

Should RVs be as reflective as possible? Since RVs are used under a wide variety of conditions, this is a valid question. Sometimes RVs get used in cold weather. When parked in sunshine in cold weather, any heat absorbed and dumped into an RV is likely to be welcome. However, in hot weather, any extra heat absorbed from the sun and dumped into an RV will be unwelcome. Because it is much easier to artificially add heat an RV on a cold day than it is to artificially cool an RV on a hot day, my answer is: *Yes, RVs should be as reflective as possible.*

II.) Insulation:

Insulation is material that retards the flow of heat. In an RV, the primarily locations of insulation are between its outer roof and ceiling, between the outer sides and interior walls, and between the RV bottom and its floor. Insulation should also be located around interior mounted heat producing appliances (ovens, hot water tanks).

Is good insulation beneficial for RVs? In most instances it is. When outdoor temperatures are cold, furnaces don't need to work as hard in well insulated RVs as they do in poorly insulated ones. When outdoor temperatures are hot, air conditioners don't need to work as hard in well insulated RVs as they do in poorly insulated ones. Also, when outdoor temperatures climb to uncomfortable levels through midday, well insulated RVs retain their cooler comfortable morning temperatures longer than poorly insulated ones (especially if the windows are closed), cutting back on or delaying the need for air conditioning. Good insulation can be a drawback only in moderately warm temperatures if heat producing appliances (stoves, ovens, gas mantle lights) are used indoors, because it helps trap heat inside. However this is not a critical problem--- opening windows, roof vents and doors can usually rid most of the unwanted heat quite well. Overall, the benefits of good insulation far outweigh any possible benefits of poor insulation.

How are values of insulation measured? There are three values that describe the abilities of materials or samples thereof to impede the flow of heat: 1) thermal resistance, 2) resistivity (also called *R value per unit thickness* and *specific thermal resistance*) and 3) R values. Let's cover each. 1) Thermal resistance describes a specific *sample* of material, taking into account its width and depth. Thermal resistance is directly proportional to a sample's thickness (measured in the same direction as heat flows) and inversely proportional to its surface area (measured perpendicular to the direction that heat flows). This means: A) Doubling the thickness of the insulation doubles its thermal resistance. B) Doubling the surface area of the wall (or other object) cuts its thermal resistance in half. 2) Resistivity (also *R value per unit thickness*) is a value that describes the *material only*, thus is *independent of a sample's thickness and surface area.* (The term *conductivity* is used to describe its reciprocal--- a material's ability to *conduct* heat). The formula that relates these two values is:

(Resistivity) = (Thermal Resistance) × (surface area) / (thickness).

3) R value is independent of the surface area, but does take into account the thickness. The formulas that relate R values to the other two quantities are:

(R Value) = (Thermal Resistance) × (surface area)
and
(R Value) = (Resistivity) × (thickness).

The formula that relates heat flow to the temperature difference between the two sides of a wall/object and the R value is:

$$\Delta T = \dot{Q}_A \times (\text{R value})$$

where

ΔT describes the temperature difference between the two sides of the insulation

(measured in degrees Kelvin or Celsius: ^{o}K or ^{o}C)

$\dot{Q}_A$ describes the flow of heat divided by the surface area of the wall

(measured in Watts per square meter: $\frac{\text{W}}{\text{m}^2}$)

and the R value is measured in Kelvin meter squared per Watt: $\frac{\text{K m}^2}{\text{W}}$.

Note #1: Above formula is in metric units. R values in the U.S.A. can be obtained by multiplying the metric R values by 5.678. Note #2: This heat flow formula is very similar to Ohm's law for electricity: (Voltage) = (Current) × (Resistance).

Which values are most important for designing/evaluating/describing RVs? The R value is perhaps the most important quantity when insulating buildings/RVs, because thickness matters but surface areas do not. (If you stand in the shadow of a sunlit wall that has two layers of identical insulation instead of one, you feel noticeably cooler. The two-layer wall has twice the R value as the one-layer wall. If you stand in the shadow of a 100 square foot sunlit wall, you feel exactly the same as you would standing behind a 50 square foot wall, having identical insulation of the same thickness. Both the 100 square foot and the 50 square foot walls have the same R values.) Resistivity (also *R value per unit thickness*) is the quantity used when evaluating the insulation material alone. Using thicker layers of any insulating material always increases the final R value, so it is theoretically possible to attain any R value using any type of insulation. But beware: The amount of space available for insulation (in this case, between an RV outer skin and inner wall/ceiling/floor) is always limited, thus so too are the theoretical attainable R values for any particular type of insulation.

What types of insulation are possible and available for RVs? RVs can (and do) use the same insulation types that get used in regular stationary houses and other buildings. The types are: 1) rolls and batts [batts are flat pieces precut to specific lengths], 2) loose fill, 3) rigid foam sheets, 4) foam in place and 5) vacuum panels. Rolls/batts, rigid foam sheets and vacuum panels are installed during construction. They may not be applied to existing RVs later, unless a wall/ceiling/floor or a large piece of the outer siding/roof gets removed. Loose fill is blown into place. Foam in place, as its name suggests, is applied as a gooey liquid at its final destination, where it solidifies.

Detail on Major Insulation Choices:

1) Fiberglass rolls/batts: This is probably the most common and most familiar of all the insulation types. It's cheap, it does the job well and it can be easily and quickly packed to fill any and all voids. (Liquids are harder to work with because they can ooze in uncontrolled directions. Solid boards need to be cut to fit unusual spaces--- this takes time.) Here is an interesting tidbit of information about all fiber insulation: Solids (like fiberglass) generally have lower resistivity than gasses (like air) do (meaning solids conduct heat better than gasses do). If this is true, why use any insulation at all? Wouldn't just plain air be the best insulation? No, because air moves around and carries heat with it when it does. Fibers greatly prevent the motion of air between them. This effectively creates hundreds (maybe thousands) of individual stagnant air pockets. Heat must migrate from one air pocket to the next, all the way through the fiber matrix, in order to fully penetrate the insulation layer. Yes, it is not the *fiberglass itself*, it is actually the *empty space* between the fibers that does most of the insulating! What does this mean? Compressing fiberglass gets rid of some of that valuable air space, thus actually lowers a sample's R value. However, compressing does allow more insulation to be packed into a space of limited thickness, so some compressing and stuffing is still a good idea because it can raise the overall R value of a wall/ceiling/floor. (In other words, compressing fiberglass to half its width, and stuffing in twice as much *does* raise the overall R value but *does not* double it.) If solid panel insulation is used, fiberglass still can and should be used to fill voids.

2) Loose fill might be alright for stationary buildings but probably shouldn't be used for RVs, because during travel, substantial wind speeds develop. If there are any air leaks between the wall space where the insulation is located and the outside world and RV interior (a probable situation), those winds could blow insulation material into RV interiors. There, if visible, it could be messy, or if microscopic, it could be a health hazard, make people feel uncomfortable and make them cough/sneeze (as the place would get dusty).

3) Rigid foam sheets: These are easy to cut. Polystyrene (brand name: Styrofoam) and polyurethane have excellent resistivity--- about twice that of fiberglass, so rigid foam sheets are a great choice for RVs in construction. Styrofoam can gradually lose some of its R value over time, because the actual insulating material is chlorofluoro-carbon gas, which is trapped within tiny bubbles called closed cells. These cells can gradually leak out the gas and become filled with regular air, which doesn't insulate as well. However the overall performance of this sheeting still remains very good for long time periods.

4) Foam applied in place starts off as a frothy liquid, then solidifies. It has a substantially lower resistivity than rigid foam panels (about half). The resistivity of foam in place is about the same as fiberglass, so it would seem to offer no clear advantage over fiberglass. However, foam in place could be a good choice because it is likely to infiltrate small nooks and crannies, preventing air circulation, which is a major inhibitor to the effect of insulation. Foam in place has another great advantage: it can be applied to poorly insulated existing RVs after construction (through holes drilled in the walls and ceilings), without necessitating removal of walls/ceilings/floors or the outer skin, possibly sparing those RVs from becoming "hot boxes" in summer and "cold boxes" in winter.

5) Vacuum panels: Perhaps the best material for insulation isn't any specific material, it's actually *no material at all*, in other words, a vacuum. Thermos bottles use vacuums, and wow, they work so well for keeping coffee hot or ice cold, that they almost seem to defy common sense. A vacuum insulation panel is basically are two adjoined rigid sheets, forming an air-tight cavity, in which the air within that cavity gets removed. Vacuum panels have superior insulating ability--- their resistivities are 4 to 6 times as great as the best of the other conventional insulation types (which are rigid foam boards). Vacuum panels would seem to be the answer to heat infiltration of RVs with rubber roofs and fiberglass/plastic siding. However they do have several problems: 1) Air can eventually leak into the cavity of an insulation board, rendering its superior insulation properties useless. At that point, at best the panel becomes no better than fiberglass of equal size, and at worst may become almost like no insulation at all. 2) They are very expensive [$3 to $5 per square foot]. 3) They cannot be cut, since cutting one would destroy the vacuum within, so some pieces must be made-to-order. This further drives up prices.

What type of insulation is best? For RVs with metal highly reflective skin, I think that rigid foam panels would be the overall best first choice, since they have about twice the resistivity as fiberglass rolls/batts, and the space within RV walls/ceilings/floors to put insulation is limited. But for RVs with unreflective rubber roofs and fiberglass or plastic siding, vacuum panel insulation might be the only way to prevent those RVs from becoming unbearable "hot boxes" in the sun. But vacuum panels are expensive, and they would need to be manufactured well enough to retain their vacuums over time. Any spaces and cracks not filled should be either tightly packed with fiberglass or foam in place to prevent air currents.

How much insulation should be used? I don't know enough about insulation to give specific recommendations on R values. I do however propose a very general test: Stand in an RV on a hot day when the sun is shining on it. If you can feel an abnormal amount of heat radiating from the walls and ceiling, then either the RV skin reflectivity or the R value of its insulation is too low (or both). Some heat will radiate from the ceiling and walls, but remember that when you are inside, you are in the shade--- it should feel like the shade.

Other considerations: 1) The entire space between the inner walls/ceiling/floor and outer sides/roof/bottom should be filled with some kind of insulation. Fiberglass is cheap. Empty space serves no purpose. 2) Insulation should not be changed/degraded or become matted by exposure to water. Sadly, RVs sometimes leak. They aren't supposed to, but life isn't perfect. [Fiberglass and foam are waterproof. This recommendation applies to some other insulation types.] 3) A thicker roof might be a good piece of engineering. Thicker walls/roofs have more space between their outer and inner facing surfaces to stuff insulation, hence can have substantially greater R values. Having thicker walls probably isn't a good idea, because RVs can only be a maximum width to legally drive on the roads, so if the walls are made thicker, the expansion must be inward, intruding on their already small interior living spaces. However making a roof thicker would cause outward and upward expansion (the interior ceiling height could be the same) and only make an RV a couple inches taller. This would be unlikely to make driving noticeably more difficult and wouldn't noticeably affect gas mileage, yet the roof's R value could be greatly increased.

How do reflection and insulation work together to keep an RV cool? Ideally, both high outer surface reflectivity and good insulation (with high R values) should be used to passively keep an RV cool. If the surface reflectivity is near 1, then good insulation isn't so important (but insulation is important for other reasons already discussed). If the reflectivity is poor, it may still be possible for good insulation to spare an RV from becoming a "hot box," however I haven't run any tests, so I don't know whether or not RVs with rubber roofs and plastic/fiberglass sides can be made to adequately withstand being parked in direct sunshine on hot days.

Problem with RV insulation: The insulation in an RV is located where it can't be seen. Unless a prospective RV buyer has the good sense to dig through and understand those technical R value specifications in the owner's manual, or actually enters the RV when parked in the sunshine on a hot day when the air conditioner hasn't been used, that buyer probably won't know whether or not that RV properly blocks heat until after buying and using it, at which point, it's too late. Sadly, RVs are sometimes poorly insulated. Insulation can't be seen, therefore skimping on it is a great way for RV designer/manufacturers to cut corners in order to boost profits.

III.) Ventilation:

It's one of the oldest tricks in the book: If indoor temperatures are too warm, and the outdoor air is cooler, open the windows to let the warm air out and the cooler air in. This procedure works because air stores heat. Thus, if air is allowed to escape, any heat stored within it also escapes.

Ventilation in RVs: RVs are not like regular houses. RVs have small air volumes, with windows located close to all interiors points. Windows are plentiful and often located on all four sides. Thus, RVs ventilate very well and quickly. (Often, only few minutes of everything-open ventilation are needed to radically change indoor temperatures or remove nasty odors--- even in windless conditions.) Most windows on an RV should be capable of opening, in order to ventilate it well.

Screens: Most windows and all roof vents should have screens. Also a full-screen-door is a must. Many areas have mosquitoes, black flies, and other large non-biting insects [like moths], that are likely enter as unwanted guests, without screens. Unfortunately, tiny insects can fit through screens. "No-see-'ems" are tiny bugs that bite and leave rather large swelling/itching areas. Other tiny non-biting insects fit through the screens and are attracted to artificial interior lights at night. They can get behind electric light lenses and die in large quantities in RV lights. They can also die by flying into hot gas mantle lights, at which point they fall onto the table/couch, or even into food. This gets disgusting. Hence, on some nights, RV users might need to choose between having the lights on and the windows closed or having the lights off and the windows open. Special "no-see-'em screen" is available. It is often used in the construction of tents. It is readily available at camping stores that carry gear for people who camp in tents. But its holes are so tiny that ventilation through it is limited. I wouldn't recommend having "no-see-'em screen" on every window all the time, but some RVers might want to have it on one or two windows, or even have a second set of "no-see-'em screens" that they could exchange with the regular screens, to allow some ventilation in heavily infested areas.

Roof vents: Roof vents are especially good for removing hot air near the ceiling from stoves/ovens/ gas mantle lights, and also from sunshine on roofs on hot days. They are important to have because hot air tends to accumulate near ceilings. Removing hot air from cooking breakfast indoors early in the day reduces the need for air conditioning later in the day. *Recommendations:* 1) Do not block all roof vents with air conditioners. If all are blocked, cut and install additional unobstructed roof vents. 2) Roof vents need to be closed when it rains! Else when it rains outside, it also rains in your RV. [It is permissible though to leave them open just a crack.] Do not leave the roof vents open when you vacate your RV for any time length, like when you shop or go sightseeing. 3) Roof vents need to be closed during travel. The high winds from traveling through the air at high speeds can break a roof vent right off or crack its plastic. 4) Plastic roof vents need to be replaced occasionally. Plastic deteriorates with time and becomes brittle and can shatter. Plastic roof vent lids can break just with age but are easily replaceable. 5) Do not have/use roof vent covers. They are absolutely useless. A roof vent cover is a rear-vented enclosure over an entire roof vent. It allows the enclosed lid to remain open in rain or during travel [the enclosed vent can be opened completely--- the cover doesn't obstruct lid extension], yet theoretically still allows air flow. In practice though, roof vent covers reduce air flow to such an extent, that having them over roof vents is almost like having no roof vents at all (although they still let in light). Also, since the covers are plastic [sometimes clear plastic], heat from sunshine concentrates under them, making any air that works its way down into the RV on sunny days actually *warmer* than the interior RV air. My mother's RV had two covered roof vents. I removed both covers. Discipline yourself to close those roof vents when you vacate or store your RV, and you won't need to contend with rain indoors.

left: roof vent cover over roof vent
right: opened roof vent with no cover

Accidental ventilation: Windows/ roof vents/ doors can be opened and closed at will. However, closing windows does not seal them completely (especially louver windows), so some air leakage is inevitable. Air leakage is not a problem when evaporative air conditioners are used, since some open-window ventilation is necessary. Air leakage is not much of a problem when compression air conditioners are used either, because these air conditioners are powerful and often used for just a limited number of hours during afternoons, often at times when outdoor temperatures do not drastically differ from indoor temperatures. Air leakage is however a problem when heating an RV in very cold weather, especially when windy, because air leakage can make an RV drafty.

IV.) The Greenhouse Effect:

The greenhouse effect is a process that heats an enclosure up. It does so by trapping a lot of extra heat within an enclosure that seemingly doesn't belong. (The process is also sometimes called *greenhouse heating*.) Since the topic on hand is keeping cool, understanding and reducing its occurrence is important. There are three ingredients needed for greenhouse heating to occur: 1) an enclosure [either sealed or with limited ventilation], 2) a strong light source [usually sunshine] and 3) a selective filter [usually glass]. The greenhouse effect gets its name from greenhouses, which are buildings made for growing plants, constructed almost entirely of glass. Greenhouses intentionally use this effect to heat them.

What kinds of things are effected by the greenhouse effect? Cars, vans, RVs, greenhouses and entire planets (Earth and Venus).

> Have you ever wondered why the inside of a car parked in the sun with the windows shut gets so much warmer than the outside of that car? There seems to be a whole lot of extra heat in that car that doesn't belong. This extra heat is caused by the greenhouse effect.

How does the greenhouse effect work? Consider a common object greatly affected by the greenhouse effect: a car. If a car had no windows and were parked it in the sun, the sunshine entering and striking its interior would cause *solar* heating but not *greenhouse* heating. Now, let's consider a car with windows, also parked in the sun. The window glass is a selective filter--- it lets energy from the sun in easily but allows that energy to escape only with great difficulty. The energy flow out always tries to reach equilibrium with the energy flow in (meaning the energy escaping tries to equal the energy entering), and since it is more difficult for the energy to escape, a whole lot of extra energy builds up inside, before that equilibrium can be reached. That extra energy takes the form of heat and rising temperatures. (Analogy: Hang a piece of flypaper in a barn. Come back later and you are likely to see a few flies on the walls but many flies on the flypaper. Why is the density of flies so much greater on the flypaper than on the walls, when the flies probably land everywhere equally? Because it's easy for the flies to land on the flypaper but very difficult for them to leave it. On the wall, it is just as easy for the flies to leave as it is for them to land.) Now let's address another question: Why does the glass admit energy in easily but make it difficult for that energy to leave? Glass is almost completely transparent to light. When light rays enter the car, they pass through the glass almost as if it weren't there. Once inside, some of the light rays reflect back and can exit through the windows just as easily as they entered. But some of the light rays get absorbed by the upholstery, where their energy heats the material. Soon, that energy gets reemitted as far infrared radiation (heat rays), some of which heads back toward the glass. Glass is transparent to light but is *not* transparent to far infrared radiation. In fact, if you could trade in your light-seeing eyes for far-infrared-seeing eyes for just one day, and looked at that same car, the glass would not look clear, it would look deeply dark, maybe like the brown glass of some beer bottles. When an infrared ray heads toward the glass, instead of passing through it, it often gets absorbed. There, its energy heats the glass up, after which it is likely to be reemitted right back into the car. Eventually, the absorbed energy from the infrared ray might work its way through the glass and be reemitted to the outside, or the infrared ray might pass through the "dark" glass to the outside without being absorbed--- but only after several attempts. Either way, the energy contained within the infrared rays escapes the car with difficulty, even though that same energy entered the car with ease as light rays.

What factors determine the extent of the greenhouse effect in vehicles? 1) <u>The number of windows</u>. The greater the surface percentage are windows, the greater the greenhouse effect. 2) <u>The interior volume</u>. Larger containers need to absorb more energy in order for equal temperature rises to occur. Therefore, larger vehicles are less effected by the greenhouse effect than smaller vehicles. 3) <u>Ventilation</u>. Besides trapping heat, windows also trap hot air within their enclosures. If windows are open, the greenhouse effect is equal, but some of that extra greenhouse heat is allowed to escape. 4) <u>The color of the interior materials</u>. Darker materials absorb light more efficiently than lighter materials. Therefore, vehicles with light colored interiors get less effected than vehicles with dark colored interiors. 5) Sunshine. Since sunshine is the energy source that drives the greenhouse effect, there is very little greenhouse heating if a vehicle is in the shade (usually not enough to even be noticed). Cloudy days can however still subject a vehicle in the open sky to enough greenhouse heating to be noticed, just not to the extent that sunny days can. 6) The glass filter (both its chemical composition and its thickness). Greenhouse filters cause the most heating when they have a combination of being transparent to light but opaque to infrared rays. They don't work if they are transparent to both light and infrared rays or opaque to both light and infrared rays.

The most-to-least order in which vehicles are effected: 1) Cars are most effected, since their interiors are completely surrounded by large windows and they have small interior air volumes. 2) Vans have less percentages of windows to their surface areas and also larger interior volumes. Hence, they are substantially less effected. [I can often leave my dog in my conversion van with the windows open on sunny days---especially with a good wind. It is actually quite comfortable in there. I wouldn't dream of doing the same thing with a regular car.] 3) Motorhomes have still less percentages of windows and even larger interior volumes than vans. Hence, they are still less effected. But they still have big windows on the front so their drivers can see what's around them. 4) Camping trailers tend to have very few windows. They don't need many windows, because when on the road, drivers of their tow vehicles see around them not through them. Hence, camping trailers are the least effected of all vehicles with windows. 5) Vehicles without windows are not effected at all.

What can be done to counter the greenhouse effect in RVs? 1) [Best plan] Cover the windows exposed to direct sun on hot days. Covering can be accomplished either with awnings that hang above and near the windows [in positions where they block sunlight from entering the RV] or directly on the surface of the window glass. A) Covers should be metallic, not plastic. Metal reflects light; plastic or vinyl often just absorbs light then reemits it as infrared radiation. [If you position yourself in the shade of a black plastic umbrella on a hot day, you get so bombarded by scattered invisible infrared radiation, that you feel like you are still in the sun.] Aluminum foil works very well as a temporary window surface cover. B) Covers should be on the outside of an RV, not the inside. If on the inside, and not 100% reflective, any absorbed light becomes heat and has great difficulty escaping. Furthermore, some infrared rays do make their way through the glass, even though it's difficult for them to do so, and once on the inside, if they are reflected back toward the glass by an interior cover, they have great difficulty escaping. Exterior covers allow all absorbed heat to easily dissipate to the outside, either as infrared radiation or as heat transferred to the outside air. 2) Alter the chemistry of the glass. I don't know how possible this is, but if some type of glass were to be more transparent to infrared radiation, the greenhouse effect would be less prominent. [However, using a glass type more transparent to infrared radiation would cause valuable furnace heat to be lost through those windows on cold nights, so this might not be worth the tradeoff.] Tinting windows [to light] may also reduce the greenhouse effect. 3) Have less or smaller windows. This is a viable option since huge windows aren't needed in RVs and usually aren't desired because some privacy is wanted. However, windows do allow natural daylight to enter an RV, which allows occupants to see better and provides psychological uplifting by brightening the place up [especially when parked in dark forests]. Also, having too few windows that can open limits ventilation, and thus stifles air cooling, so this option shouldn't be taken to an extreme.

Recommendations for window awnings: If an RV is probably going to be used frequently in hot areas: 1) Have an awning for every window. Each awning could be small. 2) Four positions should be possible for each window awning: A) *Retracted* [rolled up above the window], for cooler days when sun blocking is not needed or when the RV travels on the road. Rolled up awnings would be inconspicuous. B) *Extended horizontally* [with a slight taper downward--- maybe about 10°]. This position would block sunshine from high in the sky but might still admit it when the sun is at low angles. The slight downward taper would allow rain to run off the awning. C) *Extended at a 45° downward angle.* This position would block sunlight from lower angles or from the side better than when extended horizontally (B). This position would still allow the windows [even louver windows] to be opened to air-cool the RV. D) *Vertical hanging downward over the window.* This position would allow complete blocking of the sun but would not allow louver windows to be open and would only allow limited ventilation from windows that open by sliding sideways. 3) For motorhomes, use RV skirts that cover the front [driving] window and side passenger door windows. These already exist and are readily available. They are intended mostly for privacy but could also block sunshine. They need to have highly reflective metallic particles in them though, if made from flexible plastic, in order for them to block heat and light properly.

The good news: Because RV windows tend to cover only small percentages of RV surfaces, the greenhouse effect usually isn't terribly significant for them. The real culprits that turn RVs into uninhabitable "hot boxes" are usually poor surface reflectivity and poor insulation.

Entire Planets Are Effected by The Greenhouse Effect

One of the components of the atmosphere of Earth is a small amount of naturally occurring carbon dioxide, which is a greenhouse gas. Sunlight from space passes through this carbon dioxide easily, but infrared radiation from absorbed sunlight is somewhat blocked from leaving Earth by this carbon dioxide. A certain amount of this carbon-dioxide-caused greenhouse heating is good for Earth, because without it, Earth's average temperature would only be about the freezing point of water. Climatologists worry though that extra unabsorbed carbon dioxide in the atmosphere from burning fossil fuels could raise Earth's temperature a few degrees and wreak havoc.

The atmosphere of the planet Venus is made almost entirely from carbon dioxide. This atmosphere is very dense (about 92 times that of Earth's). Consequently, the surface temperature of Venus is usually about a sweltering 870° F, which is as hot as the inside of a toaster oven. I like traveling, but I don't think I'm going to be visiting Venus anytime soon.

V.) <u>Heat Storage Within Objects</u>:

All objects are like containers that store heat. (This includes samples of liquid and gas.) The capacity for heat storage within an object or given sample of material is called its *thermal mass* (also called *heat capacity*). Objects also appear to store coldness as well, but this is an illusion--- "coldness" is a relative term--- it means that the object or material has less heat stored than the objects or materials around it. Stored heat/coldness can be retained for quite some time--- it can take quite awhile for that stored heat or coldness to diffuse into the environment around it.

> When Grandma makes a big pot of soup on the stove, you might think the pot stores only soup. But after she cooks it, it must remain out for a long time before going into the fridge, because that container of soup also stores heat, which takes time to dissipate. Also, a solid chunk of soup fresh out of Grandma's freezer stores coldness, which also takes time to dissipate.

What does increasing the thermal mass of a dwelling do? It doesn't change its average temperature but it does keep temperatures closer to that average, preventing those temperatures from going to extremes.

Using thermal mass to advantage in houses: Some very well designed regular houses make good use of thermal mass by placing chunks of concrete or rock inside them where the sun hits during the daytime in winter. (Concrete and rock chunks have very high thermal masses.) The sun warms the concrete/rock during the day, then that stored heat gets released into the house at night. The concrete/rock chunks have another purpose: in the summertime, they store the cooler temperatures of night, then keep the house cooler during the day.

Using thermal mass to advantage in RVs: At first thought, it might seem like a good idea to increase the thermal mass of an RV interior to keep it cool during the day and warm at night. However, the only way to do this would be to add weight somewhere inside. Adding useless weight to an RV is of course unthinkable, because RVs need to be light and interior space is precious. However bear this in mind: all interior RV objects, even though they have other purposes, also have thermal mass (thus store heat/coldness). This means the walls, the table and counters, the pots and pans, the cloths, those unopened cans of soup, the toolbox, and all other objects in the cupboards/closets have thermal mass and thus store heat/coldness. Because RVs are small, storage compartments tend to be well packed, so thermal mass relative to the interior space tends to be substantial. If you don't have an air conditioner (or you have one but would rather not use it), here is how to use the interior thermal mass to keep an RV cool when nights are cool but days are hot: 1) If temperatures don't get too cold, don't use the furnace at night. 2) In the cold of morning, open all the windows for awhile, to get the interior as cool as possible. 3) Don't cook breakfast indoors, instead cook outside. Also don't use gas mantle lights in the morning. 4) When climbing outdoor temperatures exceed the indoor temperature, close all the windows, the roof vents and the door, to store that coldness inside. 5) Don't open the windows until the indoor temperature catches up to the rising outdoor temperature. Often, this equalization does not occur until about noon or 1 PM. Note #1: Sometimes I have closed the windows of my trailer early in the morning, then driven it to a hot sunny parking lot. Even though the outdoors are sweltering, the interior is surprisingly cool and comfortable. Note #2: This procedure works well for highly reflective well insulated RVs (such as my aluminum trailer). It is not likely to work for heat absorbing rubber/fiberglass/plastic poorly insulated RVs.

The one acceptable way to increase RV interior thermal mass: There is something that you may be able to add to an RV to increase its thermal mass: *water*. Water has a very high *specific heat* (which is the capability to store heat divided by its mass), so if you fill your freshwater tank, you add significant thermal mass. If your filled freshwater tank is centrally located, the water can really retain the coolness of night and keep the RV cool for quite awhile during the day. By nightfall, the water has absorbed daytime heat, and is warmer than the outdoor air, so it can often keep your RV cozy without furnace use well into the night (maybe the entire night). Of course, freshwater takes up zero effective extra space, since it is stored inside a tank specifically made for it, and that tank occupies the same space whether empty or full. Filling the tank does add weight, however an RV is designed to carry that extra weight. (A full tank has a second advantage: an RV sways less and is less affected by wind when the tank is full, so road travel is a bit safer--- yes, I actually do notice a difference when driving with my ten gallon tank full as opposed to empty.) Note: This filling-the-tank trick works well in my trailer because my gravity-feed tank is located near the ceiling and quite centrally, and my trailer is rather small (12 feet interior length). A tank located on the floor near a wall might not have quite such a noticeable effect, especially if the RV is big and especially if there is a barrier of thermal insulation between the tank and the RV interior. However, a bigger RV is likely to have a bigger water tank, which translates into a greater thermal mass when that water tank is full.

VI.) The Chimney Effect:

Consider the structure of the walls of RVs and other buildings: vertical wooden beams encased by an outer side and an inner wall. Inside might be insulation or just plain air. Suppose: 1) only air is present in the empty space, 2) the empty space starts near the ground and extends to the roof unblocked [no windows or other openings], 3) holes exist somewhere up top and on the bottom of the vertical shaft [allowing air to enter and exit] and 4) some heat from the sun shinning on the side transfers to that interior air. *Q:* What do you have? *A:* A chimney. Think about this for a moment. You've got an unobstructed vertical air tunnel with upper and lower ventilation and a heat source warming the air within. Naturally occurring convection currents will cause quite a bit of that air to rise and be vented out at roof level, taking much of the unwanted heat from the sun with it, while replacing that warmed air with cooler air taken in near the ground. Convection is of course very reliable, and like a fireplace chimney, works when the air is heated and shuts down when the air is not heated.

Walls that self-cool by "breathing" are not my own idea. I first read of this design in a book on vintage RVs, in an archived ad for a trailer manufactured at least 60 years ago, which actually used this design. It almost made me want to sit and cry. Wow. At some time in the past, somebody used some intelligence when designing and constructing RVs. Then, as usually happens, time passed and the people who had good ideas aged and died. Metal got replaced with plastic and rubber, craftsmanship got replaced with corner cutting, concern for others got replaced with the drive for maximum profits, and people who read books got replaced by people who don't read books (because they're too good for that stuff). The present-day-humans then inherited the Earth and declared themselves the final perfect end-product of a couple billion years of evolution. (But they aren't.)

Advantages of the chimney effect: It's 1) simple and 2) reliable.

Disadvantages of the chimney effect: 1) It cuts into interior space. Walls still need insulation for times when outdoor temperatures are cold and the occupied RV interior gets heated, so replacing insulation with empty space filled with plain air might help cool an RV on hot days, but it would leave no decent plan for retention of furnace heat on cold nights. The only real solution to this problem would be to double the wall thickness, using half the thickness for traditional insulation and the other half for the shafts of these wall "chimneys." Since RVs can only be a maximum width for legal road travel, this design would cause encroachment and loss of maybe four inches of precious interior RV width. 2) Water might get into the empty wall spaces, causing rot. Water might enter this space through the necessary upper vents during rain, or might enter through the lower vents when diving on wet roads or puddles during rainy days or even from raindrops bouncing off the ground. Perhaps chemically treated wooden beams could be used to prevent rot, but all enclosed panels would also need to be rot resistant. If even a small amount of water got in, it might take a long time for that water to dry, since the shafts have only limited ventilation. 3) Insects or even rodents could get into the empty wall spaces through the vents. Rodents could be kept from entering the vent openings by blocking those vent openings with rustproof animal cage wire, however wasps might decide to fly in and build a home inside your RV home. Wasps could be eradicated by spraying insecticide, but administering it would be a pain. 4) This design would also cool a heated RV on cold nights. On cold nights, air conditioning definitely isn't needed nor wanted. [In this case, the heat source driving convection wouldn't be sunshine coming from the *outside*, it would be furnace heat coming from the *inside*.] The only solution to this problem would be to design both the upper and lower vents so that they, like windows, could be tightly closed at will. Closing the vents when not needed could also keep rainwater out. 5) Maybe a second layer of good insulation in that empty space of an expanded wall would work just as well or better than a self cooling wall chimney. I don't know. I haven't run any tests to determine this.

Conclusion: While solar-powered self-cooling walls that breathe sounds like a fantastic idea, this design might present too many problems to build into RVs. However, additional vertical "shades" that extend and cover the entire RV sides, leaving air spaces of a couple inches between the shades and the sides, might work. They would need to retract to small spaces, have special compartments to store them when not in use, and anchor to the RVs so the wind wouldn't take them away. Solar-powered self-cooling shades might also be a great idea for regular houses in hot sunny climates (especially for southern facing walls), where space and size aren't issues.

Best Overall Comfort-Cooling Plan

Air conditioners are not an absolute necessity for RVs. However, all RVs should be designed to keep unwanted heat from sunshine out. The roof and sides should be highly reflective, preferably shiny metal. There should be good insulation between the walls and outer sides and the ceiling and roof. The RV should have adequate windows that open for ventilation, but not too many. Windows should have retractable outdoor shades/awnings, to keep direct sunshine from entering the RV. If the users do want an air conditioner, a compression unit, an evaporative unit, or one of each (in big RVs) are acceptable solutions. If a compression unit is selected, a generator is needed to run it. If an evaporative unit is selected, a generator is not needed to run it (users may choose to not even have a generator), but a good sized solar panel should be included to power its fan. Finally, it is a good idea to have a solar powered fan near the ceiling as well, to remove warm interior air when the RV is temporarily vacated, so that users don't come home to an oven.

Water Systems

RV water systems are both multi-component systems in and of themselves and also usually components of electrical and gas systems as well. An RV water system is part of its electrical systems because water is often moved using an electric pump. (In many systems the pump is absolutely necessary--- if the pump fails, no water comes out of the faucets.) An RV water system is also part of its gas system if it includes a hot water tank (which uses gas to heat the water).

Components

An RV water system typically consists of one freshwater storage tank, one or two wastewater storage tanks, water pipes, faucets, drain tank valves, a fill hose (for clean water), a special sewer hose (for emptying wastewater) and a pump for moving the water (unless gravity fed). More complex systems may also contain a hot water tank, a toilet, additional portable tanks for transporting wastewater to dump stations, and additional piping/valve for allowing a direct connection to city water (bypassing the freshwater tank). Simpler systems for simpler RVs may not have these last four things.

Freshwater tanks may be made out of plastic or metal. Tanks may be designed to store water only at normal atmospheric pressure (these are usually plastic) or they may be designed to store water and air at high pressure (these are usually metal). A tank that stores water at normal pressure has an air vent, to allow air to enter and exit the tank as water gets removed and added. A high pressure tank is sealed from the outside environment during use, therefore has no such vent, instead it has special (normally sealed) openings for admitting water and air under controlled conditions.

Freshwater tank drains: The drain of any freshwater tank should be as close to the bottom of the tank as possible. Water in any portion of the tank lower than the drain can't exit and get to the faucets, therefore that water and tank space get wasted.

Freshwater tank air vents: The air vent of a low pressure freshwater tank should be as close to the top of the tank as possible. Any portion of the tank higher than the air vent can't hold water, therefore that tank space gets wasted. This is because: 1) any water there would just drain out through the air vent and/or 2) a pocket of trapped air forms there, which prevents further water from being added. An air vent is a hole in the tank that lets it breath, but it's more than just that. That hole should have a watertight airtight connection to a hose, and the other end of that hose should lead to the outside of the RV. This prevents water from spilling inside the RV when the tank gets overfilled or when water sloshes around during road travel. The "top" of an air vent is actually the highest point of the air vent hose, which may be well above the tank itself. Having the highest hose point well above the tank is a good idea because it discourages water from leaving the tank and spilling on the road during travel. The air vent should be separate from the manual filling intake opening, so water may enter and air may leave the tank at the same time. (Using the same pipe for both the air vent and filling causes bubbling and burping during filling and also slows the process.)

Water pipes may by made out of either P.V.C. (plastic) or copper. P.V.C. pipes are by far more common today; decades-older RVs sometimes used copper. P.V.C. pipes are flexible; copper pipes have much less of an ability to bend (copper pipes retain their shapes permanently when bent, until bent again). P.V.C. piping and its associated fittings and valves are standard and readily available. I personally see no reason to use copper piping instead of P.V.C. for today's RV water systems.

Faucets for RV sinks and showers are the same as faucets for regular houses, unless they also contain hand pumps. They can be purchased at any store that carries home building supplies. RV toilets have special water-conserving foot-actuated pedals for flushing. (RV toilets are covered later in this section.)

The fill hose: It may be very tempting to just get a regular garden hose to fill water tanks and hookup to an RV. But this is a bad idea because: 1) garden hoses add toxic bad tasting chemicals to water and 2) regular garden hoses take up quite a bit of storage space. Hoses made for direct RV hookup/filling are made out of special materials that are non-toxic and do not decompose too rapidly, therefore do not add bad tasting compounds to drinking water. Storage space is always an issue with RVs. Some RV hoses are made out of very flexible plastic that flattens up and can be wound onto a spool, making the hose (and spool) very compact and easy to store. Garden hoses for regular houses/yards usually take up unnecessary storage space, because for houses, storage space is rarely an issue. Always use a fill hose specially made for RVs.

Wastewater tanks: An RV with no toilet only needs one wastewater tank to hold the wastewater from the sinks and the shower (if it has a shower). This water, called *graywater*, usually has a low level of pollution, consisting of dissolved soap, food particles and dirt. The tank that stores this water is called the *graywater tank.* An RV with a toilet needs a second wastewater tank to hold the wastewater from the toilet. This water, called *blackwater*, has a higher level of pollution, consisting of solid and liquid human waste, toilet paper and special chemicals which RVers add to control odor. In addition, RVers may have one or two secondary detachable tanks which are intended more for transporting wastewater rather than storing it. These are necessary when an occupied RV with permanently attached storage tanks sits in one location for a long time. (If the only wastewater tanks are permanently attached to an RV, legal emptying of them is impossible without moving the entire RV.) The secondary tanks, called tote tanks or "blue boys," have wheels and a hook which attaches to the hitch ball of car, so an entire heavy tank may be towed a few hundred feet (or even a couple of miles) to and from a dump station. Some people have even devised special platforms and lifting equipment, to allow heavy storage tanks to be lifted and anchored to trucks/cars, allowing them to be safely transported many miles to and from dump stations.

The wastewater hose is a short pleated expandable very flexible hose that allows wastewater to empty from wastewater tanks and into the drain holes of dump stations. This hose is only used for this one purpose. It stores underneath the RV when not in use. Dumping can be done all at once (emptying a substantially full tank) or wastewater can trickle out slowly on an as-needed basis (when the hose directly connects an RV to a sewer hookup at an RV park/campsite). Do not let the blackwater trickle out slowly. Solid human waste and toilet paper can dry and solidify, permanently blocking pipes. (I read of one case where somebody did just this. His pipes got so badly blocked that he had to hire a plumber to saw them off with a hacksaw before replacing them. Hope that plumber got paid a lot for his efforts.) Instead, let the blackwater tank get substantially full, then empty it all at once periodically, immediately follow by emptying the graywater tank (to help flush out and clean the sewer hose), then close both drain tank valves and let some wastewater accumulate before repeating the procedure. Letting wastewater accumulate is good procedure even when at RV parks with direct sewer hookups.

<u>Note on dumping</u>:

Never dump blackwater by a roadside or on the ground anywhere else. Toilet paper takes a really long time to disintegrate and looks disgusting. There is little more revolting than a large amount of fresh feces in a concentrated location. Toilet chemicals don't belong on the ground draining into streams either. Graywater can sometimes be dumped on the ground in remote locations, but always check with Park/Forest Service personnel to determine environmental impact. Many camping places that are even moderately visited by humans severely discourage or forbid the dumping of even graywater on the ground. If a dump station exists nearby, always use it.

Winterization

Normally, when objects or liquids warm, they expand, and when they cool, they contract. Water however has an anomalous range of temperature for several degrees about freezing, in which ice expands as it cools and contracts as it warms, thus cooling ice takes up more space than freshly frozen ice. Expanding ice can stress and crack almost anything that contains it, which in RV water systems, includes all components of such systems, such as pipes, tanks, valves and the insides of water pumps. RVs that are not heated during wintertime must therefore have all the water removed from their water systems, if there is a chance or certainty that temperatures will dip below freezing. The process of removing all water from water system components is called *winterization.*

Removing all water from RV pipes and tanks might sound like a daunting task, but in practice it is not that difficult. There are several strategies:

Non-toxic antifreeze: One common and easy RV winterization method is to flush the water in the system out and replace it with a non-toxic antifreeze specially made for this purpose. This antifreeze is readily available at RV parts stores. The antifreeze mixes with the small amounts of water remaining in the system and the entire mixture doesn't freeze. The antifreeze can often be admitted into a system using only an inexpensive hand pump, which sucks the antifreeze out of its jug and puts it into the piping. One problem with this method is that the antifreeze chemical does have some taste, so the following spring/summer, after water gets reentered into the system and the antifreeze gets flushed out, some people can still taste some of this residual chemical in their drinking water. (For this reason, I personally don't use antifreeze.)

Use compressed air to blow the water out: This is done by connecting a source of compressed air to the system through one end while opening up the other end. Air from a tank or directly from a powerful compressor can be admitted into a system either in a forward direction (enters the RV through the end of a garden hose at the coupling where water normally enters the RV, exits at open faucets) or in a reverse direction (enters through open faucets, exits at beginning of water line, which users temporarily disconnect from the entrance coupling). If you own the air compressor or tank (as opposed to borrowing it), and can thus leave it in place for awhile, it may be best to blow air through the pipes several times, waiting several hours or days between applications, to get out every last bit of water. It's good to get most of the water out, however small water droplets in the lines won't brake the pipes when freezing. This method works well.

Bypassing tanks: The freshwater and hot water tanks store quite a bit of liquid, even when nearly empty. It is often best to temporarily disconnect these tanks from the system and drain them completely, rather than adding antifreeze to them, because: 1) too much antifreeze may be needed, 2) if too much water mixes with the antifreeze, its antifreeze properties might not be effective and 3) if antifreeze is put in the tanks, it may be very difficult to completely rid those tanks of that antifreeze the following spring or summer when drinking water gets reintroduced into the system. Fresh and hot water tanks have valves on their bottoms which allow complete draining. Leave the drains open the whole winter. Bypass kits are available which allow hot water tanks to be disconnected and bypassed from the rest of the water systems. Non-toxic antifreeze or compressed air can then be used to flush the water out of the rest of a system excluding its hot water tank.

Take the system apart: Simpler systems may be winterized by just disconnecting all couplings then blowing all the water out of all the lines, either with a simple hand pump or even with your mouth. After you're sure all water is out of the lines and tank, reconnect all couplings. (This is the strategy I have used for three decades. So far, it has worked well for me--- nothing busted.)

Wastewater tanks also need to be winterized. Drain them completely then put a little antifreeze in them.

How do you know if you haven't winterized properly? If you neglect or forget to winterize or don't do it properly, you may find mysterious leaks or tears in your water lines or joints the first time in the spring or summer that you try to use your water system. You also might discover big ruptures in one or all of your water tanks that make them look like they have been exploded by terrorists.

Methods of avoiding winterization: 1) In some geographic areas, wintertime temperatures may stay warm enough such that winterization isn't needed for RVs stored there. However, in the United States, the only such locations are most of Florida and possibly some parts of southern California, and even these areas can undergo exceptionally cold snaps. For cold snaps in these areas, RV owners choosing not to winterize should have a plan. Heaters should be ready to be put in action--- the RV's propane heater could be turned on [be sure not to run out of propane] or if stored with access to an electrical outlet, electric space heaters could be used. [Usually, only a little heat is needed to keeps the pipes and tanks from freezing.] 2) Winterization can be delayed, ended early in the spring or even sometimes avoided when an RV is continuously occupied [therefore heated] rather than being in unoccupied storage [therefore not heated]. If an RV interior is heated, so too are its pipes and tanks. But beware--- water tanks are often located in low compartments, and pipes tend to run in compartments near outer walls, so even if an RV is heated, some pipes and tanks may still get cold enough to freeze in very cold weather. 3) RVs don't need to be winterized when stored indoors in heated storage facilities. However, heated garages cost lots of money to rent. It costs much less to hire someone to winterize an RV then dewinterize (flush out the antifreeze and reintroduce drinking water in the spring), than it does to store an RV in a heated garage. Finally: 4) If water doesn't get reintroduced into a previously winterized RV during the spring or summer, then it doesn't need a subsequent winterization the following autumn. If an RV isn't to be used in a particular summer, don't reintroduce freshwater into the system that year.

There is much written on the subject of winterization. Videos also exist which demonstrate these procedures. Some videos even demonstrate winterizing procedures for specific RV makes and models. Use these other sources. The information in this book should only be considered a guide.

Water Pumps

Types of Water Pumps

I.) Demand Pumps:

The demand pump is an electric water pump that is sensitive to pressure. When first turned on, it pumps until pressure builds in the downstream water lines (the lines connected to the faucets). After the pressure builds to a certain degree, the pumping stops. When a faucet gets turned on, water comes out instantly and lowers the line pressure. This lowered line pressure causes the demand pump to start and continue pumping water until all faucets get turned off and the pressure in the downstream lines rebuilds. The demand pump works when it is needed and shuts off (but remains ready) when not needed.

II.) Compressed Air:

In this strategy, compressed air is introduced into a special airtight water tank capable of withstanding high pressures. When a faucet gets turned on, air pressure in the tank causes water to move through pipes upward and out that faucet in a direction that gravity does not want to take it. Air compression may be achieved: 1) by an electric air compressor, 2) by a hand or foot operated air compressor, 3) from pressure in a city water hose [when filling the tank] or 4) from pressurized water forced into the tank from a demand [water] pump.

III.) Hybrids:

A hybrid system uses both a demand water pump and compressed air. Two freshwater tanks are used: a large low pressure tank upstream of the pump and a small high pressure tank downstream of the pump. When the demand pump forces water into the second tank, that water compresses the air. When a faucet is opened, the compressed air forces the water out.

IV.) Gravity:

In this strategy, a freshwater storage tank is located above all faucets. When a faucet is opened, gravity causes the water to naturally drain out from the tank through that faucet.

V.) Manual Pumps:

These work by the mechanical action of people's hands or feet. A hand pump is part of a special faucet. It sucks water from below. A foot pump pushes water from below up through a regular faucet.

VI.) Municipal Water:

This is not a wilderness strategy, it occurs only at RV parks or regular houses. A hose from a municipal water supply gets connected to the RV and takes over and bypasses the RV's wilderness water system.

Which Water Pump Types are Best Suited For RVs?

I.) Demand Pumps:

The demand pump is *very appropriate* for RVs. It is by far the most frequently used strategy for moving water in today's RVs. In fact, this strategy is the one almost exclusively used in today's RVs. It lends itself well to complex water systems, but can also be used for very simple water systems. (However other methods also work well for very simple water systems.)

Advantages: 1) Demand pumps are cheap. Most models cost less than $100, and many are available for $60 to $90 [2015 prices]. 2) They are reliable. The one in my mother's RV never failed [when electricity was available]. So far, I have never heard a single complaint from another RV user about a failing demand pump. However, if one does fail, it can be easily replaced for the next two reasons. 3) They are easy to install. I have read internet accounts of people installing them in minutes. [I don't know how reliable these accounts are.] A pump requires only two connections to water lines [an entrance from the freshwater tank and an exit to the faucets], a connection to each of the two terminals [+ and –] of a 12 Volt power supply, and anchoring screws/bolts. Of course, it must be located somewhere accessible to human hands and tools, in order for installation to be easy. 4) They are readily available at all RV parts stores, also from internet vendors. 5) They are not very big. A typical size might be 8×4×4 inches. A typical weight might be 4 or 5 pounds. Size and weight are always issues for RVs, both in designing new ones and customizing/improving old ones. 6) They use only small amounts of electricity. According to manufacturers, they use 4–7 Amps when pumping [I've seen estimates as high as 9 Amps.]. This might sound like quite a bit, however water pumps do not spend much time running each day, because freshwater and wastewater tank spaces are limited, and filling/emptying is a hassle, so RV users naturally learn to conserve water [and thus pump run time]. If the pump runs for 15 minutes each day, this only drains the battery by about 2 Amp hours or less per day. 7) They can be used with low pressure water tanks. Low [normal atmospheric] pressure water tanks have the following advantages over high pressure water tanks: A) Low pressure tanks can be made out of plastic [polyethylene] instead of metal, therefore are cheaper, and provide clean tasting water. B) Low pressure tanks can be made rectangular [instead of elliptical or cylindrical], therefore store more water in less space. C) Low pressure tanks are readily available from RV parts stores and internet vendors. 8) Systems using demand pumps can replace older compressed air water systems in vintage RVs. Many owners of usable antique RVs have chosen to convert to newer demand pump water systems, when existing systems fail, rather than try to get parts for and fix the older systems. 9) Demand pumps can actually be used to compress air indirectly in a pressurized water tank. This secondary freshwater tank is called an *accumulator*. This water pump / compressed air system is covered shortly under its own subheading [III. Hybrids].

Disadvantages: 1) Demand pumps are noisy. This rarely bothers anyone during the day, however demand pump rattling from faucet use of a fully awake RV user can awaken other sleeping RV occupants at night. 2) Water sometimes pulsates when coming out of faucets/showers. I don't know why this is considered a problem; it shouldn't be when filling any container such as a sink or coffeepot or flushing a toilet, but some people find the pulsating machine-gun-like water spurting out of the showerhead to be annoying. Some pump models claim to minimize or eliminate this spurting. Other systems use an accumulator tank, which completely eliminates this uneven water flow. 3) Demand pumps occasionally pulse at night. Water pressure in the downstream lines naturally drops very slowly, even when all faucets are off, presumably due to backward leakage through the pump. This can cause an occasional single pump "thud" cycle at night, which can awaken people. [My mother was really bothered by this. She solved the problem by simply turning the pump off at night. It's very easy to temporarily turn the pump switch back on at night, when water is needed, so this solution worked well.] 4) Water only flows when the pump is turned on. There is no noise-free way [except with an accumulator tank] to get water out of the faucets or flush the toilet at night. And [very important]: 5) Demand pumps are absolutely dependent upon electricity. As stated in Chapter 2 [in "The Octopus Effect"], RV electrical systems are unreliable, as there are many situations that can cause them to fail. Since a demand pump is only as good as the 12 Volt DC power supply that runs it, the entire water system may be unreliable. Keeping several gallons of spare freshwater in jugs, and having backup procedures for extracting water from the tank without the pump, is highly recommended.

Priming: Dry pumps need to be primed (moistened with water) before they start pumping water right. Most pumps today are designed to be self priming. When first connected to a water tank and turned on, a pump sucks up air for a while until water hits it, then it works. Because both the freshwater tank and the pump are likely to be on the floor, at about the same level, gravity usually provides water to self prime a pump. If this doesn't work, then a water line needs to be temporarily disconnected from the pump, then water needs to be poured into the pump (usually backward). The water line gets reconnected after the pump gets primed. Always be sure to open a faucet when priming (self or manual) and first filling RV lines with water. One needs to do this in order to allow air in the water lines to escape and get replaced by water.

Running dry: Most pumps made today are designed to run dry without damage. (They run dry briefly when they self prime.) I even read of a case where somebody put an RV into storage plugged into electricity and accidentally left the pump on, running dry, for over a week before shutting it off. After he added water to the system, he turned the pump back on and it worked just fine. Still, it's not a good idea to leave a pump running for no reason. It cycles endlessly, trying to build pressure downstream, and with air, it just can't build that pressure. Running any motor endlessly, for no purpose, can wear out that motor. Also, some pump models may not be designed to run dry, so they might be damaged.

Conclusion: Although I have done quite a bit of complaining about many aspects of current RV designs, I don't have any complaints about demand pump water systems. Overall, these systems work quite well. This is a case in which design/manufacturing trends of RVs coincide with technological integrity.

II.) Compressed Air:

Compressed air is a great strategy for moving freshwater in RVs. Compressed air water systems were widely used in RVs until about the 1970s, when they became obsolete, not because they didn't work, but only because cheap demand pumps became available, and water systems using demand pumps worked so well. A special high pressure water tank is needed for a compressed air water system. An electric air compressor should be used, because although manual pumping of air is possible, it is tedious and often must be done frequently. Compressed air lends itself well to both simple and complex water systems.

How are pressures measured? A common measurement in America is the *P.S.I.*, which stands for "Pounds per Square Inch." The metric measurement is the *Pascal* (Pa) or the *kilopascal* (kPa). One Pascal equals one Newton per square meter. The *atmosphere* is perhaps the easiest measurement to understand. It describes the pressure as a multiple of the normal atmospheric pressure of the Earth at sea level (which is 1 atmosphere). Pressures are also described in terms of their abilities to lift columns of liquid mercury or water. [1 mm. of Mercury (Hg) is also called 1 *torr*.] Finally, pneumatic pressures are usually given in terms of their functional abilities, as *pressure differences* between the tanks/components and the outdoor air. The *gauge pressure* is the difference between the pressure in a pneumatic tank or component and the outdoor air. The *absolute pressure* is the pressure difference between the tank or component and zero (an absolute vacuum). (Values for outdoor air pressures are also usually given as absolute pressures.) For handy reference (rounded to three decimals):

1 atmosphere = the atmospheric pressure of Earth's air at sea level.

1 atmosphere =
14.7 P.S.I. =
101 kPa =
760 mm. Hg (also 760 torr) =
33.8 feet water column.

(Gauge pressure) = (Absolute pressure) – (1 atmosphere).
[at sea level]
or
(Gauge pressure) = (Absolute pressure) – (outdoor air pressure)
[at some other altitude].

Note: For most equations, the outdoor air pressure is approximated to be 1 atmosphere, even if not at sea level. This is close enough for most practical applications.

Typical water system pressures: Typical water pressures from municipal water supplies start out at 50 to 100 P.S.I. (gauge), arriving at the houses at 50 to 70 P.S.I., with 60 P.S.I. a good healthy average. RV plumbing is usually tested to high levels of 100 to 125 P.S.I., however to prevent problems and to keep warranties valid, 40 to 50 P.S.I. with 45 P.S.I. a good healthy average and the standard for RV plumbing. A pressure regulator should be used to lower excessive upstream pressures from municipal water connections to the downstream RV plumbing.

Using a compressed air system: If the tank is to be pressurized only once during each drain cycle, it should be filled (or refilled) only about halfway with water. This is because in order for it to work, both water and air must be present in the tank. Air is very compressible whereas liquid water is almost completely incompressible. If too much water gets put in a tank, there isn't much room for the air, so after pressurization, that small amount of air decompresses very quickly as water gets released, and soon, there is no pressure left and water stops flowing. If a tank is to be repressurized several times during each drain cycle, then the tank may be filled more completely, however there should always be some empty space left in the tank for air.

<u>Why Should a High Pressure Tank Be Only Partially Filled With Water</u>
<u>If Only Pressurized Once</u>?
(Putting Some Numbers On Air Pressure)

A) Suppose a 40 Liter (about 10 gallons) tank were initially filled almost completely with 39 Liters of water, leaving only 1 Liter for air, compressed to 45 P.S.I., which is about 3 atmospheres (gauge) or 4 atmospheres (absolute). That 1 Liter of 4 atmosphere (absolute) compressed air can and will only expand to a volume of 4 Liters before equalizing with the 1 atmosphere (absolute) pressure of the Earth, at which point it loses all ability to push any more water. Since that 1 Liter of air can only expand from 1 to 4 Liters, only 3 Liters of water could be pushed out of the faucet before all pressure gets lost and water flow stops. (Actually, because the faucet is high, water would stop flowing slightly before the release of the whole 3 Liters. Toward the end, the water pressure would be hopelessly low.)

B) Suppose the same 40 Liter tank were initially filled halfway with water, leaving 20 Liters for air, also compressed to 45 P.S.I., which is about 3 atmospheres (gauge) or 4 atmospheres (absolute). When the tank runs dry, that 20 Liters of 4 atmospheres (absolute) compressed air would expand to the entire 40 Liter volume of the tank, thus doubling its volume and finishing at half its absolute pressure, at 2 atmospheres (absolute) or 1 atmosphere (gauge). 1 atmosphere of gauge pressure is still quite strong, it is capable of lifting water about 34 feet up a pipe. Nobody has an RV 35 feet tall, so every last droplet of water would be easily pushed out.

C) Suppose the same 40 Liter tank were filled with only 1 Liter of water before pressurizing. Yes, there's plenty of room for the air, but you don't need to do the math--- only 1 Liter of water would be pushed out before air would start coming out of the faucet.

Methods of achieving air compression:

1) Electric air compressors: A 12 Volt DC air compressor can be used to pressurize the tank. An inexpensive compressor can be used, but inexpensive models work slowly, so they may take quite a while to build the desired pressure. Inexpensive models are also light-duty--- they are designed for only occasional use, so they might break with regular frequent use. Faster heavy duty models may be more reliable. Faster models use more current, however they don't need to work nearly as long as slow weak models do to do the same job, so theoretically there should be no difference in the amount of battery discharge between fast or slow models. Since slow models only cost $20 to $30, and can also be used to inflate tires, air shocks and volleyballs, it may be a good idea to have an inexpensive spare air compressor around just in case the main one fails.

2) Manually operated pumps: Air compression doesn't need to be done with an electric pump, it can be done with either a hand or foot operated pump (the type often called "bicycle pumps") by persons in reasonably good physical shape. However this task is time consuming, somewhat physically demanding and a major inconvenience. Many times RV users just might not feel up to the task, and may have to be without water pressure for awhile, or they may become neurotic about water conservation. (The good news is: you can save money by cancelling that gym membership if you pressurize your water tanks by hand. Just kidding--- when you're in the wilderness, you don't need a gym membership because you're nowhere near the gym.)

> My father had a high pressure water tank in the first trailer that our family had in the 1960s. Dad brought the water from home in a big plastic jug, unscrewed the pipe cap from the opening, poured the water in (this task took awhile itself), screwed the airtight pipe cap back on (which had a small air check valve on top of it, identical to the valve on car tires), connected a foot pump to the valve, then started pumping. It took him quite awhile to pump the water up, but once he did, water came out of the single sink faucet like magic. I always was fascinated by both Dad's ability to make the water come out of that faucet and by his patience, even I was only around seven years old at the time.

3) City water pressure: Water from a municipal water supply already has substantial pressure. Pressure from city water can be used to compress air in the tank, if at the time the tank gets filled, it gets filled from a pressurized city water source. For this to work right, both the tank and the hose coupling need to be watertight and airtight. There are several problems with this method: *A) The tank pressure drops to its initial filling pressure when the tank reaches empty.* If the pressure is zero when water first gets added, that pressure drops back to zero when the tank reaches empty, so the last of the water comes out sluggishly slow, then not at all. This problem can be solved by putting some compressed air into the tank (with and electric or hand pump) before beginning to fill it. *B) This setup doesn't allow water to be added from a non-pressurized source.* Bringing water to an RV that's been stationary for awhile is a common occurrence. Water in portable jugs is not pressurized. An air compressor is still thus needed. *C) The tank and all the water lines in the system must be able to withstand city water pressure.* If they can't, city water must pass through a pressure regulator before entering the RV, in order to prevent pressures from getting too high and bursting something.

Question:
Could city water pressure cause air in an empty hose to pre-pressurize a water tank?

Answer:
It could, but it would have only a small effect. A typical 50 foot long hose with a 5/8 inch inside diameter only contains about 3.0 Liters of air when empty. If one end gets connected to a city water source and the other gets connected to a water tank, that 3.0 Liters of air gets forced into the tank. However, adding 3.0 Liters of air to even a small tank of 40 Liters, only increases the pressure by 1.1 P.S.I. by the time the first of the water starts entering the tank. [(40+3 L)/(40 L) × (14.7 P.S.I.) = (15.8 P.S.I.), which is 1.1 P.S.I. greater than the outside air pressure of 14.7 P.S.I.] 1.1 P.S.I. of gauge pressure is only enough raise water about 2.5 feet up. The faucet is probably higher, so this couldn't push the last of the water though that faucet (though pressure from city water could effectively push the first of the water through at the beginning of the drain cycle). Even if 1.1 P.S.I. were at the faucet, the water would still come out sluggishly slow then stop just before the tank ran empty. One would need a very long or very fat hose in order for this scheme to actually work. Get an air compressor to pre-pressurize the tank.

4) Air pressure from a water pump: A demand pump can force water into a secondary tank called an accumulator and thus build air pressure. These systems are covered in the next heading (III. Hybrids).

Which method should be used? Electric air compressors are really the best single method of pressurizing tanks. Pressurizing water tanks by hand/foot is impractical; this method is only for fitness fanatics, masochists or people with very simple water systems who only use them occasionally and don't mind the work involved in pumping them. Manual pumping however is a great backup method for emergencies, if an electric pump or supporting electrical system that powers it fails, so each RV user with a compressed air water system should still have a manual pump. Electric air compressors have another advantage over hand pumps: because they are so easy to turn on and off, tanks can be easily repressurized several times during each drain cycle. This allows tanks to be filled more completely. Repressurizing a tank several times during each single drain cycle manually is really tedious. Finally, city water pressure can and probably should be used to pressurize tanks when city water is available. However, this shouldn't be the only method, because it doesn't allow water to be added under zero pressure, and tanks should be pre-pressurized before adding city water, else they have zero pressure at the end of their drain cycles. Also, some air may leak from hoses connected to city water, so all pressure may be lost before tanks completely empty. I believe the best overall plan should use all three air compression methods: electric air compressor, manual pump (emergency) and city water pressure (when available).

Advantages and Disadvantages of Compressed Air RV Water Systems:

Advantages: 1) Air compressor pumps can be used at different times from faucets. This differs from water moved by the standard popular demand pumps, which must be used simultaneously with faucets. Compressed air is an energy storage medium. Electrical or muscle energy from a pump gets transferred into compressed air, which then can sit in a tank almost indefinitely, so it is not just normal for faucets to get activated at different times from pumps, it is likely. This is important for two reasons: *A) Electric pumps make noise, whereas water coming out of faucets is rather quiet.* Water can thus be drawn from a faucet at night by one awake RV user without disturbing other sleeping RV occupants. *B) Sometimes the functional battery drain of an air compressor is zero.* All electrical pumps [both compressed air and demand water] draw current, so they should discharge RV batteries, but this is only true when an RV's only electrical source is its batteries. When a motorhome or tow vehicle/trailer combination gets driven, its alternator makes electricity, so if an air compressor gets operated while driving, no battery drain occurs at all. When a generator gets run for some other reason, any air compressor operated simultaneously also doesn't drain the battery at all. Finally, running an air compressor from a battery located in a tow/spare vehicle, just before that vehicle takes a ride, has no functional discharge to that battery, because the subsequent ride usually fully charges that battery. 2) Compressed air is extremely reliable. As long as there is water and sufficient compressed air in a tank, and there are no leaks or blockages in the system (due to ice), water comes out of any faucet when opened. 3) Pressure can be achieved by other means, if the event of an electric pump failure. Water can be pressurized manually. In a pinch, even a bicycle pump can be used. Water can also be pressurized using city water pressure, when available. Thus, compressed air water systems can still be used in electrical blackouts [both temporary and indefinite]. 4) Water flow is very smooth and quiet. This differs from water moved by the standard popular demand pumps, which have pulsated flow and make some noise.

Disadvantages: 1) A high pressure tank is required. This creates several problems: *A) New tanks are generally unavailable.* Since compressed air water systems are generally considered obsolete, new tanks generally aren't manufactured today. Therefore: *B) New tanks probably need to be custom fabricated* or *C) old tanks need to be used.* One big problem with both custom making new tanks or using old tanks is: *D) some tanks are toxic.* Toxic new tanks can result from fabricators being unfamiliar with [or unconcerned about] which materials are suitable for food and water. Toxic really old tanks probably exist because decades ago, there may have been less concern for and knowledge about safety, and safety laws may have been more relaxed. [More on tanks next subheading.] 2) Adding water to a pressurized tank from a non-pressurized jug is a real bother. This situation occurs frequently, when an RV water tank still has some water in it, the RV is stationary for a long period of time, the users bring water to the RV and want to add that water to the tank. At this point, they have two choices. They can: *A) Use some sort of water pump to pressurize the water before sending that water into the tank through the city water input coupling.* A standard RV demand pump could do this job. But at this point, both a water pump and an air compressor are needed. Having two pumps adds complication and expense to a system and lessens reliability. *B) Unscrew the cap on the opening of the tank, pour the water in, then screw the cap back on.* But wait--- opening the tank lets all the pressure out. One can do this, and one must do this in order to add non-pressurized water, but one must then repressurize the tank from zero every time. Quite a bother. And if the air compressor happens to malfunction just after adding water... 3) High-pressure tanks take up more space than low pressure tanks. There are two reasons for this: *A) Because of their need to withstand pressure, high pressure tanks must be cylindrical or elliptical [egg shaped].* This wastes space. Low pressure tanks are usually rectangular, so they fit snugly in compartments. Of course, an experienced packer [such as myself] could make use of the space around a cylindrical or elliptical tank by jamming lots of small objects there. *B) If pressurized only once, high pressure tanks should only be filled halfway.* This further limits the maximum water storage capacity of these tanks, thus wasting more space and necessitating still bigger tanks to equal the storage capacity of low pressure tanks. This problem can be alleviated by filling the tank almost completely, but one must repressurize several times during each drain cycle. 4) Water pressure varies as the tank empties. This is especially true at low pressures. Varying pressure could have an effect upon showers, but probably wouldn't adversely effect faucets or toilets. This varying pressure problem can be somewhat less annoying if higher pressures get used, since even at the low end, water flows nicely. 5) If a pressurized tank springs a leak, things get messy. Pressure exists at all times in a tank that's in use, so if a leak develops, water doesn't just ooze out, it sprays out with force until all the pressure or water is gone. Things can get really wet. This situation has been known to happen with older tanks in vintage RVs. These older tanks have been known to rust, deteriorate and spring leaks.

High pressure water tanks:

Not just any tanks will do for compressed air water systems, special tanks need to be designed to withstand the substantial pressures of both RV and municipal water systems.

Composition: Metal is generally used because metal is often stronger than plastic. Decades old high pressure tanks were usually made out of galvanized steel or aluminum. Galvanized steel tends to be inexpensive but has definite disadvantages. Although galvanization greatly inhibits the rusting process, it does not stop it completely, so tanks can and do eventually rust. Rusty water tastes bad. [Also, a rusting tank is a disintegrating tank, and disintegrating tanks can spring leaks.] Furthermore, the galvanization process inhibits rust by sacrificing a surface coating of zinc, which ionizes and dissolves into the solution, which in this case, is drinking water. In other words, galvanization works by *putting zinc into the water*. Although zinc is necessary for the human

Don't Eat or Drink Batteries

Once, at a large public gathering, large amounts of grape juice got served. The hosts decided to store this grape juice in and serve it out of brand new clean galvanized garbage cans. Bad idea. The acid in the grape juice along with the two different metals (zinc and iron) actually turned each can into a short circuited self discharging battery. (Yes, you *can* make a battery by accident.) The zinc ionized and dissolved into the grape juice, leaving some electrons behind in the zinc, which migrated to the iron, which when on the surface then combined with the hydrogen ions in the solution (acids are defined as solutions with lots of hydrogen ions) to form hydrogen gas. The process continued, gradually increasing the concentration of zinc in the juice. Several people got sick from drinking this tainted grape juice.

I read about this incident decades ago in an article entitled: "Galvanized Garbage Cans Not Meant For Food." While fresh water isn't very acidic, it can be *slightly* acidic, so this same process could gradually occur in a galvanized freshwater tank.

body as a trace element, too much of it is poisonous. Aluminum is a much better metal choice for freshwater tanks than galvanized steel, because aluminum is much less likely to poison the water. However, there is widespread disagreement as to whether or not aluminum tanks are suitable for drinking water. (The American Boat & Yacht Council recommends that aluminum not be used--- but only in newer codes. Many older boats have them. Some boaters who have had and used aluminum tanks for decades report no problems with sickness or senility.) One problem is that aluminum and chlorine do not mix well. Chlorinated water starts reacting with the aluminum, forming tiny pieces of oxidized metal, which can break off into the drinking water. These can clog a system or might pose health risks if ingested. The solution to this is to put a filter in the water lines or faucets. Doing so however slows down water flow. Furthermore, filters need to be cleaned or changed once in a while. And as with any metal, there are many different alloys. The aluminum metal of a tank might have something else mixed with it that taints the water. Stainless steel seems like a really good choice for freshwater tanks. However, stainless steel comes in many qualities, and from many manufacturers, some of which are dubious, so some low quality "stainless" steels actually do rust. Some plastic fiber reinforced composites can be made very strong, so they could easily withstand the required pressures, so they might be suitable for high pressure freshwater tanks. Many plastics too however are toxic. One possible solution might be to coat the inside of a super strong tank (plastic or metal) with polyethylene, which is considered non-toxic. Whatever the case, food grade material should be selected for a freshwater tank.

Some people with vintage RVs are so adamant about keeping these older water systems alive and historically intact that they are willing to hire welders to custom make high pressure tanks. Some can-tinkering RV individuals who are diehard supporters of compressed air do the same. But if one doesn't know what one is doing, or one hasn't done the proper research, this is risky.

Low pressure systems:

An alternative to standard RV water systems (using demand pumps or compressed air) are low pressure compressed air systems. Standard RV water systems use high pressures--- about 45 P.S.I. and can sometimes go as high as 60 to 85 P.S.I. However, this amount of pressure isn't necessary; it actually only takes about ½ P.S.I. at a faucet to make water come out. (Don't believe me? Ever take a bath with the water 1 foot deep? What happens when you're done and you pull the drain? The pressure at the mouth of the drain is only about ½ P.S.I. when the drain first gets pulled, and even less as the bathtub empties. Yet water flows down at a considerable rate.) Large pressures aren't needed for lifting water either. It takes only 1 P.S.I. to lift water 2 feet up in the air. The total minimum tank pressure needed to make a system functional is thus 2 P.S.I. (1½ P.S.I. to lift water to a faucet 3 feet up from a freshwater tank on the floor and another ½ P.S.I. to make the water come out). Low pressure compressed air water systems behave much like gravity fed water systems, except that air needs to be recompressed frequently.

Advantages: 1) Low pressure systems can be used with standard RV polyethylene freshwater tanks. These tanks are not designed to withstand pressure, however they can withstand a few P.S.I. Tanks are readily available and non-toxic. These tanks do however have air/water vents, which need to be sealed, in order to withstand any pressure. 2) Low pressure systems encourage water conservation. Yes, it may be a little annoying to patiently wait for water to fill that pan or coffeepot. But one wastes far less water doing tasks, such as washing hands or a dish, when the water flows slow. The time saved by not needing to fill and empty water tanks as often exceeds the time lost waiting for slow water flow.

Disadvantages: 1) Can't take a shower, flush a toilet or use a hot water heater properly. Pressure is needed to take a proper shower or flush the RV toilet. And the hot water heater would need to be rather close to the faucet, or else the water might cool too much by the time the water comes out and the container fills. Alternative procedures for using these are covered two headings down ["IV. Gravity"]. 2) Compression needs to be done very often. Low pressure air expands and equalizes with the surrounding air quickly, therefore it cannot push very much water. Thus: 3) Compression should be done simultaneously with faucet use. This means frequent pump use. Also, if the air compressor works too slowly, one might need to wait a couple minutes after turning on the compressor before using the faucet. This could be somewhat of a bother.

Could a low pressure tank could be modified to withstand moderate pressures? Low pressure tanks are polyethylene, which is not very strong. They begin to visibly deform with only about 1 P.S.I. gauge. But they are also rectangular. It would seem to be easy to surround and encase an entire tank with six pieces of strong plywood, all securely fastened together. (The side with the water inlets would need to have holes cut for pipes to be connected.) The tank would try to expand when pressurized, but the plywood surrounding it would prevent it from doing so. I don't know how much pressure such a modified tank could hold. Somebody would need to be willing to experiment a lot and burst a few tanks in order to experimentally determine this information.

Using a separate compressed air tank: To avoid problems of tank unavailability and toxicity, perhaps a compressed air tank separate from the water tank could be used. The high pressure air tank would need a pressure regulator between itself and the low pressure water tank. The regulator would feed low pressure air into the water tank on an as-needed basis whenever water got drawn from a faucet. (This is the same way a regulator feeds low pressure propane into the propane pipes from a high pressure propane tank on an as-needed basis.) This would eliminate the need for frequent pumping and repressurizing. It would allow water to flow out of faucets at a steady but slow rate. And if a valve shutting off the compressed air were added, (with valve off) the water tank could be opened and depressurized, losing only a little pressure, then zero pressure water from a jug could be added, then the water tank could be closed and (valve then opened) repressurized.

Conclusion: Compressed air could work very well as a mechanism for moving water to and through RV faucets, but compressed air water systems are considered obsolete, so those wishing to use them must be willing to can-tinker and experiment. Many people feel that since demand water pumps work so well and are inexpensive, it is not worth the bother of trying to use compressed air.

III.) Hybrids:

A hybrid system is an indirect setup: high pressure water from the demand pump compresses the air, then the compressed air forces water out of the faucets. A hybrid system offers the best of both the worlds of demand water pumps and compressed air. However such a system adds expense and complication and takes up more space, as two freshwater tanks are needed (not including the hot water tank). This setup is definitely only for upscale RVs with complex water systems.

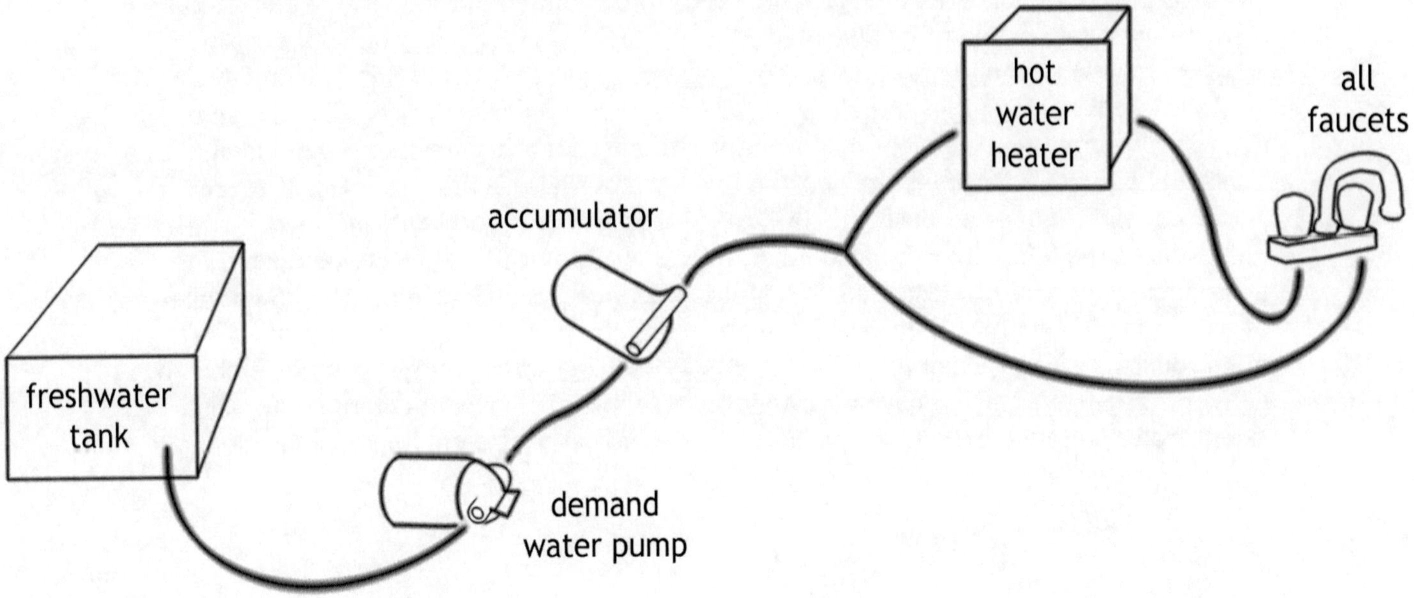

The tanks: The main freshwater tank is the tank that holds most of the water, and does so under normal atmospheric pressure. It might hold 20 to 40 gallons or more. The second freshwater tank, called an *accumulator*, is the tank that holds water under high pressure and also holds compressed air. This tank is much smaller; it holds maybe only 2 gallons of water. The accumulator is located downstream of the low pressure water tank and demand pump, and upstream of all the faucets, the hot water tank and the rest of the plumbing.

Advantages: 1) Water flows smoothly and steadily. When the demand pump works simultaneously with faucet use, the compressed air in the accumulator acts as a shock absorber and takes out the pulsations. [The accumulator acts like a capacitor filter in pulsating electronic circuits.] 2) Some water can be extracted when the pump is off. This is because compressed air exists in the accumulator at all times. The demand pump may be shut off at night, allowing a couple flushes of the toilet, without causing that noisy pump, probably located under the bed just underneath one's sleeping spouse's head, to turn on and wake that sleeping spouse up. Water can also be extracted for short durations during electrical outages. 3) Existing RV demand pump water systems can be converted into hybrid systems. All one needs to do is install an accumulator downstream of the demand pump and make sure everything is airtight. There must however be room for this additional tank.

Disadvantages: 1) Hybrid systems are more complicated. In addition to an accumulator, a hybrid system probably should also have an air compressor, to restore air pressure in case it gets lost. Accumulators are often pre-pressurized [maybe to 20 P.S.I.] before any water gets added. They will work without pre-pressurization, just not as well. Pre-pressurization could get lost from air leakage when an RV is in storage or from opening up and completely emptying an accumulator tank during winterization. However, restoring lost pressure probably needs to be done so infrequently that even a manual foot pump would probably suffice for this task. 2) Complication adds cost to any system. 3) Accumulators take up additional RV storage space. Space is precious in all RVs. 4) Winterizing is more difficult. Getting water out of every nook and cranny of an RV water system becomes increasingly difficult as more components get added. Tanks often get disconnected, drained and bypassed during this process. If antifreeze gets used, all of it must be flushed out of every nook and cranny when water gets reintroduced the following spring/summer, else the water may become contaminated by the antifreeze or may just taste bad. Complicated water systems increase the probability of contamination from antifreeze and demand more complete flushing when water gets reintroduced. 5) The accumulator may not be needed, as some pumps claim to be able to deliver ripple free water. I haven't tested these claims. I also can't determine to what extent water pulsating bothers people, as all RV users have different tolerance levels for annoyance.

IV.) Gravity:

Gravity fed water systems can be very appropriate for simple RV water systems but are not appropriate for more complex water systems. This is because in RVs, gravity cannot create the pressures needed for the fast water flow required for some of the water appliances of these complex systems.

Installation: One could be an idiot, yet still be able to install a gravity fed RV water system. The main thing one must do is *install the freshwater tank higher than the faucet.* (Actually, the bottom of the tank needs to be higher than the top of the faucet. It doesn't even need to be that much higher--- 4 inches will do.) When opened, the faucet acts as a drain to the tank and water comes out. The other main requirement is strong physical support for the tank, so the tank doesn't fall to the floor.

Advantages: 1) Gravity is reliable. So far, I have never known gravity to fail. Any heavy object or liquid on Earth, not physically supported, falls down. 2) Gravity is cheap. The only way gravity could be cheaper than its current price [of zero], is if somebody paid you to use it. 3) Gravity is in abundant supply. Unlike water or gasoline stored in a tank, or oil buried underground, gravity doesn't run out or even run low from being used. It always seems to be there. 4) Gravity is found everywhere on Earth. Its strength is approximately equal at all places on the surface of the Earth. 5) Gravity is safe for the environment. The pollution created from using gravity to move water is zero. And: 6) Gravity water systems can be used with low pressure polyethylene water tanks. These tanks are readily available and non-toxic.

Disadvantages: 1) In RVs, gravity can't create the pressure needed for traditional RV water systems. For every one foot the top of the water level in the tank is above the faucet, only about ½ P.S.I. of pressure gets created. It is possible to create substantial pressures by gravity; in fact, municipal water systems use gravity to create their pressures using water towers. [After purification, electric pumps move the water into water tower tanks high above the ground, then gravity pulls the water back down to people's houses.] To create the 45 P.S.I. of pressure usually used in RV systems by gravity, a water tank would need to be about 103 feet higher than the faucet. [14.7 P.S.I. raises water 33.8 feet.] Nobody has an RV 106 feet tall--- that's as tall as a ten story building! The maximum pressure one can hope to achieve in an RV by gravity is about 2 P.S.I. [This is achieved from a full freshwater tank just below the roof of an RV whose roof is 7 feet above the floor and 4 feet above a faucet that's 3 feet above the floor.] Because gravity only makes low water pressures in RVs: 2) Special procedures are needed for gravity fed water systems. The next subheading covers using low pressure water systems. 3) A large full water tank high in an RV can make the RV top-heavy. Having too much weight too high can make a vehicle prone to tipping over when driving on the road. However, in my own personal experience I've found this not to be a problem. I installed a ten gallon water tank high in my trailer and found that when full, the trailer is actually *more* stable on the road. [I've never "almost" tipped it.] I think this is due to two principles of inertia: *Spinning inertia ["egg spinning principle"]:* You can spin a hard boiled egg because it's solid. If you try to put a spin on a raw egg, it just doesn't want to accept the spin. A raw egg is gooey and liquidy on the inside, so friction in this liquid absorbs and thus resists the spinning motion. Water in an RV tank may actually work like the raw egg--- absorbing sudden spinning torques, therefore making the RV less prone to fishtailing [sway]. *Translational inertia:* [In physics, "translation" refers to motion in a straight line.] Weight tends to stabilize motion, and is especially good at resisting wind. If you tow a heavy RV, it may hold the road better than if you were to tow a light parachute, which would probably try to blow all over the place. *The final verdict:* Adding water weight high in an RV can make the RV less road stable due to top-heaviness, however advantages of inertia somewhat offset this instability. The advantages due to inertia should work equally well whether a tank is high or low. 4) Freshwater tank size of high mounted tanks should be limited. I wouldn't want to make an RV too top-heavy, even though there are certain advantages to adding weight. I suggest that a 10 gallon tank be the maximum for a 15 foot travel trailer. [Maybe a big heavy RV could safely have a bigger water tank, because the weight proportion would be the same as a smaller RV with a smaller water tank. Somebody else needs to test this.] 5) Adding water from a jug is tricky. Because of the height involved, the manual water intake opening must be very high up--- almost at roof level. Unless you're 18 feet tall, you're probably not even going to be able to reach it, let alone hold and pour from a jug higher than it. You could climb a ladder, but standing on an unstable ladder several feet off the ground while holding a heavy jug and attempting to feed water into a tank is risky. The best option is to place the filling jug on top of the RV roof, climb up onto the roof, and feed the water in from there. The water intake opening must be easily accessible from the roof. A person must be kind of light weight and in good enough shape to climb up to and down from the roof safely. [I'm both.] Also, the roof must be able to support the weight of the person on it without caving in. Filling a high mounted tank for a gravity water system from a pressurized water hose [instead of jugs] poses no problems. Unlike the zero pressure manual water intake, the high pressure RV intake hose coupling can be very low. Water pressure will easily force water up to the tank near the RV roof. 6) A large heavy tank of water needs to be physically well supported. Constructing proper support can be a bother but it must be done. You certainly don't want to find the remains of your full water tank, its contents and part of your cabinetry on your floor after taking a ride.

Special consideration for air vents: The air vent hose for any freshwater tank should have a high point somewhat above the tank level, to discourage sloshing water from exiting the tank and spilling onto the street during road travel. Putting an air vent substantially higher than the freshwater tank is impossible to do for an RV with a gravity fed system, because the water tank needs to be very close to the top of the RV in order to work right. The solution to water loss through sloshing is to put a valve in the air vent hose that allows the air vent to be shut off (blocked) during road travel and opened when parked. In my gravity system, I use a long vertical hose as an air vent. One open end is inside the tank near its top, facing up. The other open end is outside the trailer below its floor, facing down. The first time I took a ride with a full water tank, without a valve, half of the tank's contents spilled onto the road. That was unacceptable. Since installing a valve in the hose, and closing it during travel, spillage doesn't happen anymore. I can even use the faucet for awhile when the valve is closed--- the tank bends inward, absorbing some negative pressure created by the exiting water.

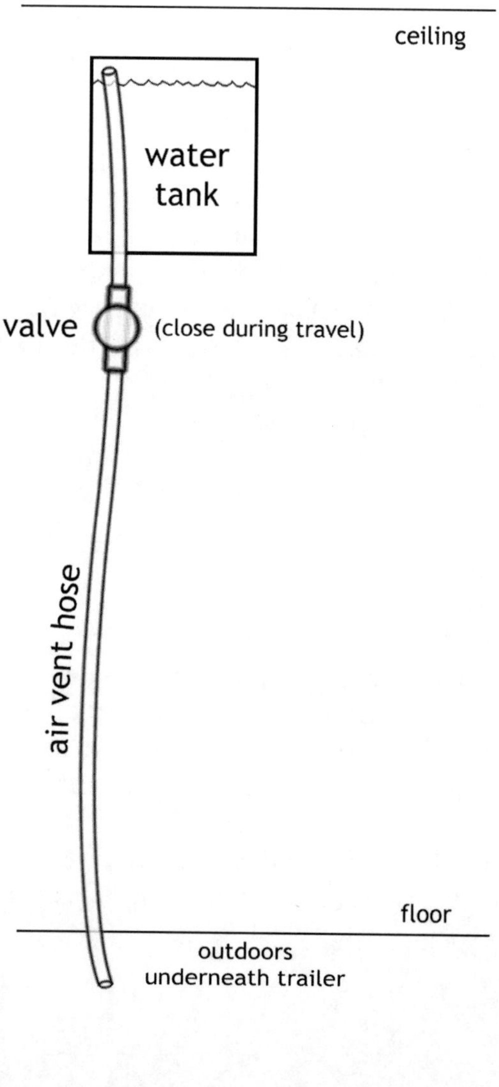

This air vent hose design with a valve works well in my trailer. I installed such a system in 1985. I still use it today in 2017.

Low Pressure Water Systems:

Low pressure water systems do not work well with hot water heaters, showers and toilets, so special procedures must be used.

Hot water heaters: These are likely to heat water just fine. Getting the hot water out of the hot water tank with a gravity system is the challenging part. The water is likely to flow so slow that it probably will come out of the faucet unacceptably cool. Also, it may take quite awhile to fill a large container, during which time the "hot" water in that container cools unacceptably. Fortunately, a hot water heater is not a necessity. Cold water can easily be heated on a propane stove to do dishes, cook with, even bathe. (I don't have a hot water heater in my trailer. I don't feel the least bit deprived.) If one absolutely needs a hot water heater, it probably is a good idea to put a faucet right on the bottom of the hot water tank or very close to it. (Large party-size coffeepots have valves on their bottoms that work by gravity. Their coffee/water comes out hot.)

Showers: Water is likely to come out of a showerhead way too slow with an RV gravity system to be acceptable--- especially if water needs to flow through piping of considerable length. Fortunately, one can take a decent shower by simply holding a container over one's head and spilling warm water onto one's body. (Coffeepots and plastic gallon-size water jugs work well for this purpose.) I've taken multiple showers with this method over prolonged camping trips. I get just as clean as I do with regular "house" showers. If one's arms are too weak to lift and hold a container overhead, or if one prefers, one can just hang a dishpan with a hole in it overhead. When warm water gets added, it flows out the hole and can make a person clean.

Toilets: One does not need to use water in a toilet; one can use a dry toilet. One can urinate into a jug then dump the liquid outside. One can take a dump into a plastic kitchen garbage bag, dig a hole then bury the contents of the bag in that hole. (These are methods that I use. They work just fine.) If one insists upon using a water toilet with a gravity system, one will probably need to flush by dumping a bucket of water in that toilet. More on toilets in an upcoming section.

Conclusion: Despite the inconveniences associated with RV water systems, gravity can work great in RVs. It's cheap, safe, abundant and one of the most reliable mechanisms known to man.

Gravity
It's not just a good idea, *it's the law.*

V.) Manual Pumps:

Manual pumps are primarily for very small and simple RVs (like pop-ups). They are usually used only with single-faucet water systems. They are appropriate for people who don't want to use electricity at all (other than portable batteries). They may also be useful in more upscale RVs as emergency backups in case main water pumps fail.

Advantages: 1) Manual pumps work without electricity. Therefore, they are unaffected by the unreliability of RV electrical systems. They can be used in campers that don't have electricity at all. [Some simple RVers are diehard electrophobes. They want to go out into the boonies for good long time periods and they don't want to be bothered by anything electrical. They want candles as lights, catalytic furnaces which are non-electric, and water that pumps by hand or foot.] Also, some people with electric demand pumps actually install manual pumps as well, to conserve on electricity while out in the boonies. 2) Manual pumps are inexpensive. They are less expensive than demand pumps, but not by a huge margin. [A manual hand pump seems to cost about half what an electric demand pump does--- but the price range of both pump types is considerable. Cheap hand pumps run from $27 to $40.] 3) Manual pump water systems are simple. These systems are simple to install and can have a minimum of components. No need to mess with electric demand pumps or air compressors. The freshwater tank can be a low pressure polyethylene tank [which is readily available and non-toxic]. The tank can be mounted low on the floor.

Disadvantages: 1) Mechanical pumping is somewhat of a bother. It's much easier to turn on a faucet and have the water come out by itself, which is what happens with demand pump, compressed air or gravity systems. [#2–#5 apply only to hand pumps] 2) Air may accumulate in the line. This occurs from the water level dropping back down to the tank between faucet pump usages [which occurs from air leaking backward through the faucet]. The first few pumps may only pump air out of the faucet, which is annoying. Installing a check valve [which prevents water from backing up] can eliminate this problem. Some faucets don't need a check valve, as they are pretty good at being airtight and holding the water level high. 3) When using a hand pump, only one hand is free. Thus: 4) Hand pumps make washing things difficult. Usually, two hands hold an object [dish, cup, pan] when washing it. Usually when one washes hands, two hands go under the faucet at the same time and wash each other. People have found ways to deal with this. Some keep a bowl of water for washing. Others fill an empty dishsoap bottle with water and keep it at the sink for washing. The only other option is to have a very cooperative spouse or camping buddy. 5) Hand pumps may dry and lose their prime. This can occur if the water level drops down after pump usage and a long time passes before the next attempted usage. 6) Manual pumps are not appropriate for showers. Showers connected to a water system need both hot and cold water. The logistics of mixing hot and cold water in the right proportions and sucking the warm water through several feet of water lines without it cooling, using a single pump, seem formidable. Also, how many hands would a person need? One hand to hold the showerhead, one hand to crank, one hand to soap and massage one's body. How many people have three hands? Do you have three hands? I don't.

Hand pumps verses foot pumps: If I had to choose between a hand pump and a foot pump, I would choose a foot pump. Foot pumps offer several advantages over hand pumps. 1) Foot pumps allow both hands to be free. Thus, washing hands or kitchen objects directly under faucets is much more possible with foot pumps than with hand pumps. 2) Foot pumps push water from below whereas hand pumps pull water from above. It's better to push water than to pull it. This is because pushing creates positive pressure whereas pulling creates negative pressure. Any tiny leak below a hand pump will suck up air into the line due to the negative pressure created by the hand pump. Negative pressure then causes that air to expand, dropping the water level down toward the tank. A foot pump creates positive pressure in the water line, so in the event of a pinhole leak, air won't get sucked into that line. 3) Foot pumps are much less likely to dry and lose their prime than hand pumps. Since a foot pump sits at or below water tank level, it is likely to remain wet just due to gravity, as long as there is some water in the tank. A hand pump sit high. If it leaks even slightly, the water level in the line drops, so if a leaky pump sits awhile without use, it can dry out.

Manual pumps in upscale RVs: I have always advocated an emergency procedure for freshwater access in upscale RVs. Some RVs cost a pretty penny. Fuel to drive these big rigs costs a lot too. And for many people time is precious, so for these people, it might not be practical to abort a trip due to water pump failure then start again at a later date (or continue the trip miserable and "dry"). RV users out in the boonies need their water. Water systems can fail for several reasons (faulty pump, faulty electrical system, water leak in the pipes). A manual pump/faucet, drawing water from a point in the system upstream from the electric pump (therefore bypassing it), would allow access to all the water in the event of such a failure. It wouldn't be optimal, but it would be much better than having no water access at all. Yes, I do believe that manual pumps do have a place even in more upscale RVs.

VI.) Municipal Water:

Connecting an RV to city water essentially turns a whole RV water system into a really big hose attachment. City water is only appropriate when taking a "vacation" from the wilderness at an RV park (for whatever reason) or when the RV is parked at one's regular home. When available, one should probably use it. (However I personally don't. On those rare occasions when I stay someplace with a water hookup, I prefer the low pressures from my gravity fed tank.) City water may arrive at pressures that are unacceptably high for RVs. This high pressure transmits into the interior RV lines and can burst them. Using a pressure regulator between the city water output and the RV input prevents this from happening. Always use a pressure regulator when connecting to city water.

Hot Water Heaters

The typical RV hot water heater is the only RV appliance that is part of all three of its life-blood systems: electricity, gas and water. Obviously it is part of the water system. It is part of the gas system because propane gas gets used to heat its water. The very common D.S.I. type is also part of the electrical system because electricity gets used to make decisions, open its gas valve and light its burner.

Types of Hot Water Heaters

I.) Pilot Light (Gas):

These work the same way that regular household gas hot water heaters and passive convection RV furnaces do. Each unit has a pilot light and a thermostatically controlled main burner. The pilot light always stays lit when the tank is ready for use. When the water temperature dips below a critical low, this triggers the main burner valve to open, and the burner is lit by the pilot light. When the water temperature exceeds a critical high, this triggers the main burner valve to close, causing the burner flame to be extinguished. If the pilot light goes out, a heat sensing thermocouple shuts off the gas to both the pilot light and the main burner. This prevents unburned gas from accumulating and presenting an explosion hazard or poisoning RV occupants. Ignition of the pilot light may be accomplished from a flame (match or butane lighter), a piezoelectric igniter or a battery powered sparking igniter.

II.) D.S.I. (Gas):

"D.S.I." stands for "Direct Spark Ignition." When lighting, the gas valve opens and sparks get used to ignite the propane. There is no pilot light; when the burner valve closes, all flame gets extinguished. Although the actual heating of the water is done with gas, the opening and closing of the valve, all temperature monitoring and decision making (via a circuit board), and the making of the sparks for ignition are all done with electricity (from a 12 Volts DC power supply).

III.) Electric:

These work by using electric heating elements to heat the water. Electricity is also used to monitor temperatures and make decisions. Most electric RV hot water heaters are actually two separate water heaters in one (one electric and one gas) which share a single hot water storage tank.

IV.) Tankless:

These don't store much water, instead they heat water that passes through them on an as-needed basis. Almost all RV models heat the water with gas but use electricity for decision making and sparking.

V.) MotorAid:

A MotorAid system uses waste heat from the engine of a motorhome to heat freshwater while driving. Some of the hot engine coolant circulates through sealed pipes that pass through the hot water tank, transferring that heat into the freshwater.

VI.) Hot Water Heaters Combined With Furnaces:

These are comfort heating furnaces (diesel and electric) that also have a second function: heating freshwater for faucets/showers.

Which Hot Water Heater Types are Best Suited For RVs?

I.) Pilot Light (Gas):

These are very appropriate for RVs. They have a more primitive design (than the D.S.I.) and are more likely to be found on simpler or older RVs. Despite their simplicity, they are still manufactured today, and are quite common, though probably not nearly as common as the D.S.I. gas water heaters.

Advantages: 1) Gas burners, valves and controls are simple and reliable. 2) The general valve and burner types of these heaters are very widely used. If something fails, one should not have difficulty finding either parts or a gas technician to do repairs. 3) The pilot lights can be lit with matches or butane lighters, if piezoelectric igniters fail.

Disadvantage: Pilot lights sometimes blow out during travel. This necessitates frequent relighting, which can be somewhat of a bother. However, some models have automatic 12 Volt battery powered relighters, which sense the absence of flame, then quickly create sparks to relight their pilot lights. These greatly reduce failures due to pilot light blowouts. Even if all flames go out and stay out, water heats up within about 20 to 30 minutes after relighting (in 6 gallon tanks), so users aren't likely to need to go for too long without hot water.

II.) D.S.I. (Gas):

These are very appropriate for RVs. They are very commonly used in RVs today.

How does such a unit work? An internal sensor and an electronic circuit board monitor water temperatures. When the water cools below a certain critical temperature (or when the unit first gets turned on), a decision is made to fire the burner. The gas valve opens and sparks are made to ignite it. The flame is electronically sensed, so as soon as the gas is lit, the sparks stop. If not lit within a certain time span (usually 15 seconds), the valve closes and the unit gives up trying to light the burner. (The burner won't light if the main gas valve is shut off, the propane tank is empty or there is air in the gas lines because the gas has just recently been turned on.) The user must then turn the unit off, allow a short pause, make sure gas is present in the gas lines, then turn the unit back on again. This causes the burner to once again attempt to light.

Advantages: 1) They do work. 2) They don't use much electricity. If an RV's batteries run dead, it's not likely to be due to D.S.I. hot water heater use. 3) No need to deal with pilot lights or relighting. The pilot light can't blow out when traveling, because there is no pilot light. If somehow the burner does blow out, the unit will sense this and reignite the flame quickly. 4) It's very easy to turn the water heater on and off. A switch fundamentally identical to a light switch is used. [How difficult is it to turn a light on and off, using a wall switch?] If the burner blows out and does not reignite, simply switching the unit off then on again a couple minutes later causes another ignition attempt.

Disadvantages: 1) Electricity is required. No doubt you have noticed that I favor appliances that can work with no electricity at all [lights, furnaces, refrigerators], because these can still work in power outages. The same might seem true of gas water heaters [pilot light models still work without electricity, whereas D.S.I. models don't], however electricity is still needed for the water pump, so if the electricity fails, no water of any kind comes out of the faucets. What good is having a [pilot light] hot water heater that works in a blackout, if there's no way to get the water out of the tank during a blackout? 2) Slightly noisy. My mother was a little annoyed by the frequent lighting of the burner at night. [The unit was under her bed, right underneath where she put her head. She didn't need hot water while sleeping, so she liked to turn the water heater off at night, then back on the following morning.]

III.) Electric:

Electric water heaters are completely inappropriate because anything that uses an electric heating element doesn't work in any RV that even occasionally goes off grid. However, if a tank also has the apparatus for heating water with propane gas while off grid, the unused electric heating element poses no disadvantages.

Advantages: None--- as far as I can discern.

Disadvantages: 1) These gobble up massive amounts of electricity, far more than RV batteries typically store, therefore: 2) they can't be used off grid. 3) Having an electric only water heater might cause someone to *think* that he/she has a working hot water heater, when in fact that person does not.

IV.) Tankless:

I have no experience with these, but internet reviews suggest that they work well. There seem to be many satisfied customers. Beware though--- a few RV models only run on electricity. Always choose a unit that runs on propane gas. Anything with an electric heating element won't work in an RV that even occasionally goes off grid.

Advantages: 1) Hot water is available continuously. These don't run out of hot water and go cold, the way water heaters with tanks do. However, despite manufacturer claims of "endless" hot water, remember that an RV off grid only stores limited freshwater (wastewater tank space is limited too), thus the amount of hot water is limited by the amount of freshwater in the tank, so "endless" hot water isn't possible. 2) These use less gas. This is because they only produce hot water upon demand. Hot water heaters with tanks need to occasionally light to maintain the temperature of their water [which is always cooling], even when that water isn't being used.

Disadvantages: 1) These are expensive. Indoor tankless water heaters can run about $1000, whereas those with tanks typically range from maybe $260 to $800. 2) Tankless water heaters need to be drained and winterized before taking a ride when temperatures dip below freezing. This is a big problem. [Yes, if you RV extensively you *will* occasionally find yourself in subfreezing temperatures, even if you don't plan on encountering them.] What a bother! Perhaps this problem could be avoided by connecting a parallel water line through which one would pump some of the water downstream of the water heater in a backward direction into the line upstream of the water heater. [As cold water passes through the water heater, this causes it to fire up to heat the water. If it fires up, its contents won't be able to freeze.] 3) Beware of excess electricity consumption. I can't seem to pull up data on electricity consumption. From the descriptions, it sounds like these water heaters work a lot like forced-air furnaces, which use quite a bit of electricity. However, hot water tanks aren't run to nearly the extent that furnaces are, so they might be alright. Always be sure that the estimated daily electricity consumption of any appliance is within your electricity usage budget. 4) Some tankless water heaters are for outdoor use only. Don't buy one of these unless it is your intention to use the tank only outdoors. These are meant to be portable and shouldn't be permanently mounted to any RV.

V.) MotorAid:

MotorAid systems could supplement the propane gas hot water heating capability of motorhomes, but the added costs and complication associated with such systems probably outweigh any benefits. These systems are definitely only for upscale motorhomes occupied by full-timers who spend a lot of time driving. Weekend campers or RVers who park for extended time periods (hosts or maintenance staff) without driving aren't likely to get much use of MotorAid systems.

Advantages: 1) These work while driving. There is zero chance for a pilot light or main gas burner blowing out during travel, because gas flames aren't used. [Such a system while in use temporarily replaces the need for a gas burner]. However, pilot lights and main burners can be effectively shielded from wind, so that they don't blow out from driving. Blown out pilot lights can also be automatically reignited with the right sparking equipment. Thus, blown out pilot lights may not be much of an issue. 2) These use waste engine heat. Anything that finds a second use for an otherwise useless byproduct of some other machine or function, seems like it should be a good idea.

Disadvantages: 1) These add cost and complication to overall systems. Gas burners work really well and don't consume very much gas, so why bother with a supplemental system? 2) These *only* work while driving [a motorhome]. Therefore: 3) MotorAid systems do not replace the need for gas burners. A MotorAid system, if installed, must be *in addition* to a gas burner water heating system, not *in place of it*, because most of the time that hot water is needed, the motorhome is likely to be parked. 4) What happens if the system springs a leak? If automotive radiator fluid gets into the drinking water system, it might contaminate the drinking water components to the extent that most of the freshwater plumbing system might need replacement. If the radiator fluid lines leading to or from the freshwater tank spring a leak, hot radiator fluid under pressure could spray into interior RV compartments. What a mess this could be!

VI.) Hot Water Heaters Combined With Furnaces:

This sounds incredibly complicated and expensive. Diesel and electric? I don't even want to know the logistics of these! Why make things complicated and expensive if they can be made simple and cheap? Passive convection furnaces work very well. So too do RV gas water heaters (both pilot light and D.S.I.). What is the point of combining both these appliances into one mega-multiunit? Still, there might be can-tinkerers or engineers who can build and market such a unit and have it actually work, so if somebody does, and you want to buy it and give it a try, go ahead. (In the past, some regular household furnaces also heated water. I've heard that some of these worked really well, so I'll keep an open mind.)

Water Heater Notes

The standard temperature of hot water in RV tanks is 140° F. This is often not adjustable. However, adjustable replacement thermostats are available for some models.

The standard capacities of RV hot water tanks are 6 or 10 gallons. The 6 gallon models require 20-30 minutes to fully heat the water from cold; the 10 gallon models require almost an hour. Larger 16 gallon equivalent models are available. These do not actually hold 16 gallons, instead they hold 10 gallons of superheated water, which mixes with some cold water on the way out, resulting in the equivalent of 16 gallons of standard temperature hot water.

Do not ignite hot water heaters when there is no water in the tanks. They can crack. Tankless models do not have this problem; they are designed to fire up only when water actually passes through them. Tankless models are unaffected by the water running dry; they simply shut their heat down when the water stops flowing.

Suggested Improvement

Hot water recovery valve:

When any faucet designated as "hot" first gets turned on, the initial water is room temperature because that water has been sitting in the water lines between the tank and the faucet for a while. In houses, this intermediate cool water is often just spilled down the drain. Water is so cheap and abundant, that nobody gives much thought about wasting this water. In RVs, both freshwater and wastewater tank space are very limited, so wasting nearly a pint of freshwater every time one turns on a hot water faucet is not such a great idea. For demand pump systems, I propose a solution: a special faucet design that has an extra valve on the back, which I call the "hot water recovery valve." This extra valve would intake from the hot water line, and exit into an additional water line heading back to the RV's freshwater tank (which is upstream of the demand pump). Just before turning on a hot water faucet, the users would turn on the hot water recovery valve for a few seconds, circulating the line water back to the tank. Then, the "hot" water would come out of faucet immediately hot, with no waste.

Advantage: This would cut down on water waste.

Disadvantages: 1) This would add complication and expense to the overall system. However, valves and pipes are inexpensive. 2) Extra water lines would require greater effort during winterization. 3) The intermediate water can be saved by other simpler means. Some people catch this water in containers and save it for future uses, such as dishwashing or cooking. [This topic has actually been written about in other books.] Also, this water could be caught and then manually transported and poured back into the freshwater tank, without needing additional pipes and valves. 4) RVs with multiple faucets would require multiple hot water recovery valves and lines. This would further increase cost and complexity.

The "hot water recovery valve" might be most useful for the shower, where the mixture of hot and cold water might be preset to achieve some specific desired temperature.

Final Thoughts On Hot Water Heaters

Of all the appliances that RVs usually have, the hot water heater is the least needed. It is so easy to heat water on cooking stoves, for use in washing dishes, general cleaning, even taking baths. Sure, it's a bit nicer to spontaneously wash one's hands under a faucet stream of warm water rather than cold water, but even cold water works well for this purpose and is usually not that cold nor uncomfortable. The only real "need" for hot water delivered directly from a faucet is probably for showers--- especially when families or guests camp who are not accustomed to RV "roughing it." But even a "shower" can be taken by simply pouring pre-stove-heated water onto one's own body. If you want a hot water heater and got the bucks and the room for it, get one. If your RV already has a working hot water heater, use it. Personally, I don't see much of a need. I don't have a hot water heater in my trailer. I do not feel the least bit deprived.

Toilets

Water Toilets Versus Dry Toilets

terms I use:
"water toilet" means any toilet that uses water
"dry toilet" means any toilet that does not use water

At the beginning of this chapter I stated that some appliances that work well in regular houses work poorly in RVs. No example is more classic than the water toilet. In regular houses and skyscrapers, water toilets work great. Water is cheap and all the freshwater and wastewater pipes are permanently in place. When freshwater and human waste get mixed, the resulting sewage is very liquid, so massive amounts of human waste from high concentrations of people can be safely removed and transported miles away to sewage treatment plants, where the stuff can be dealt with responsibly (hopefully). Pumping of the wastewater is often not needed (unless going uphill), because new wastewater introduced at the beginning of sewer pipes and gravity pushes the wastewater already in the pipes further along, creating a natural flow. In RVs, mixing water with human waste might not bc such a great idea. Water is heavy and significant quantities of it takes up lots of space. Adding weight and removing storage space are both really bad things to do in RVs, which are inherently small. Unlike buildings with permanent pipes, there is no possible way (even theoretically) to permanently attach any pipes to an RV, because it moves frequently and often goes into remote areas where pipes don't exist. Thus, when an RV has been parked in the wilderness for awhile, both freshwater and wastewater must be transported to and from the RV, which is a real bother. What is the purpose of mixing water with human waste in an RV anyway? In stationary buildings, it is to transport that waste many miles away. After one flushes a toilet in an RV, the waste moves what--- maybe two feet to the holding tank? Finally, what's so great about carrying a tank of fermenting human waste in one's RV all the time? Dry toilets offer a very viable alternative option for any size RV.

Types of Toilets

I.) Folding Toilets, Bags & Jugs (dry)
This is one of the simplest and cheapest methods known to man. It is the "state of the art" method that I use. The title lists the components used.

II.) DryFlush (dry)
This "toilet that will change the world" (as the manufacturer calls it--- I'm less enthusiastic) is a permanent toilet fixture that when "flushed," uses a vacuum to wrap the mess in a plastic bag, which then gets stored in a larger plastic bag (with other individual smaller mess bags), for eventual disposal in a dumpster.

III.) Composting (dry)

These toilets separate solid human waste (and toilet paper) from liquid human waste and store each in separate tanks. The liquid waste gets emptied often. The solid waste gets mixed with organic material (usually peat) and allowed to decompose slowly. The solid waste only gets emptied occasionally. A vent with a fan creates a slight constant negative air pressure within the solid waste tank and moves and deposits that air to a point somewhere outside the RV/boat/cabin, to control odor.

IV.) RV Water Toilets (wet)

These are the typical toilets that almost all RVs (with toilets) have. Such a toilet uses freshwater from the RV tank (or sometimes reclaimed greywater) and stores the water-waste mixture (blackwater) in a holding tank separate from the greywater tank.

V.) Small Portable Chemical Toilets (wet)

These are basically scaled down versions of regular RV water toilets, except these: 1) are portable, 2) do not connect to their RVs physically, 3) do not connect to their RVs' water systems and 4) tend to use harsher chemicals. These toilets have their own separate before and after tanks.

Which Toilet Types are Best Suited For RVs?

I.) Folding Toilets, Bags & Jugs (dry)

These are extremely well suited for RVs (especially simple ones) and possibly the best overall system, because of their simplicity. These can also be used by people who camp in tents. I use this system.

To take a leak: Urinate into a gallon-size plastic jug. Sometime before the jug fills, take the jug outside and empty it onto the ground (if in a remote area) or into a flush toilet, outhouse or dump station (if in a well used area where such amenities exist nearby). By the way--- don't use thin plastic jugs, like those in which spring water are sold. They can spring leaks. [I've used the thin jugs to store drinking water (and only water), and usually do so in my van. When they spring leaks (and they do), parts of the van get wet, but soon the water dries up, leaving no residue.] If female, find a jug with a wider mouth and a tight cap. To avoid spillage when traveling, be sure to jam or tie the jug in such a manner to ensure that it rides upright without tipping. To control odor, a little floor cleaner/disinfectant, rubbing alcohol or white vinegar can be added, if desired. To avoid polluting, don't dump on the ground if floor cleaner or any other chemicals get added (except alcohol, which evaporates, but can still kill grass).

> I bought my pee-toilet in a department store auto parts section for less than \$5. It came with a free gallon of window-washer fluid, which I transferred into my car/van before using the jug for its real purpose.

To take a dump: Do your thing into a plastic kitchen garbage bag opened and placed over a folding toilet that resembles a cheap folding chair. (Don't use plastic grocery bags as toilets--- they often leak.) Put a bucket underneath the seat and bag, to contain any possible leaks (but these kitchen garbage bags never seem to leak). There are two basic variations on this procedure, depending upon whether one disposes the waste immediately or stores it for a brief time for later disposal. *1) If disposing immediately:* Many remote places do allow the burying of human waste. Burying is the best thing to do if legal. Do not bury toilet paper; it takes a really long time to decompose and animals may dig it up and bring it to the surface, where it looks disgusting. Instead, put the toilet paper into a second garbage bag. Take bag #1 outside, dig a hole at least a few inches deep (to discourage animals from digging up its contents), empty its contents into the hole then bury. After emptying, place the empty bag #1 into bag #2 with the toilet paper, then fold and tie closed. Store this bag under a rock (so it doesn't blow away) until you can take it to a dumpster. *2) If storing for awhile:* Some remote places do not allow the burying of human waste, because these areas are very environmentally sensitive. It is best to obey these laws. Also, sometimes one needs to do one's thing in a parking lot of a well populated city or a rest area without toilets. Put both the toilet paper and body waste into a single bag. Close. Store this bag in a sealed container, such as a coffee can, nut can, empty milk/juice carton (these work well--- tape them shut) or a large disposable plastic drink cup from a fast food restaurant (these also work very well, can be reused many times, and garbage picked for free--- be sure cans/cups have undamaged lids). Store somewhere outside the RV until one can take to and empty the contents of each bag into a toilet or outhouse then throw the empty bags into a dumpster. One can make a special container outside the RV for storing these "toilet cups" during travel, however I have found that inside the double propane tank cover on the front of my trailer between the propane tanks there is ample space for these toilet cups. Several cups fit snugly in there and do not tip during travel. Once closed and sealed, there is not much outdoor odor--- still, dispose of their contents as quickly as practical.

This folding toilet cost me $10 in 1985. I still use it to this day (2017). The bucket cost $1.

Advantages: 1) <u>All dry toilets drastically reduce the need to deal with water</u>. One still needs freshwater for other purposes and produces wastewater from other purposes, so transporting water and filling/emptying tanks is still necessary. However, toilets [even those used in RVs] use lots of water, so eliminating their water share saves substantial water moving work. 2) <u>Weight gets saved</u>. Water is heavy. Zero water for a toilet weighs nothing. 3) <u>Space gets saved</u>. A permanent toilet and blackwater tank already installed in an RV take up the same space whether or not they get used. But a folding toilet collapses and becomes nearly flat [only about 2 inches thick] when not in use. Also, one has the option of making a folding toilet out of a folding chair and a standard toilet seat purchased from a home building supply store. 4) No electricity is needed for a water pump. Not only does this method save electricity, it also works if the electricity fails. 5) One can use this plastic bag strategy even if one has a regular RV toilet. Simply put the unfolded plastic kitchen bag over the existing RV toilet before use. 6) This strategy is super cheap. I bought my folding toilet for around $10 in 1985, and still use it today in 2017. Suitable urine jugs can be garbage picked for free. Plastic kitchen non-tie garbage bags cost around 15 cents apiece. If I ever find myself broke and homeless, it won't be from using too many bags for my RV toilet. 7) This is a clean system. Burying human waste only minutes after producing it is a much cleaner option than retaining and carrying it in a tank with water where it ferments. Also, because each bag covers the toilet and is used only once, the toilet itself never comes into contact with human waste, therefore never becomes dirty and does not stink. After one buries, the waste is gone. One can't see it, smell it or step in it. It is organic natural material, so it decomposes without pollution. Even if it is necessary to store several bags of solid waste for later disposal, once the bags are closed and stored in sealed containers with lids, it is surprising how little odor escapes. A urine jug can be emptied as often and washed as often as one likes. Finally, the interior of an RV airs out quickly if all the windows are open, even with no fan. After I take a dump and remove it, the RV is usually stink free within fifteen minutes if the windows are open--- often by the time I'm done burying the mess.

Disadvantages: 1) Using folding toilets & jugs could become chaotic if several people share the RV. This is especially true if waste must be stored for awhile before disposal. 2) People are accustomed to having water in toilets. Some people [especially children and teens], if asked to do something outside the norm, might flat out refuse or have brief psychiatric episodes if forced to do so. 3) One must remove solid waste from the RV very quickly after producing it. Obviously. This is only a problem if it's raining or there are bitter cold temperatures outside. [In heavy rain, it is permissible to just drop the used bag outside by the door, then deal with it later after the rain stops.] Using folding toilets & kitchen bags is more work initially after taking each dump, however in the long run, work is saved later by not needing to empty the non-existent blackwater holding tank. 4) Plastic bags take up space in landfills. Some people complain that throwing away lots of plastic bags pollutes the environment. This is true but there are worse offenders. Grocery shopping bags take up landfill space too. Taking used plastic grocery bags back to the store for reuse (until the bags tear or become contaminated by food) or for recycling can save on landfill space. I often repurpose old plastic grocery bags as regular trash bags. This almost eliminates the need for regular plastic trash bags. Landfill space can also be saved by other means, such as reusing (or not using) disposable plastic fast food drink cups or picnicware.

Feces and dumpsters: In many places, it is illegal to dump feces into regular garbage containers. It is often considered to be medical waste and a potential hazard to sanitation workers. (However, many people put dog poop, used cat litter, and used disposable baby diapers out with the regular garbage--- few get arrested. I have yet to come upon a dumpster or trash can patrolled by an armed guard, asking me what I'm tossing in there.) It is best to dispose of feces in toilets or by burying and place only the bags and toilet paper into the garbage. Wrap bags up well and contain them in other garbage, to minimize both smell and the risk of exposure to others. *Note:* Chemical toilet bags also exist for human feces disposal. Their chemicals sanitize, deodorize, break up and contain the mess. As far as I know, it is legal to toss these bags into the regular trash.

II.) DryFlush Toilets (dry)

I don't know much about these toilets and have never used one, so my advantage/ disadvantage list is largely guesswork. These possibly could work well in RVs because they are waterless.

Advantages: 1) All dry toilets drastically reduce the need to deal with water. This saves time from not needing to transport water to/from the RV for toilet use. 2) Weight is saved. Zero water for a toilet weighs nothing. Having no water mixed with the waste not only lightens the RV when traveling, it also lightens the waste to the extent that it can be easily lifted and moved by a reasonably healthy person for disposal. 3) The bag can be emptied often. The less time the waste remains with the RV, the less chance exists for odor.

Disadvantages: 1) As a permanent fixture, it takes up space. A folding dry toilet on the other hand takes up very little space. 2) This may be expensive. I'm sure that the folding toilet method that I use is much cheaper. 3) Electricity is used for each vacuum flush. How much, I don't know. 4) How reliable is the flushing mechanism? [I don't know.] How easy or quickly is it to find a technician to fix a problem, should a problem occur? 5) One is condemned to buy special bags from the manufacturer forever (as long as the toilet gets used). If users run out of bags in some remote area and it takes time to ship more bags… If the manufacturer goes out of business and stops producing these special bags and they become unavailable… I've heard the bags can get expensive… 6) Plastic bags take up landfill space. 7) It is illegal to dump feces into dumpsters in many areas. People do it nonetheless.

This seems like it might be a complex device to fulfill a simple need. Still, who knows… Maybe this type of toilet will actually catch on. I like to keep an open mind.

III.) Composting Toilets (dry)

These toilets could be useful in RVs. Apparently some RVers use them. There are some rave customer reviews posted on a manufacturer website--- but they may be only telling one side of the story. I don't have enough familiarity with these toilets to recommend either for or against them. I am extremely concerned as to what happens if the electricity runs out and the fan stops working and air stops circulating.

Advantages: 1) These toilets supposedly have very low odor. One of the main contributors of odor arises from mixing liquid with feces. This is because liquid hastens decomposition and decomposition causes odor. The liquid can be either water or urine. By directing urine into a separate container, the feces is kept relatively dry, thus slowing decomposition and lowering the odor. [This may be true. I've noticed that when picking up my dog's poop in my backyard, there isn't much odor when the poops are dry--- and I have a big dog.] 2) Storage space is saved. The holding tank on a composting toilet can be made smaller than a comparable holding tank on an RV water toilet, due to the lack of need for water. 3) Weight is saved by not adding any water to the human waste. 4) The solid waste holding tank does not need to be winterized. Freezing poses no problems, except that decomposition does not take place below 55° F. When temperatures rise above 55° F, slow decomposition resumes. 5) The solid waste holding tank doesn't need to be emptied very often. In fact, if such a toilet is used during the summer, one manufacturer recommends that the tank not be emptied until the following spring! If the feces/peat mixture is allowed to decompose for a long time, it supposedly comes out with only an earthy smell. It apparently stinks more if it comes out too soon. Of course, how often one needs to empty the tank depends upon how much the toilet gets used, therefore how quickly the tank gets filled to capacity.

Disadvantages: 1) These are expensive. Any fancy toilets such as these are not likely to be cheap. One model that claims to be about half the price of the competition, is priced just below $1000. 2) [very important] A vent fan must run constantly! Even a very low power fan will eventually run a battery down if that fan must run 24 hours per day. What happens if the fan stops working? Batteries don't store all that much electricity, equipment [and procedures] for proper battery charging is expensive, RV electrical systems are unreliable--- blah blah blah… I've already covered these points before, I'll probably need to make them again--- so many times, maybe I risk getting carpel tunnel syndrome from typing them into my computer keyboard--- but here's the deal: so many people just don't get these points; I feel compelled to keep repeating them. The toilet vent tube could be designed to move air by convection instead of a fan, if a small propane gas flame [maybe the size of a butane lighter flame] burned within it and the tube were well thermally insulated. I don't know if methane from decomposition could cause an explosion hazard though. If it did, the flame would need to be encased within its own separate adjacent tube airtight from the toilet vent tube. Putting the heat from even a very small flame within the vent tube would turn that vent tube into an air sucking chimney, thus eliminating the need for a fan and any electricity used for a fan. Perhaps the best vent tube could circulate air either from an electric fan or by convection from the heat of a small flame.

IV.) RV Water Toilets (wet)

These are very appropriate for RVs, especially considering that most RVs (except the very basic ones) already have them built in, so there isn't any need to add something. RV water toilets do work--- they're not great, but they are alright. They are somewhat different from regular house toilets in several respects.

Differences between RV toilets and house toilets: 1) RV toilets are lightweight. Unlike regular house toilets, which are made of thick heavy porcelain, RV toilets are made of thin lightweight durable plastic. 2) RV toilets are smaller than house toilets. But they are not that much smaller. 3) An RV toilet has no tank in its rear. A regular house toilet flushes by rapidly releasing the water from its storage tank. An RV toilet is instead connected directly to the water lines. When it flushes, water squirts from those lines directly into the bowl, and the horizontal flap at the bottom of the bowl temporarily retreats, allowing the contents to fall into the blackwater tank. 4) RV toilets use far less water than house toilets. Rather than filling the bowl entirely, only a small amount of water remains upon the toilet's horizontal flap after each flush. This horizontal flap should be watertight with the rest of the toilet when not flushing. The small amount of water at the bottom of the bowl creates a "water seal" which prevents really stinky air within the blackwater tank from moving up, entering and mixing with the RV living space air. 5) RV toilets flush with foot pedals. Some models have only one pedal, others have two. On single pedal models, pushing the pedal slightly usually results in only water being added to the bowl [but no flushing]; pushing the pedal all the way down results is flushing [in addition to water being squirted into the bowl]. On two pedal models, one pedal is for squirting water and flushing, the other is for water only. 6) RV toilets have different procedures for #1 and #2. When only urinating, one pedal uses only a very small amount of water when flushing. When taking a dump, the other pedal causes more water to get used. Just prior to taking the dump, one usually fills the bowl with more water than gets used for just urination. This helps prevent the mess from sticking to the bowl, and also helps to move it and the toilet paper downward during the flush. 7) Chemicals are usually added to the blackwater tank. This is done to inhibit odor and also to help physically decompose both the feces and the toilet paper into a sludge that is liquid enough to move through the wastewater hose when being emptied at a dump station. There are apparently many different chemicals to choose from.

typical RV water toilet
(note the foot flush pedals on right front and the lack of water tank in back)

Advantages: 1) RV water toilets are very common. Therefore, you will find no shortage of literature or friendly talkative campers explaining how to use them. You are also likely to easily find repair parts and RV technicians who can fix any problems that might occur. 2) RV water toilets do work. An RV toilet can be flushed as long as there is freshwater in the system, space in the tank and electricity for a working water pump. The flushing mechanism is mechanical, so if there is no water in the freshwater tank or the water pump fails, the toilet can still be flushed by manually pouring water into the bowl from a container while depressing the flushing pedal. 3) Emptying the wastewater tanks and refilling the freshwater tank is easy when the RV is on the road anyway and passes by a dump station. The entire process only takes minutes. However, making a special trip with an entire RV to a dump station for the sole purpose of emptying/refilling is neither easy nor fast.

Disadvantages: 1) Raw sewage is retained in a tank for awhile. The best thing to do with solid human waste is to get rid of it immediately, by burying, whenever possible. 2) Mixing water with feces causes fermentation and odor. Chemicals can be added to the blackwater tank to control odor, but they are never completely effective and they have their own perfume odors, which can be offensive. 3) The blackwater tank needs to be emptied often, perhaps every few days. Freshwater needs to be brought to the RV equally as often. 4) Water is heavy. It adds weight to the RV. 5) Electricity is used to pump freshwater into the toilet when flushing. This puts a drain on the battery. 6) When parked, dumping/refilling is a real bother. Transporting freshwater and wastewater to and from an RV that's been parked awhile takes time and effort. The other alternative is to drive the entire RV to and back from a dump station, but this requires packing [to ride-stabilize], de-leveling [taking jacks out from underneath], possibly retracting slideouts and awnings, driving what might be a lengthy route with a gas-guzzler, then repositioning the RV when parking, possibly extending slideouts and awnings, releveling, then finally unpacking items from their "driving" places and putting them into their "parked" places. Quite a bother. 7) If freshwater or holding tank space runs out, the toilet is useless. What is "Plan B?"

V.) Small Portable Chemical Toilets (wet)

For most RVs, these are not appropriate, simply because most RVs already have their own toilets. Small portable chemical toilets may have some use for some very small RVs not already having toilets, or as replacements in very old RVs whose in-house toilets have broken. They also have use for tenters.

Advantages: 1) They are small, portable and relatively lightweight. 2) Compared to other toilet types, they are inexpensive.

Disadvantages: 1) These toilets have limited storage capacities, therefore: 2) they must be emptied frequently. They are definitely not for frequent prolonged use. 3) The chemicals needed for these toilets stink. Harsh chemicals are needed for disinfecting and deodorizing. These chemicals can have objectionable perfume smells and some are toxic. Do not omit chemicals because: 4) the toilet tanks can explode! This is because decomposition releases gasses [such as methane and hydrogen sulfide]. Adding gasses to an airtight container causes pressure to build. Building pressure is what can cause a tank to "blow." 5) The toilets themselves can stink as they get old. This is because the plastic absorbs odors deep within itself; so deep that these odors can never be washed away. 6) These toilets need to be winterized, to avoid freezing and bursting.

Emptying: These toilets should be emptied into septic systems. They may also be emptied into indoor flush toilets. Generally, they should not be dumped at RV dump stations because of their chemicals. Emptying into outhouses is also not a good idea, because their chemicals slow down decomposition, which is a normal expected and engineered process in outhouses.

Final Thought On Toilet Complexity

It never ceases to amaze me how some people cannot do anything simply or cheaply. They take the simplest task, and by the time they get done engineering a solution, they've turned it into something that must cost thousands of dollars. Relieving oneself is a very simple task. Cleaning up afterward is only slightly more complicated. Animals relieve themselves outdoors wherever they stand, usually nature takes care of it. Humans like a bit more privacy, and sometimes people cluster in high concentrations, so some sanitation engineering is necessary. Still, the task of cleaning and disposing of human waste seems simple--- I could probably devise ten RV methods within a couple hours that would be clean and acceptable. Sometimes I think that some people have too much money or are overeducated.

Final Thought On Water in RV Toilets

Why do so many RV users and designers feel a need to mix water with human waste? My best guess is: *habit*. People are so accustomed to having water in a toilet, that they think the world will stop rotating, crack open and pull them into a grave of hot lava if things are any other way. I've said a truth earlier in this book, I feel compelled to repeat it here: *Familiarity is a lousy barometer for determining suitability, but familiarity is what many people use.*

Washing Machines & Dryers

All right, let's get real. You need a washing machine and dryer in your RV almost as much as you need a hole in your head. These things are highly impractical in RVs. But I'll cover them anyway.

Washing Machines

Suppose the following situations occur when it's time to do the laundry:

Case #1: You are parked in some wilderness area quite some distance from civilization. You could: A) pack the cloths into the car or tow vehicle, drive to town, and wash them at a public laundry facility or B) wash the cloths in the washing machine in your RV, if you have one. Possibility (B) sounds easier and rather appealing, until two questions get asked: Which weighs more--- the cloths or the water required to wash them? And: Which is easier to transport--- the cloths or the water required to wash them? If (A) you drive to and back from town to do the cloths at a public laundry facility, the *cloths* will make a round trip to and back from town. And so will you. If (B) you wash them in a machine in your RV, the *water* will need to make a round trip back from and to town. And so will you. (Yes, the water makes the round trip in the reverse direction, but this is irrelevant. In practice, the trip would entail hauling away wastewater and returning with freshwater.) Assuming the dump/fill station is the same distance from your RV as the laundry facility, and assuming the cloths require a sufficient quantity of water to warrant one extra trip to that station, no time nor effort could be saved by doing laundry "home" at the RV. And extra time and effort would be needed to do the wash "home" at the RV because: Cloths are lightweight and easy to move, as they can be put into flexible bags or even haphazardly plopped on a seat. The water required to wash the cloths weighs much more, requires two tanks to carry (one for the wastewater and a second for the freshwater), may require special lifting equipment/procedures (due to weight) and requires connecting/disconnecting of hoses and opening/closing of valves. And if a wastewater tank somehow leaks in your tow/spare vehicle, things could get messy. Even if the dump/fill station were considerably closer than the public laundry facility, even one extra trip to that station while fussing with heavy clumsy water tanks seems harder than just tossing some cloths into a car and taking a drive. The only situation where dumping and filling water is relatively easy occurs when the whole RV gets driving anyway and just so happens to pass very close to a dump/fill station along the way. If the cloths were washed in town, time would be required there, but that time could be put to good use for shopping, dining, reading (or my favorite use--- using my laptop to write a book about RVs). Of course, a washing machine is not an absolute necessity. You could always take the third option: C) throwing the cloths into a lake or river to get them clean. You would be surprised at how often this option is viable.

Case #2: You are parked at an RV park, hooked up to water, sewer and electricity. In this case, the washing machine is likely to work with ease. However, many RV parks have their own laundry facilities right there. In these places, your washing machine/dryer would be redundant.

Case #3: Your RV is permanently parked on a piece of private vacation property with a well and septic tank. This is the only case in which the washer could come in handy. However, if you only use the RV on weekends or for short time spans, it would probably be easier to pack enough cloths to last the entire durations of your stays, then transport and wash all the dirty cloths back at your regular permanent home after your camping stays end. It's not difficult to go a week without needing to do a wash.

Dryers

To grasp how badly you need a dryer, consider this: One year, the dryer in my regular traditional house broke in February. I didn't get around to replacing it until August. Cloths can almost always be dried outdoors on clotheslines. And during rain, or when the weather is cold, moderate amounts of cloths can be dried indoors, using the heat from gas lights or furnaces.

Disadvantages of RV washers & dryers: 1) They take up precious RV space [perhaps two cubic yards]. 2) A washer & dryer would cost quite a bit [hundreds of extra dollars]. And: 3) Extra money would need to be spent on fuel while driving, because these appliances would add weight to the RV.

above photos: gas mantle lights make lots of heat, thus are great for drying cloths

right: cloths drying in updraft of RV furnace

Conclusion

Yeah, a washer & dryer might seem nice, but in an RV it's just not practical. It's just too easy to go to a Laundromat, or dunk cloths in a lake then dry them indoors.

TVs, Stereos & Other Electronics

120 Volts AC Appliances In RVs

In regular houses, nearly all appliances run from 120 Volts AC. (Some run from 240 Volts, but these are too heavy duty to be considered for RVs.) In RVs, 120 Volt appliances can be run indirectly from 12 Volt RV or automotive batteries, using power inverters, which as covered in Chapter 1, are transformers that convert 12 Volts DC into 120 Volts AC. There are disadvantages and advantages to using regular 120 Volts AC household appliances from 12 Volt batteries and power inverters. As a general rule, it is best to choose 12 Volts DC appliances if they are available.

Disadvantages of using 120 Volts AC appliances: 1) Appliances made for regular houses often aren't designed to be energy saving. In regular houses, cheap electricity is available 24 hours per day 7 days per week in seemingly limitless quantities. [Consider this: In 2013, the average household electrical usage was 30 kWh per day. This is **40 times** the energy that a typical 100 Amp hour RV battery stores, when discharged from full to 40% full with an average of 12.5 Volts, which is 0.75 kWh.] Therefore, household appliances often don't need to be energy saving, thus they aren't. Although many such appliances could be run from 12 Volt batteries, they would run the batteries down too fast to make such operation practical. 2) Power inverters consume energy. In addition to taking a cut when running something, power inverters waste electricity even when nothing is connected to them, so they should be left off when not running anything and not on standby. 3) Power inverters make small amounts of noise because they are usually air cooled by fans. The noise is not terribly objectionable, however it can be easily heard by their immediate users in remote quiet areas, especially at night. 4) Most electronic internal components run from DC. A power inverter converts low voltage DC to higher voltage AC. The first thing an electronic appliance [such as a stereo] does, is convert that higher voltage AC into a lower voltage DC. [Believe it or not!] Each conversion wastes a bit of energy. Logically, it would be most efficient to use gadgets designed to run directly from a battery. However, DC voltage regulation is often still needed, and this throws some energy away.

Advantages of using 120 Volts AC appliances: 1) Many more appliance choices are available. This is because most people live in regular houses hooked up to the grid, whereas very few people live in RVs or remote cabins off grid. Manufacturers legitimately need to target their products to the largest population of potential buyers. Whether or not an appliance could *theoretically* run better if designed to run directly from 12 Volt batteries is completely irrelevant, if nobody actually makes a 12 Volts DC version of that appliance. 2) Some AC appliances are energy efficient. LED lights, TVs, laptops, generally don't use very much electricity, therefore are well suited to be run by batteries and power inverters. 3) Sometimes AC appliances get used for only very brief time periods. Kitchen gadgets, such as blenders, mixers, food processors and juicers, all work well from power inverters, because they operate so briefly [usually no more than a few minutes], that their power consumptions are negligible. The same is true of power tools, such as drills, saber saws and sanders. I can easily use such power tools for short time periods, with no fear of running my batteries down. 4) Power inverters don't use that much electricity. They only use about an additional 20% over that which the gadget uses. This waste, while slightly annoying, is unlikely to run down battery powered electrical systems easily. 5) Power inverters are cheap. 100 or 450 Watt output models can be purchased for less than $50 [2016 prices]. If a power inverter fails, simply buy another one. They are readily available.

In an early draft of this book, I advised RVers to avoid using 120 Volts AC appliances wherever possible. I have since revised my position. While avoiding 120 Volts AC appliances is a good piece of advice theoretically, it's just not practical. There are too many nice household gadgets commercially available that have no 12 Volt DC counterparts. And while electricity conservation is usually a good idea, there comes a point at which one should allow oneself to use gadgets guilt free. Still, many 12 Volt DC appliances do exist and are available at RV parts stores or online, so do check them out before automatically grabbing 120 Volts AC appliances.

TVs:

There are many considerations that go into choosing a TV for your RV. Most notably, do you really even "need" one? After all, shouldn't you spend most of your RV living time seeing and experiencing the wilderness? TVs are nice on rainy days though. I've got one in my RV and I certainly like it. If you do decide that you want a TV, how fussy should you be? Are you willing to be content with just about any old "Good Will" model? Would just watching occasional movies on your laptop be good enough? Or do you want a halfway decent TV? Getting a decent new TV for your RV is worth considering because small high quality TVs don't cost that much. There is a vast selection of TVs available today, in many different sizes and qualities and from many different manufacturers (and more keep coming out). In general, any TV buyer should: 1) view the TV in use in the store, to determine best picture quality and 2) consider price. For the RV user, there are two more special considerations: 1) How much power [including the power inverter] does the TV consume? [Include in this equation how often and for how long one intends to use it.] And: 2) Is there room in the RV for a TV that size? [I will let you answer question #2 for yourself.]

RV TV power sources: TVs can be designed to be powered from either 120 Volts AC or 12 Volts DC. While 12 Volts DC models might seem like a better choice for RVs, hunting for them may be a bit impractical, because regular household 120 Volts AC TVs are so much more available and in so many more styles and sizes. Also, if RV users need to run power inverters anyway, to operate DVRs or DVD players for feeding signals into those TVs, then there isn't much point to having a 12 Volts DC TV instead of a 120 Volts AC TV.

> I used to have a 12 Volts DC TV built into my conversion van. I really liked it. After it broke, I really missed it, so I went shopping for another TV. I found one at a van dealership, but it was $600. I told the dealer that I would need to "go home and think about it." I then found a small 120 Volts AC TV at a home electronics store for just over $100. It didn't use much power. So I opted for the $100 TV. I figured that with a savings like that, I could easily buy an additional car battery, if this TV used too much electricity. It didn't. I never needed to buy an additional battery; my new TV worked fine.

"Old" TV type: Cathode Ray: These are the "picture tube" TVs, manufactured until about halfway through the first decade of the 2000s. They were appropriately referred to as "ion bombardment tubes" by the "Coneheads" of the 1970s satire-comedy "Saturday Night Live." Each unit uses a magnet and a large vacuum tube, through which beams of electrons are directed toward a phosphorescent screen, which lights up in proportion to the flow of electrons hitting it. These TVs were appropriate for RVs not too long ago, when they were the only type available. Today, they are considered obsolete. They pose no advantages. They use between 3 and 4 times as much power as the newer LCD TVs, their picture quality is inferior to all TV types made today, they weigh more and are take up more space than the newer TVs. People are now putting them out to the garbage, even though they still work, or selling them real cheap at garage sales. If you want to garbage pick a cathode ray TV or get a cheap one at a thrift store for your garage workshop or basement, that's alright, but don't get this type for your RV. If your RV already has one, I recommend replacing it with a newer type.

Current TV types: 1) Plasma: These contain many gas filled gels, which glow in different primary colors when exposed to ultraviolet radiation, created by electric excitation of gas. 2) LCD [acronym for Liquid Crystal Display]: Such a flat screen TV uses a lit background over which light gets selectively blocked, in different colors. The default backing is fluorescent light [CCFL--- "Cold Cathode Fluorescent Light"]. These are also called LCD [CCFL] TVs. 3) LED TV: Such a flat screen TV uses the same LCD blocking screen as the LCD fluorescent, except that the background light comes from LEDs instead. These are also called LCD with LED backing. 4) Organic LED TVs [OLEDs]: These use thin films of carbon based diodes, that emit light when an electric current passes through them. [This process is called "electrophosphorescence."] Note: Plasma TVs and OLEDs actually *make* (emit) light. LCD (CCFL) [type #2] and LCD (LED) [type #3] start off with a lit background, and their screens *remove* light.

Relative efficiencies of TVs (rated from worst to best--- compared to LCDs): 1) Cathode Ray: These typically use 3 to 4 times as much power to make their pictures as LCDs [CCFL]. Wasting this much energy is why you definitely should consider a newer TV for your RV. 2) Plasma: These use 1.5 to 3 times as much power as LCDs [CCFL]. 3) LCDs [CCFL]: These are good efficient TVs. 4) LED TVs [LCD with LED backing]: These are right now the best choice in terms of efficiency, as they use 20% to 30% less power as the LCDs with CCFL backing. 5) OLED TVs: These are kind of new as of 2015. Their average efficiencies are even better than LCD LED TVs, but not by a wide margin--- only maybe 10%. Because they *make* light and use energy to do so, they are very efficient [using 0.4 times the energy] if the picture is dark, but can be inefficient [using 3 times the energy] if the picture is light. Because picture brightness fluctuates, it is difficult to determine with precision their efficiencies. Note: Efficiencies are only general approximations. TV sizes also affect relative efficiencies.

Practical Considerations: 1) Take your DVR from home. If you have a home DVR or a TIVO, and nobody else is going to remain living in your house when you live in your RV, record as many shows as your unit's memory will hold and take the unit with you. It will supply you with TV shows to watch for quite awhile. 2) Take a DVD player too. DVDs are available cheap at many grocery stores and Wal-Mart, and used DVDs are available real cheap at Good Will. Also, some stores still do rent a wide selection of DVDs at very reasonable prices. You may not get good antenna signals when deep in remote areas. 3) You may want to patch the sound into your RV stereo. This results in much better quality sound. If you do, the efficiency of your TV could be slightly increased if you can disable and completely turn off the TV's sound system.

Stereos:

Music from a stereo can be a great way for an RV user to take a break from all that time in the woods and have a bit of "civilization." Music also has a way of uplifting the soul.

Stereo power sources: Like TVs, stereos can be designed to be powered by 120 Volts AC or 12 Volts DC. Small stereos can also be powered by batteries. Stereos that run from 12 Volts DC are readily available and generally referred to as "car stereos." RV users shouldn't have any trouble finding or replacing 12 Volts DC stereos, considering that car stereos have been around longer than I have. I see no reason to choose a 120 Volts AC regular household stereo over a car stereo, however if you do, consider its power consumption as well as how long and how often you intend to use it. Also, if you choose a regular home stereo, find a way to mount it so that it doesn't break loose and become a missile or become damaged due to shaking during transit.

Built-in versus portable: Built-in stereos have the advantage that they are always there in the RV--- they don't need to be taken out for use then put away when done. Some RVers may opt for just small portable stereos (sometimes called "boom boxes" or "ghetto blasters"). These have the following advantages: 1) They are often small [so they can fit in many places]. 2) They are readily available. 3) They don't need installing, so they are ready for use immediately after buying. 4) They can be taken and used outside [where they are great by campfires]. 5) Some of them with subwoofers have an awesome quality of sound that is superior to that which was available just a few years ago. 6) The same units can usually be powered from a choice of either 120 Volts AC or batteries. Boom boxes can also be easily converted to be powered by the RV 12 Volt batteries instead of by their internal disposable battery packs, using inexpensive components.

As with TVs, how fussy one wants to be probably depends upon how often one intends to use the stereo, including how much time one plans to spend actually living in the RV. As with TVs, it is a very good idea to actually listen to a stereo in the store before buying it.

Computers:

Regular laptops are very usable and appropriate for RVs, as they are small and don't use much electricity. In RVs in the field, they can be used to write and e-mail letters to friends and family, to edit/alter digital photographs, get wireless internet through cell towers (some of which work for miles into wilderness), play movies or patch music into stereos. Even though a laptop runs from DC, electronic experts recommend that power inverters be used to convert the 12 Volts DC from RV batteries into 120 Volts AC, even though the computer turns that AC right back into DC for its own internal use. This is because strict voltage regulation is needed, as even slight voltage variations can wreak havoc computers.

Printers:

Printers can be used to print digital photographs or written documents for the purpose of mailing to friends. Having a printer in your RV might seem appealing at first, but if you think about this a bit, it probably isn't. Printers are very precise machines, so they could easily be damaged by the extraneous jostling of road travel. Then there is dust--- yes, you will encounter tiny dust particles in your RV, from either driving on unpaved roads, or silt storms. While driving, wind blows these particles everywhere. (Airborne particles of river silt entered my parked trailer once, even though all my windows were closed. My camera equipment has never been the same since.) Silt could murder a printer. Then there is the possibility of objects, food or liquids falling onto it. Also, printers can be so sensitive to improper voltage fluctuations, that some of them can't work right with certain power inverters. If you insist upon taking a printer in your RV, you're on your own. Instead, if you want physical prints made, I recommend taking any digitally stored recent photographs or text documents to a photography department at a drug or department store or an office supply store. Their printing machines are very advanced and produce outstanding quality at low prices--- even printed on ordinary paper, color photos look fantastic. Unfortunately, computers sometimes can't read each other's files. If your photos are stored as JPEG images and text documents are stored as PDFs, you should be alright, as anybody's computer can open these. If you are photographically inclined, when you go to a store that can make prints, you may feel like a kid turned loose in a candy store.

Small electronics with rechargeable batteries:

12 Volts DC chargers that plug into cigarette lighter type plugs are available for many small electronic gadgets, such as cell phones, G.P.S. devices or i-pads. Don't neglect to keep the batteries for these items charged in the field. If you don't have 12 Volts DC chargers for everything, do one or both of the following: 1) When not in use, keep these items connected to their regular household chargers and plugged into the 120 Volts AC receptacles of your power inverter. Then, when you turn your power inverter on for some other reason [watching TV, running LED lights, running kitchen gadgets], your little e-gadgets will automatically start charging--- even if you've completely forgotten about them. These tiny batteries use so little power, they hardly drain your deep-cycle batteries at all when charging. 2) Put these gadgets in your car, and plug them all into the small cigarette lighter power inverter that you plug in or turn on when you take a ride. Your little e-gadgets will charge as you drive with absolutely no battery drain.

Electronics then and now…

I bought my trailer in 1985. I used it for several initial trips between 1985 and 1991. Back then, letters home were written with pencils and pens on paper and sent through the U.S. mail. Now, people send texts or e-mails. Back then, I used public pay phones to call home. Now, I use a cell phone. Back then, photography was done with film, photosensitive paper and chemicals. When I took a photograph, I didn't get to see it until months later, when I finally got home from a trip and developed the film in my basement (which was lots of work). If I wanted to send a really nice picture home, I bought and mailed a postcard of a photo somebody else took. Now, travelers take pictures digitally, edit them on laptops while way out in the bush, connect to wireless internet in cities, and e-mail their photos to friends/family very soon after being taken and while still out on their trips. (Or maybe they post them on FaceBook.) My, my, how the world has changed. Some of these changes are for the better.

Things That Support Appliances

Switches

Master Disconnects

RVs are seldom occupied 365 days per year. Usually, they are used as vacation or second homes, and even full-timers sometimes stay somewhere else for awhile. Thus, even a typical heavily used RV is likely to spend long time spans of several months with little human monitoring. If even one of the house appliances gets left on when an RV gets put into storage, the result can be disastrous--- a totally discharged and ruined battery (or maybe several ruined batteries). The probability of leaving an electrical gadget on increases as RV electrical system complexity increases. Complex RV electrical systems not only have more components (to leave on) than simple RV electrical systems, but many complex components have circuit boards, which can place slight (phantom) loads on RV batteries, even if nothing appears to be turned on. Furthermore, a complex multi-component system has a multitude of wires, any of which could short, and if a short is very light (high but not infinite resistance), its presence may not be detected by any means other than noticing that the battery tends to mysteriously and repeatedly go completely dead after the RV sits for awhile. A high-current super-heavy-duty switch called a master disconnect switch is the answer. When in "off" mode, this switch shuts *everything* off (it disconnects the house battery from *all* the household appliances). Fortunately, some RVs do have master disconnect switches. (I did not think this idea up myself.) I believe that all RVs, except maybe those with extremely simple electrical systems having only a few components, should have master disconnect switches. It is just too easy to accidentally leave something on and ruin a battery.

If a solar panel or 120 Volts AC trickle charger is used to prevent battery discharge during RV storage, the charger should be connected upstream of the master disconnect switch (so that the charger still is connected to the battery and charges when the switch is in the "off" position).

Timer Switches

My mother's motorhome had a feature that was both nice and not-so-nice: the cupboards and closets had their own interior lights. At first I thought this was such a great idea. Lights in cupboards allowed us to easily see the contents of those cupboards. But then I realized one reason why this might not be such a great idea: one needs to remember to turn those lights off. If one turns on a light in any compartment, not visible from the RV interior or exterior, forgets to turn that light off, then closes the door, that light can stay on for a very long time period and substantially drain the battery(s). If even one small closed-compartment light is left on when an RV (with no master disconnect) gets put into storage, it can completely drain and trash an RV battery--- a mistake that costs $100 or more. (Yes, someone made this mistake in my mother's RV.)

My proposed solution is that all lights in closets, cupboards and compartments (accessible from the inside, outside or both), not visible when closed, have special electronic switches that automatically turn themselves off about fifteen minutes after they get turned on. This way, if a user forgets to turn the switch off, the switch does it for him/her. Anyone wanting a light on for more than fifteen minutes (maybe to clean a cupboard) wouldn't be too inconvenienced by needing to hit the "on" button again every fifteen minutes. Tiny electronics do not cost much money, so such switches could be very reasonably priced.

There are RV switches that automatically turn on and off when doors open and close. This seems like it might also be a reasonable solution to the lights-on-indefinitely-in-cupboards problem, but I wonder if such switches could malfunction and leave a light on after a door is shut. (If the door is open just a crack, might the light stay on indefinitely? Could some other circumstance fool the switch into thinking that the door was open when it was actually closed, and cause the light to turn on indefinitely?) I believe that my timer-switch idea is best for all lights in closed compartments.

Timer-switches could also benefit other electrical stuff, such as power inverters (when no 120 Volts AC load is applied for awhile), bathroom fans, water pumps (which can keep pumping indefinitely when the water runs dry), TVs (in case somebody wants to fall asleep with the TV on) and spark (re)igniters (which might spark indefinitely to try to light burners that can't light if the propane runs out). Timer switches connected to gas valves might also benefit small catalytic propane heaters, by turning them off after awhile, as they shouldn't be left on indefinitely (especially if somebody falls asleep).

Connectors

Trailer connectors:

A trailer and its tow vehicle usually do not stay connected all the time. When they disconnect physically and part company for awhile, they must also disconnect electrically. When they reconnect physically, they must also reconnect electrically. The plugs that are used to make this connection tend to have a serious defect: *they are often made out of aluminum (or some alloy thereof).*

Aluminum is a poor metal choice for electrical connectors

Why is aluminum a poor metal choice for connectors? As a bare metal, aluminum is a very good conductor of electricity. Bare aluminum exists on the *inside* of all wires made of aluminum. The oxide of aluminum however, does not conduct electricity. The oxide of aluminum tends to exist on the *outside surfaces* of wires and connectors made of aluminum. Thus, where two separate pieces of aluminum meet and are in physical contact with each other (as in electrical connectors), their surface oxides block the flow of some electrons. Some electrons do get through anyway, so lights and other electrical gadgetry may work, but sometimes, enough electrons get blocked to cause complete malfunction or hindered performance of some electrical gadgetry (lights may be too dim, heating elements may not get hot enough, motors may struggle).

Where does the surface oxide come from? Every aluminum wire and connector starts its life as a piece of pure aluminum (or an alloy thereof). Aluminum however is a very reactive metal. Any surface, not encased in plastic/rubber or solder, gets exposed to the oxygen in the Earth's atmosphere, where it quickly oxidizes into aluminum oxide. The aluminum oxide then adheres to the pure metal beneath it, and impedes the flow of electrons (provides an electrical resistance) from one piece to another.

Does aluminum "rust" like iron? In some ways, aluminum does "rust" like iron, in other ways, it "rusts" very differently from the way iron does. Yes, aluminum does react with oxygen in much the same way that iron does. But both the oxidation processes and the physical oxides of aluminum and iron differ in several ways: 1) The color of aluminum oxide is white or gray, so its presence upon a surface may go visually unnoticed. The oxide of iron [rust] is an unsightly reddish brown, so it's very noticeable. 2) Aluminum is more reactive [with oxygen] than iron. 3) The oxide of aluminum stubbornly adheres to the bare metal substrate beneath it, thereby actually protecting the bare aluminum which it encases. The oxide of iron binds very loosely to its iron substrate, so it tends to fall off, exposing new bare iron underneath, which then rusts, falls off, exposing more new iron, etcetera. In this process, an unpainted iron fence or structure is likely to rust away until there is nothing left of it, whereas an aluminum structure may last a long time.

> Because aluminum oxide binds very stubbornly to the metal beneath it, thereby protecting it, an aluminum structure can last for perhaps centuries, whereas a similar iron structure might rust away completely in just a couple of decades. This is true even though aluminum is a much more reactive metal than iron is.

In outer space, aluminum is a great metal for electrical wiring and connectors, since it is lightweight (hence easier to accelerate to orbit) and there is no air in the vacuum of outer space to corrode its surface. But here on Earth, unprotected aluminum is a lousy metal for transmission of electricity, especially at unsoldered connection points.

If aluminum is such a bad metal for automotive and RV electrical connectors, why is it used so often? Damned if I know.

> Aluminum is so reactive that for a long time, it was difficult to win it from its ore. Even though aluminum is quite common (in compounds in the Earth's crust), in the 1800s, metallic aluminum cost more than gold.
>
> Aluminum is so reactive that its powder can be (and sometimes is) used to make explosives.

What can be done to improve trailer connectors? 1) <u>Electrical connectors could be plated with noble metals</u>. Gold is a good choice. It just doesn't seem to want to oxidize, therefore its surface keeps on conducting electricity as it ages. This might seem like an outrageous suggestion, considering the price of gold, however only a super-thin coating [over a much cheaper metal] is needed. Some flashlight battery connection points were once plated with gold. Trailer connectors are likely to be used for much longer time spans than disposable batteries, so if plating could be done to batteries, why

not with trailer connectors? Plated trailer connectors probably won't last forever, as the surface plating will probably wear away as the connectors get plugged into and unplugged from each other multiple times, but they probably could last for quite awhile. 2) Spring clips could apply force to connecting points. These would greatly reduce interruptions in circuits, occurring from jiggling while traveling on roads. 3) Dielectric jelly could be used. A greasy jelly made specifically for this purpose supposedly conducts electricity but impedes corrosion [by blocking air from contacting metal]. This jelly can be applied to the surfaces of connector pins and holes. Some people swear this works well. I've tried it, but found that my connectors corroded almost as quickly as they did without the jelly [I didn't notice any difference], so I've abandoned its use. Some people instead use ordinary petroleum jelly. They say this works well too.
4) The connectors can be occasionally sanded and reamed. In the absence of using a noble metal plated spring clipped connector [somebody would need to design and market such a connector, and I don't know if anybody does], this is the best option. Although the oxides of aluminum and many other metals adhere stubbornly to the metal surfaces, sanding and scratching can abrade them away, exposing bare metal beneath, which conducts electricity very well. Sanding and reaming can thus render old electrical connectors good as new. Take a small piece of fine grain sandpaper and work and agitate it between all the connector pins until they are all shiny and new looking. Take a drill with a bit smaller than the female connector holes, energize the drill to start spinning the bit, and agitate it while scratching away internal corrosion. This procedure takes maybe the better part of an hour, and should be repeated every couple of months with actively used tow vehicle / trailer combinations [as connectors corrode quickly], but it's worth doing. It certainly is a better option than driving around with trailer lights and brakes that just don't work right.

left: Corroded connector pins do not conduct electricity very well.

right: Sanding off the corrosion exposes bare metal, which does conduct electricity well. Note that for this photo, only some of the corrosion has been removed.

Appliance & wire connectors/connections:

In order for each DC appliance to work, electricity must be able to get to that appliance from the RV battery (or power converter). To do this, good electrical connections are needed. If electrical connections are poor, due to being loose or corroded, appliances will struggle or not work at all, even if batteries are fully charged, and even if that appliance is in perfect working order. Appliance connectors occur where appliances meet the wires that carry electricity to them. Connectors may also occur within the internal structures of appliances (to and from on/off switches, for example) or in the wiring outside appliances (to and from fuse boxes or at wire junctions, for examples). Appliance and wire connectors suffer from the same serious defect that trailer connectors do: *they are often made out of aluminum (or some alloy thereof).*

What can be done to improve appliance and wire connections? 1) Don't use aluminum connectors. Instead: 2) Connect the wires to each other directly. Strip the ends of the wires, twist them together, then wrap up in electrical tape [or use an electrical cap]. Electrical tape tends to seal out much of the oxygen in the air, preserving that connection's integrity. Connections made this way can last a long time. [I made several connections by this method over 30 years ago, which still work today.] Connections may still corrode and degrade, even if the wire is copper. If you really want good connections: 3) Solder twisted wires together. Properly applied solder encases metal within metal, allowing electrons to flow unimpeded from wire into hardened solder then into the other wire. The inside of the metal solder connection [where electrons flow] is completely sealed off from the outside air forever, so the connection never corrodes and thus lasts forever. Soldered connections are very strong physically too--- if you try to pull a good soldered connection apart with much force, the wire usually breaks at a point somewhere other than the point of the soldered connection. Soldered connections should still be wrapped up in electrical tape, to prevent metal from touching other stray metal or wires and causing short circuits. There is one disadvantage to soldered connections: they are so strong, that if the appliance must be removed for service, the wires need to be cut then soldered back together when the appliances gets returned. Note #1: Be sure to heat *the wires* then apply solder to those wires. This helps make the solder stick. If you instead heat *the solder* and try to apply it to cold wires, the solder tends to not stick. Note #2: Solder tends to stick only to stranded copper wires. It doesn't stick well to aluminum wires. I also can't seem to get it to stick to solid copper wires.

Testing Appliance Currents

To be successful at off-grid RVing, one needs to develop a sense as to how quickly each appliance (and the sum total of all appliances) drains the battery(s). This can be done by intuitive experimentation, which is developing "gut feelings" after noting how long various things can be left on before the battery begins to struggle. It is however much better to use an ammeter and put some numbers on the system and do it mathematically. Knowing how much current each appliance draws, and multiplying that value by the time the appliance runs (you know your own usage habits), tells you how much the battery discharges, and comparing this value with the number of Amp hours that a battery stores, allows a more precise method of gauging how long a battery can perform before running down and needing recharging.

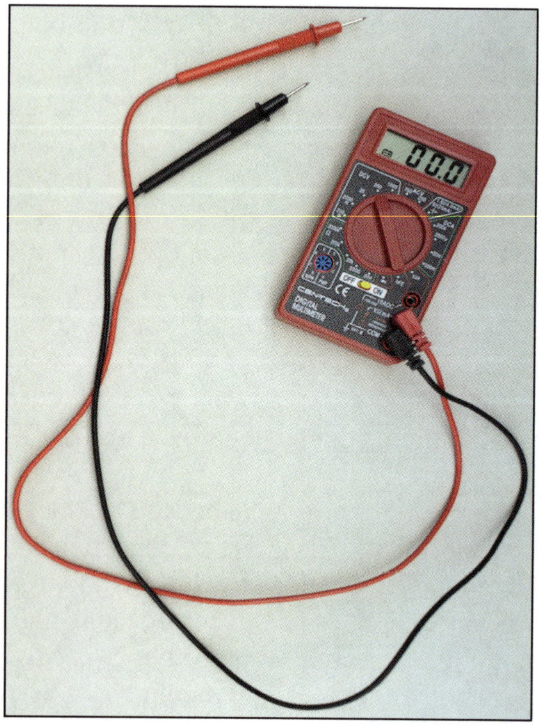

A digital multimeter allows precise measurements of currents.

You can use current estimates in RV books, but these should be used only as a starting point, as they vary widely, and even my low electricity-saving estimates may not reflect the individual idiosyncrasies of your system. You should get to know your own system. The best way to do this is to test the current draw of each gadget yourself. Some of the results may be somewhat unexpected and surprising.

To measure current, place an ammeter in series with the circuit, as shown on the next page. The ammeter can become part of the circuit in either the positive or the negative wire. Many ammeters are not supposed to accept current through them for too long (especially for higher current levels--- high currents cause heat--- prolonged times cause that heat to build up), so I like to connect a short circuit in parallel with the ammeter before turning the appliance on. I then disconnect the short (forcing all the current through the ammeter) for a brief time, long enough to take a reading, then I reconnect the short before turning the appliance off.

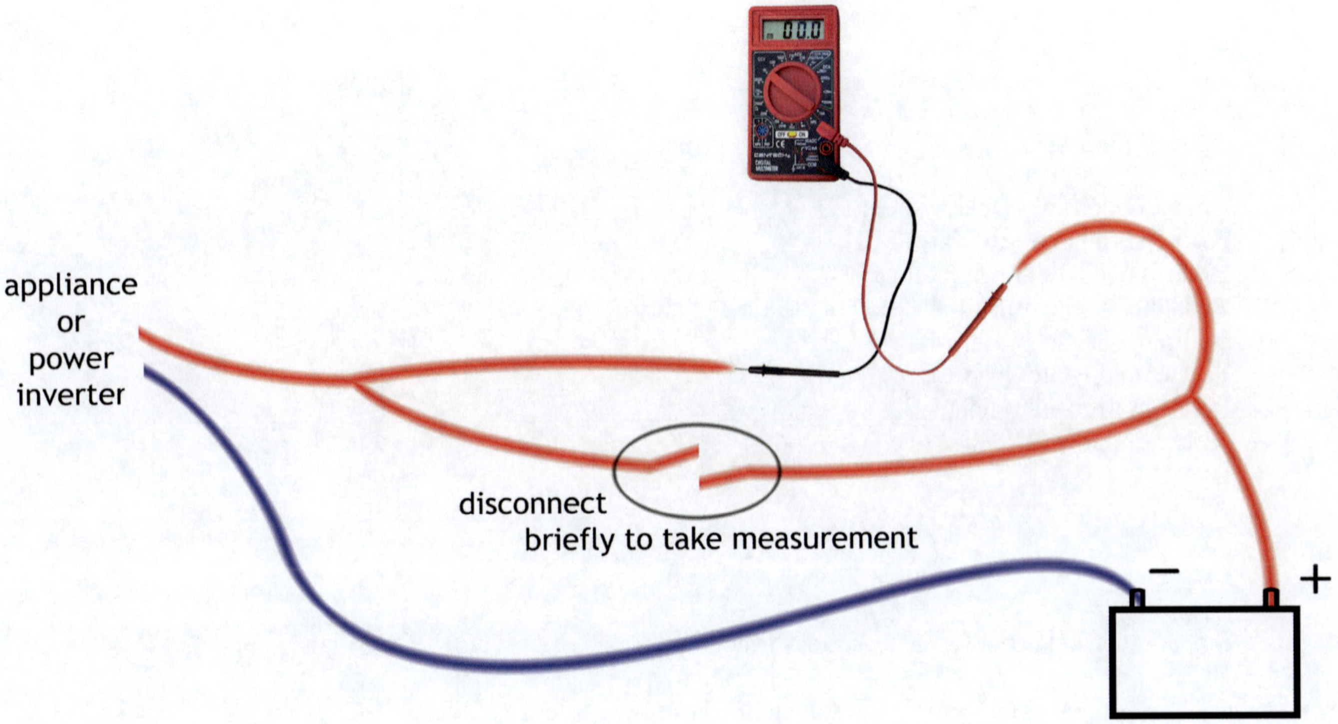

You can also use an Amp-hour meter to keep track of both currents and time, but you still should know approximately how much current each appliance draws.

On the next page is a table of the actual current measurements I took of the appliances in my own personal trailer RV. Hopefully this table can give you some insight as to what usages are possible and can be reasonably expected.

Current Measurements in Author's Trailer

12 Volt gadgets	Amps
Livingroom bright (cold white) LED	.24
Kitchen bright (warm white) LED	.09
Blue "mood" dim night light LED	.11
Red "mood" dim night light LED (a clearance light)	.07
Bathroom bright (warm white) LED	.20
Porch (cold white) light LED	.09
Porch (warm white) light LED	.09
Kitchen double (2 "bulbs") LED	.23
All nine of my above LEDs combined	1.12
Fluorescent really bright double 17 inch bulb kitchen light	2.55
Standard incandescent RV bulb (I don't use these)	1.36

(Note that all 9 of my LEDs combined use less current than a single standard incandescent RV bulb.)

120 Volt AC gadgets run by/with power inverter		Amps @ 12 Volts [1]
Inverter off but connected (phantom load)		.03
Inverter on alone (no appliances attached)		.14
60 Watt (equivalent brightness) kitchen warm white LED		1.53
60 Watt (equivalent brightness) livingroom warm white LED		1.35
Both above lights run together		2.75
60 Watt (equivalent brightness) coiled fluorescent light		1.33
100 Watt (equivalent brightness) coiled fluorescent light		2.10
Incandescent night light with small bulb (I don't use these)		1.48
TV alone (no signal fed into it)		1.80 [2]
TV and DVD player together		2.31
TV + DVD player + boom box (for sound)		2.45
CD player [3] and boom box together (my stereo)		1.16
Laptop (off but plugged in possibly charging its battery)		2.15 [4]
Laptop (on and being used)		5.6
Cell phone charger alone (while charging phone)		.29
Cell phone charger on with another device	adds	.17 [5]
Nikon charger (for Nikon digital SLR battery)		1.09

Notes:

[1] These currents are the 12 Volts DC currents leaving the battery heading toward the inverter. The 120 Volts AC currents leaving the inverter heading toward the appliances should be about 1/10th of these, but the latter are not relevant for calculating battery discharge. It is the 12 Volts DC currents (multiplied by the times the gadgets actually run) that determines battery discharge.

[2] This is a 40 Watt LED TV with a 19 inch (diagonally measured) screen. Mathematically, a 40 Watt TV should draw about 4 Amps @ 12 Volts. [(48 Watts) = (12 Volts) × (4.0 Amps). The extra 8 Watts has been added for the power inverter.] With no peripheral devices, the TV however only draws less than half of that. This case demonstrates that you should do your own measurements and not rely solely on technical specs in manuals and data in books collected by others.

[3] This "CD player" is actually my DVD player which also plays CDs.

[4] Note that this is an extraordinarily high phantom current for a device not actually on. Unplug your laptop when not in use, else your RV battery might be mysteriously draining too fast when you turn your inverter on. If you desperately need to charge your laptop battery, I suggest plugging your laptop into the inverter while driving, as no automotive battery draining takes place while driving because all the electricity is made by the alternator.

[5] Because the cell phone charger uses only a very small amount of current, if it is the only item being run, the power inverter takes a significant energy cut (41%). It is best to operate a cell phone charger simultaneously in conjunction with another appliance that would be used anyway. For this test, a 60 Watt (equivalent) LED, alone drew 1.28 Amps. When the cell phone charger was added to the light, the combined drew 1.45 Amps.

Chapter Conclusion: Choices Exist But One Must Look

Humans have an immense capacity for intelligence and creativity. Unfortunately, we also have an immense capacity for stupidity and stagnation. Our brains are somehow programmed to repeat the same behaviors over and over again. We do this with many aspects of our lives. We often select from the same small number of groceries, even though thousands are available. We often order the same food from the same restaurants, sometimes on the same days of the week, even though many menu items and restaurants exist. We hang out with the same friends, engage in the same small group of hobbies, listen to the same music over and over again (even though more music exists than any one human can listen to in a lifetime). Sadly, we are often oblivious to most of what reality has to offer us. What happens if something jolts us into realizing just how big the universe of possibilities is?

When I look at RV electrical & appliance systems, I see this same repetition and stagnation. It looks to me as though one small group of people came up with a standard RV design several decades ago, then everybody else copied it. I long for variety, yet am almost always disappointed by a clone of that same stupid design, that like a life form, somehow reproduced, jumped across brand names, and eventually proliferated and extinguished everything else. Without real choices, I can readily understand why some prospective RV buyers might get discouraged.

But don't be discouraged. Possibilities give us choices. Where there are choices there are solutions. This chapter is about appliance possibilities, choices and solutions. Perhaps you find the wealth of information presented here to be overwhelming. I hope that instead though, you take it as a constructive challenge. Your dream RV might exist only in your imagination in the present day, but remember that every bridge, skyscraper, TV, airplane and every other object that people make, started in someone's imagination before it actually became a physical object.

Sometimes it is appropriate to try new food, make new friends, engage in new hobbies, listen to new music. It is high time that new workable RV designs be tried and put into production (and older forgotten better designs be resurrected)--- in fact, such changes are long overdue. The RV world could put out good products, but they need to break away from the same-old-same-old.

Conclusion

Building Better RVs

A New Plan

One day, while camping, I started talking to a woman in the site next to me. After lamenting incessantly for awhile about poor RV designs (which I often do), and telling her about the book I've been writing, she interrupted me by saying: "Just think how good you will feel the first time ever you see an RV driving down the road with your name on it." This remark stunned me and shut me up almost instantly. Wow! What a concept! A "Thomas Childs RV." This got me thinking. Even though it is unlikely I will ever build RVs, I could still offer to design them. I could also inspect them and grant "Thomas Childs Seals of Approval" to any models that I deemed fit for the wilderness. To do this, I wouldn't need much money, social connections nor anyone else's permission, because I figure: in a worst case scenario, even if I were homeless, living under a bridge, pissing into a river, and had large lawsuit judgments hanging over me (for telling the truth), I still own all rights to my name. Nobody could compel me to grant a favorable review. Nobody could put "approved by Thomas Childs" on any RV except me.

To put this idea into practice, I realized that I would need a written set of RV standards, both for my own evaluation and for participating designers/manufacturers to strive to comply with. I decided to write a new set of codes for building and testing RVs, and also for company procedure in responding to customer feedback.

I began writing these building codes, and initially planned on including them in this book. But I ran into several problems: 1) *They were redundant and quite boring.* I've said all I need to say already. If I included such building codes in this book, the book would become hopelessly long, and I would lose many readers. My ideas too might then get lost. Also, building codes are really technical; they are like numerical measurements or a really dry wine--- they usually are of use only to people while in the process of designing, building, customizing or evaluating. 2) *Including building codes would delay publication of this book.* I've been writing this book for almost nine years now. If I don't publish it soon, maybe I never will. If any book were to be assigned a number describing its value, with 100 representing a perfect book, and 0 representing a completely useless book, well, any unpublished book is still a 0, since nobody can buy a copy and benefit from its ideas. 3) *A single set of codes can't be applied to all RVs, because RVs come in many levels of complexity and cost.* As I began writing these codes, I realized that even my own beloved trailer wouldn't pass them, because I was gearing those codes toward more upper end RVs. Several levels of code are needed, depending upon an RV's complexity. Finally: 4) *There are still many design issues of which I am uncertain.* I would need a staff to build prototypes and do extensive testing before I could state some things definitively. For that, I would need money. Some of my suggestions in this book are at this point just best guesses.

I still plan on writing such building codes, but if I do, I will publish them later separately.

Dealing With Existing Building Codes

In order to avoid breaking any existing laws, and to fulfill their purpose of fostering quality RVs, my "Thomas Childs" building codes would need to be stricter than any existing laws. They would be completely voluntary, but nobody in non-compliance could put my name/approval on any of their RVs.

The present existing codes are: 1) NFPA (National Fire Prevention Association) codes, 2) RVIA (Recreational Vehicle Industry Association) codes and 3) RV Consumer Group standards. NFPA codes (which apply to buildings and their components, including RVs) exist primary for the purpose of preventing fires. I don't want to challenge any laws that legitimately exist for safety. Apparently though, things don't need to work to pass these; things only need to not start fires. RVIA codes are a joke. They are designed by RV *manufacturers*, not RV *users*. They probably exist primarily to widen the profit margins of manufacturers, but also to project an *illusion* that reasonable standards exist, when in reality they do not. These codes are too lenient. When any type of codes or standards are too lenient, the item under review passes almost no matter what. However, passing doesn't mean it will actually hold up in the real world. Consider this: Any goofball can make an RV with all the defects described in Chapter 2, and that RV will pass RVIA codes. RV Consumer Group is an evaluation organization independent of RV manufacturers. I have much respect for this organization. It is my impression that this group focuses on different aspects of the RV than I do. (They focus a lot of attention on the physical construction of RVs.) I believe that my goals and ideas compliment the goals and ideas of this organization rather than competing with it.

Problems could arise if existing laws require specific poorly functioning equipment while prohibiting other equipment that actually works. RV manufacturers want to continue with business as usual. Also, private RV parks would not like RVs that can go off-grid, as such RVs would decrease the need to patronize their parks. There is a substantial economic incentive for both these groups to influence building codes. Also, these two groups might use litigation to cause trouble to anyone advocating RV design improvement. It may take a small army of dedicated people to break down any barriers to the cause. If anyone ever does commercially produce an RV with an electrical & appliance system that actually works, it will cause great embarrassment to those who currently make RVs.

The Appliances:

Lights would be the first difference that people would be likely to notice and they would notice this difference almost immediately upon entering the RV. I would ban the use of incandescent lights almost everywhere. Instead, I would require a combination of propane gas mantle lights and LEDs. Gas mantle lights would be the main bright light source. RVs smaller than 12 feet long would require at least one gas mantle light; longer RVs would require at least two. Most of the rest of the lights would be LEDs. A combination of both 12 Volts DC (run directly from battery) and 120 Volts AC (run from battery with a power inverter) LEDs would be used, since they have different properties that compliment each other. (120 Volt AC LEDs tend to be much brighter than their 12 Volt counterparts, but they consume more electricity and require power inverters to be on.) I would allow some 12 Volt fluorescent lights too, used sparingly. The only incandescent bulb I would allow would be one small one (10 Watts max.) in the conventional oven. (LEDs are damaged by heat, therefore are inappropriate for ovens.) Any lights in compartments, that can't be seen when closed, would be required to have timer switches, which would automatically turn those lights off 15 minutes after being turned on. Placement of lights would also be important. I would make sure that all stoves, sinks, counters and dining room tables had good artificial illumination. Finally, I would also include a couple of dim and colorful LED mood lights, for relaxing at night.

Furnaces would be the second main difference, which people might not notice until actually out in the field. I would ban inclusion of those ridiculous forced-air furnaces that don't heat. (The only way I would even reconsider a forced-air furnace, would be if it could pass the rigorous 99 day field test, detailed in Chapter 5.) Furnaces would all need to be passive (non-electric) propane burning furnaces. They could be mounted at floor level, beneath the floor, or someplace low--- of course, all flammable materials (including rugs) would need to be far enough away such that they wouldn't catch fire. Thermostatic control would be required. Electronic ignition is a good convenience, because pilot lights can be very difficult to light under high wind conditions. Electronic furnace ignition would be allowed (even encouraged), either from 12 Volt battery or piezoelectric, but only if those furnaces could also be lit by matches or butane lighters, if the power failed. Bigger RVs would have two or multiple modest size furnaces instead of one big central one. This would allow for even heating, and also *some* heating, in case one furnace failed but another didn't. Although all main furnaces would be required to vent their exhausts to the outside, bigger RVs would be allowed to have several small catalytic indoor-vented supplemental heaters in cubbyholes (low oxygen sensor and timed shut-off valves would be required). A set of storm windows, along with a special compartment to store them, would be highly recommended for all RVs and required for upper-end RVs. Storm windows are great to have in bitter cold conditions, as windows (especially louver windows) can be quite leaky. Nobody wants to heat the outside. Finally, carbon monoxide and unburned propane detectors would be required. They would be powered by the RV 12 Volt systems, but each unit would need to have and automatically revert to its own backup battery in case the power failed. I certainly don't want anybody to die in any RV approved by me, due to its design.

Electrical consumption savings with this plan: The above recommended plan for lights and furnaces would cut electrical consumption for those combined appliances by an astonishing 95% (using only one twentieth the electricity), yet the RV would still be brightly lit and warm and would even function far more reliably. Best of all, lights and furnaces are the major consumers of electricity in typical RVs, so these two changes combined would make a Thomas Childs RV a drastic improvement over a standard RV.

Refrigerators would be required in all RVs. Smaller RVs would need a refrigerator only; larger RVs would need to have freezers as well (which could be located inside the refrigerators). Fridges would be almost the same as the ones currently in use; all would be the propane fueled ammonia absorption type. They could also switch to 120 Volts AC while parked with municipal electricity, or 12 Volts DC while driving. The major requirement difference would concern circuit boards and electronic ignition. Circuit boards (which control some fridges) and 12 Volt electronic ignition would be allowed, *but only if the fridges stay working and are capable of being started (lit by match, butane lighter, piezoelectric) in complete electrical power outages.* Anything requiring electricity (even from 12 Volts RV batteries) would fail certification. Finally, 12 Volt reigniters would be required for fridges in upper end RVs and encouraged in all RVs, because flame sources sometimes mysteriously blow out, turning things off. (Reigniters electronically sense when a flame blows out, and immediately create sparks to reignite that gas flame.)

Water systems would be almost exactly the same as those currently in use. The 12 Volt water pumps in use are inexpensive, readily available and fairly reliable. They don't use that much electricity, because although they do have a high current drain while pumping (estimates range from 4 to 9 Amps), faucets aren't likely to run very long (estimated at 15 minutes per day max.), because freshwater and wastewater tank space are limited. These pumps might thus drain a battery by only 2 Amp hours per day. Propane fueled hot water heaters (either pilot light or D.S.I.) would not be required in simpler RVs but would be required in upper end RVs. Although I would often include hot water heaters, I wouldn't worry too much about their failure, because hot water from a faucet is not an absolute necessity, as it is not difficult to heat water on a stove. Finally, there is one small improvement that I would make over existing water systems: *freshwater would need to be accessible in the event of a pump/power failure.* All RVs would be required to have an outside low mounted gravity fed faucet that bypasses the pump. Upper end RVs (in addition to having regular faucets) would also be required to have a hand/foot pumped faucet that also bypasses the water pump, to get water in emergencies. Being deep in the bush, with freshwater in the tank, but not being able to get to it, could be a real drag.

Stoves fueled by propane are a necessity. Many of those currently made are fine. I would require a stove with at least three burners in each RV. Each larger RV would also need to have a propane fueled conventional oven as well, and an outdoor fold-down or slide-out stove for cooking on hot days. No electrical cooking appliances would be allowed to be built into (or sold/ given away with) any RV, except for microwave ovens. (Microwave ovens have very high power requirements, therefore they should either be powered by generators or used very sparingly from batteries.)

Air conditioners are a bit of a tricky issue because they are not a necessity but are really nice to have in hot weather. I would require air conditioners only in upper end RVs. For larger RVs, I would allow users to choose between compression or evaporative air conditioners or one of each. (Compression air conditioners cool fast and work in all weather but require generators. Evaporative air conditioners cool slowly and don't need generators but only work in low humidity. Both kinds shouldn't be used simultaneously.) One very prominent difference between the RVs currently made and those approved by me would be my requirement that RVs not get too hot when parked in the sun. The skin (sides and roof) must reflect sunshine, not absorb it and dump it into the RVs in the form of heat. Aluminum and maybe stainless steel reflect sunlight; fiberglass, plastic and rubber tend to absorb it, therefore for now, I would only approve aluminum and stainless steel for the sides and roof. If it could be demonstrated that the normally heat absorbing materials of fiberglass, plastic and rubber could be coated (painted) with light/heat reflecting material, I will reconsider my position. Good insulation in the walls would be an absolute necessity, both for keeping RVs cool in hot weather and warm in cold weather. Passive air cooling (through ventilation) is useful on moderately warm days, therefore RVs would require adequate openable windows, a full-screen door and unblocked roof vents (one for each 12 feet or less of RV length--- I wouldn't allow all ceiling holes to be blocked by air conditioners). Finally, to prevent RVs from unwanted heating due to sunshine entering through closed windows, I would include a set of outdoor highly reflective window shades to cover them, probably mounted on rollers (like awnings are) when not extended. Preventing RVs from becoming too hot in sunshine would greatly reduce the need for air conditioning, therefore reducing the electrical power requirements for air conditioning.

TVs and stereos could be built into RVs if desired, but I would not require either of these, even in upper-end RVs, because users wanting them might prefer to buy their own portable units. Any built-in stereo equipment would need to be powered by 12 Volts DC, TVs, DVD and DVR players could be powered by 120 Volts AC from power inverters. All such equipment would need to be reasonably energy efficient.

Master disconnect switches, which disconnect the entire appliance system from the RV batteries when in the "off" position, would be required in all RVs. An RV's appliance system should be disconnected when the RV is in storage and not in use, to prevent unwanted battery drain (which leads to permanent battery failure), which occurs from something being accidentally left on.

Propane is the "other" stuff that makes RV appliances work. Propane tanks would need to be large enough such that RVs could function awhile without running out. Achieving this is easier said than done, since consumption rates differ drastically in cold weather than in hot. (A tank can run out in two or three days in bitter cold winter weather due to furnace use. In hot weather when the furnace isn't used, it can take two and a half weeks for a tank to run out.) Key features of propane systems for trailers I would keep the same as they already are--- I would require two detachable tanks with a changeover valve, so that users never have to run out completely. (Having only one tank would be prohibited.) Smaller trailers would have two 20 pound tanks, larger ones would have 30 pound tanks. For people who camp in winter, I would encourage that space exist to carry additional tanks. (Three tanks could be carried on the tongue.) It would be better to carry more than two moderately sized tanks, rather than just two super heavy tanks, because heavy tanks might be difficult for some people to lift. As for the metal, I would encourage that tanks be made of aluminum instead of steel. Aluminum does not rust away the way steel does and is therefore considered safer and much longer lasting. Also, aluminum tanks weigh about half that of steel tanks when empty, therefore are easier to lift. My requirements for motorhomes would be different from current designs. They would need to have two detachable propane tanks and a changeover assembly, like those in trailers. I would prohibit single tank setups permanently mounted to the motorhomes, since permanently mounted tanks force people to pack up and drive those big monsters great distances just to fill up with propane, which is a real bother. (A permanently mounted tank *in addition to* two removable tanks would be acceptable.)

The Batteries:

There would be two major battery differences between RVs approved by me and those currently available: battery storage capacities and placement. Instead of the 100 Amp hours, that is a typical storage capacity of the single battery in many RVs, I'd go with three times that number, and make 300 Amp hours the standard for medium sized RVs. For both trailer type RVs and large motorhomes (which tow cars/trucks behind them), I'd require that the storage capacities be divided between an even number of batteries, with half the batteries located in the RV and the other half located in the tow/spare vehicle. (Very small motorhomes that don't tow extra vehicles would of course be exempt from this division.) Another barometer I would use for determining storage capacity would be time. I'd insist that the total house batteries (in both RV and car) store enough Amp hours to run an RV for five days with a moderate average electrical load, and no charging. It's just not good for any battery operated system anywhere to run down too fast. It makes such systems impractical.

For upper-end RVs, I'd consider using the lithium type batteries used in electric cars, as these batteries are considerably lighter in weight and can accept charge considerably faster than lead acid batteries.

The Charging System:

Small and medium RVs could make all of the electricity their users need, for everything except compression air conditioners, by simply charging batteries while driving. I believe that even large motorhomes with bigger needs could meet most of their needs this way, because bigger motorhomes have bigger engines, bigger more heavy-duty alternators and more space to store more batteries. Charging while driving works so well, that I would require that RV dealers offer to install half the RV house batteries in the user's tow/spare vehicle, and do the appropriate electrical wiring. On medium to large RVs, that seem to need more electricity than charge-as-you-drive can provide, I'd also require solar panels. Upper-end RVs would have enough to almost equal the RV's daily electrical needs when the sun shines. Smaller RVs would require least a maintenance-sized panel (whose output is just enough to keep the house battery from discharging when the RV is in outdoor storage). Despite the fact that I like solar panels, I would never attempt to use them as a *substitute* for charge-as-you-drive electricity, I would only use them as a *supplement* to it. This is because solar generated electricity isn't always available because the sun doesn't always shine. On large upper-end RVs, whose users wanted the most deluxe charging system of all, I'd supplement the charge-as-you-drive strategy with a standing system combination of solar panels and propane fueled thermoelectric generators (do not confuse with internal combustion engine generators). Solar panels and thermoelectric generators complement each other well. Both make electricity with no moving parts (to wear out), no noise, and can be left on for very long time durations (almost continuously), for slow and steady battery charging. They differ in that solar panels make electricity that is free but not reliable, whereas thermoelectric generators make electricity that is reliable but not free.

Now let's get to the subject of (internal combustion engine) generators. Are they even needed? Should they even be included? In my opinion, for RVs, generators are only really "needed" for compression air conditioners (which absolutely need the high electrical outputs that only internal combustion engine generators can provide). Generators are also really useful for microwave ovens, household vacuum cleaners, and "last-resort" battery charging. But are they really "needed" for these things? For heavy microwave oven use, yes. For light microwave oven use (5 or less minutes per day), no, as RV batteries (via heavy-duty power inverters) can run microwave ovens, and I believe the above battery charging strategies could keep pace with consumption. Remember that microwave ovens are not a necessity; food can be heated (or reheated) very effectively by conventional burning of propane gas; it just takes a little longer. For household vacuum cleaners, yes, generators are needed. But household vacuum cleaners aren't really "needed." Vinyl RV floors can be swept and washed. For RVs with permanently installed rugs, vacuum cleaners are needed. But if those rugs were instead installed in small pieces, and were removable, those rugs could be cleaned by being temporarily removed, beaten against trees and rocks or even submerged in bodies of water, then hung to dry, before being put back. So even for rugs, vacuum cleaners aren't really "needed." And with the above described appliance and charging system, "last-resort" battery charging perhaps may never be "needed," unless there is a major failure somewhere within that system.

So would I include generators? For all RVs having compression air conditioners, yes, I would absolutely require them. But for small to medium RVs without compression air conditioners, no, and I would strongly discourage including generators, as doing so only encourages poor RV design and user overconsumption. Any RV that did have a generator would also be required to have a halfway decent 120 Volts battery charger. Such a charger would be required to charge at a minimum rate of 30 Amps (10 times what typical RV charger/converters do). The generator would also be required to have a special "battery charging only" on mode, in which computer control, coupled with voltage sensors, would automatically shut the generator down when the batteries reached 80% charge level. (Beyond the 80% level, charging is too slow and inefficient.)

Other (Non-Electrical) Requirements:

Even though my codes would focus upon electrical and appliance systems, I would fail any RV that had a serious problem in some other area. Specifically:

1) *Roofs must not leak.* Therefore, roofs must be curved or angled, such that water puddles do not form and sit up there [where those puddles of water eventually enter the RV]. Also, the roofs must overhang the sides. If the sides overhang the roof, I'd fail it, since some water enters the RV, so matter how much sealant gets applied.

2) *The RVs must be able to withstand the shaking of road travel.* All things built into RVs would need to be properly anchored, so that they did not loosen and fly around. All cabinet doors and drawers would be required to somehow latch, so that they did not open and spill their contents.

3) *Permanently installed rugs would be prohibited.* Vinyl floors are best. Everybody likes rugs, but rugs just aren't practical for RVs because rugs get filthy really fast. If rugs are desired, they should be removable and installed in small pieces, anchored by buttons [through yarn loops], with removable metal strips covering the seams [so people don't trip]. The rugs would have nice flooring underneath them, so when those rugs got removed for cleaning, nice flooring underneath would be exposed. Also, when the rug pieces got hopelessly dirty or worn, they could be easily replaced with new pieces.

4) *RVs must be able to withstand lightning strikes.* This is not as difficult an engineering task as it may seem, but it is important, as it is surprising how often one finds oneself in a prairie or desert during an electrical storm, with the RV being the tallest and only object [therefore a target]. Cars and RVs with metal roofs and sides are naturally very lightning safe, because if struck, the metal directs the electrical current around the inside and its occupants, and the magnetic forces cancel, thus keeping insiders relatively safe. [People in cars struck by lightning frequently survive.] Plastic, fiberglass and rubber do not conduct electricity, therefore if an RV made of these materials is struck, the electricity may prefer to go through a human being [composed mainly of salt-water, which conducts electricity well], rather than through the roof and sides. Fortunately, there is a solution to this problem: the Faraday cage. This is a series of connected metal wires [resembling animal cage wire] surrounding the interior of an object to be protected [in an RV, it would be embedded in the ceiling and walls, therefore not visible]. The Faraday cage does the same thing that metal roofs and sides do during a lightning strike--- it directs the current around the interior and its occupants, rather than through the interior and its occupants. I would require a Faraday cage is all RVs not made of metal. [Airplanes, which are frequently struck by lightning, have Faraday cages in them.]

5) *Blind-spot cameras and monitors must be included.* All RVs (both motorhome and trailer) have significant blind areas, which present hazards when driving down the road or backing up. Mirrors do exist, and are always installed, but these mirrors are usually so small that they either allow only very narrow viewing angles or wide angle views that render reflections from nearby large objects (vehicles) so tiny that they might go unnoticed. In either case, the driver must take his/her eyes off the road for awhile (too long) in order to properly look at and read the mirrors. Fortunately, there is a solution. In the present day, cameras and color monitors have become quite inexpensive. A decent sized color monitor (maybe 12 inches across) and several

cameras, coupled with a voice or finger activated device, that would allow the driver to see and rapidly switch between camera views, would allow the driver to accurately and quickly perceive the vehicle's surroundings and respond accordingly. (The driver would either wear a microphone or small device on his/her hand to sense finger movements.) With today's current state of technology, vehicles shouldn't have blind-spots anymore.

6) *Automotive integrity would be required.* I am definitely not an auto mechanic or engineer. I know very little about how cars work. Fortunately, there are many experts in this field who can write books on this subject and/or properly evaluate RVs. Any serious defect in the automotive design would cause the RV to fail certification.

So there is the basic design plan for a Thomas Childs designed/approved RV. Would it work? Well, the key components of the above plan have already been implemented and tested in my simple RV and found to be successful. I can say with 100% certainty that it is possible to comfortably live in an RV for extended periods of time without plugging into municipal electricity *ever*.

Testing RVs

I know almost nothing about boats. I do however have a piece of advice for anybody who designs and makes them (which I would follow if somehow I were put in charge of a company that makes boats): Once in a while, it might be a good idea to put one of those boats you make onto the surface of a body of water, to see what happens. If it floats, you've got a good start. If it sinks, it's time to stop making those boats, and either do some serious redesigning or close up shop permanently. Unlike boats, I know quite a bit about RVs. As an RV improvement advocate, I have the same basic advice for anyone who designs and makes RVs: Once in a while, it might be a good idea to take one of those RVs you make into the wilderness and actually try living in it for awhile, to see what happens. If the RV "floats" (needs only minor adjustments), you've got a good start. If the RV "sinks" (has major shutdowns), it's time to stop making those RVs and either do some serious redesigning or close up shop permanently.

This "float or sink" principle seems like such common sense, it's silly that I need mention it. Yet I believe that those who design and make RVs often forget or don't bother to do this. Consider this: Chapter 2, the "defects" chapter, is the backbone of this book. All of the defects described there were revealed to me (most for the first time) in my mother's RV during *one single trip* into the wilderness, lasting *only three weeks*. I draw a stunning conclusion: Even though those RVs were mass produced, nobody who designed and made them could have actually taken one camping in the wilderness, even for three weeks, because if they had, they would have noticed the same defects that I did, and if they had any integrity, they would have either corrected those defects or closed up shop. I draw another stunning conclusion: Since these defects aren't the exceptions, they are the industry standards, I feel that very few RV manufacturers actually bother to properly field test their RVs.

To pass my certification, all RV manufacturer/designers would be required to do occasional three-month-live-in field tests.

The Three Month Field Test

In this test, user/testers would need to live in an RV, as actual RVers, spending nearly all of their time in the wilderness, for three continuous months.

Why three months? 1) *RV livability is effected by many variables. It takes a long time for a good representation of those variables to present themselves.* RV livability is different in hot, moderate and cold weather. RV livability is effected by the presence or absence of sunshine, wind and humidity. RV livability is effected by altitude, parking orientation (which sides face the North, South, East and West directions), lengths of time the RV stays parked before moving to new locations, proximity to vital supplies (water, propane and groceries), and the individual habits, numbers and ages of occupants. Variables can combine to produce different outcomes (hot & sun, hot & shade, cold & sun and cold & shade all produce different results). RV components can even interact with one another, much like different drugs can interact in the human body when taken simultaneously, producing disastrous results. (Sometimes one appliance actually shuts another down.) RV systems as a whole are quite complicated. Short tests just can't do them justice. 2) *Observation powers of occupants increase with longer trips.* The beginning of any trip [about the first month] is usually clumsy. After camping for awhile, users become well accustomed to their routines. This makes them better observers, better problem solvers and thus better testers.

Why wilderness? If it's designed for the wilderness, it must be tested in the wilderness. User/testers would however be allowed to take their RV through populated areas while in transit between camping locations (but must sleep without municipal electricity). They would also be permitted to have their RVs repaired at service centers in populated areas.

Why no municipal electricity? Wilderness has no municipal electricity *ever*, therefore RVs should have no electricity *ever*. (Also, batteries may not be removed and charged by municipal electricity elsewhere.) RV occupant/testers would also be prohibited from hooking up to water or sewers, except briefly at dump stations to fill/empty water tanks. (I would however also order a separate one week field test with municipal electricity, water and sewer hookups, since some people do camp with those hookups.)

Why should testers live as actual RVers? The only way to think like RVers is to actually become them. Occupant/testers should do the chores normal to RVing, including using dump stations, transporting water, propane and garbage to and back from their RVs, monitoring & recharging their batteries, doing household chores (such as dishwashing, bathing, laundry), hitching/unhitching and traveling, and preparing and eating meals at their RVs at least five days per week. Occupant/testers may vacate their RVs during days for shopping, sightseeing and visiting people, but would need to return to sleep at nights (no staying at motels). Finally, occupant/testers should strive to have RVer personalities, most notably by being friendly toward and having many conversations with other RVers. RVers are eager to share information, therefore they are great sources of RV information.

Who should do the field tests? Field tests should usually be done by people at the companies who design and build RVs. They could also however be done by the people who buy the RVs, provided those people received proper training in how to use, test, observe and thoroughly document their findings. Since most RVs are probably occupied by two people (husband and wife couples), most tests should be done with two occupants. Adult guests, children or pets should sometimes be included, since sometimes RVs travel with multiple people or with pets. (Pets however would be exempt from the requirement of reporting their findings, since most pets don't talk or write.)

How often should tests be done? This is a tricky issue, because sometimes only very minor changes might be made to an RV design. For example, suppose a refrigerator is moved six inches so that opening its door doesn't interfere with the opening of a cabinet door. Should such a minor change warrant a new three month test? I think not. For another example, suppose only the color or style of wallpaper gets changed. Should such a change warrant a new test? No. New tests are however warranted and should be performed if there are major differences in size, floor plans, addition/removal of slideouts or substitutions of appliances for appliances of different brands or designs. What if only one appliance is replaced with an appliance of a different brand, but everything else in the RV stays the same? Hmmm. I'll have to think on that one a bit. Perhaps in this case, an additional shorter one month test should be done, emphasizing the substituted appliance. Perhaps tests should be proportional to the number of RVs produced by any company. Perhaps tests should be proportional to the number of changes made in a design. Perhaps new field tests should be done whenever the manufacturers begin to feel uncertain about the overall performance of a particular RV (assuming they care). I think it should be a combination of all of these variables. But for sure, wilderness field tests should be done *sometimes*.

Should lab tests also be done? Of course. Lab test results however should be considered only *preliminary* results, not the final verdicts. Lab tests should be done frequently during various stages of development of individual RV components. After passing, those components should be installed into prototype RVs, then those RVs should be tested *as a whole* in the parking lots near the factories that make them, emphasizing known possible destructive interactions. When no further problems are known or believed to exist, then it's time to field test the RV.

Responding to Customer Feedback

Suppose somebody mailed you an envelope, containing five crisp new real $100 bills and a note saying "Here's a gift with no strings attached." Would you put those $100 bills in your wallet or would you toss them into the trash? I sure wouldn't toss free money in the trash! Now let's consider something. Suppose you work at a company that makes something, and a customer sends you a letter [or video] describing a problem, observation or suggested improvement about that product. Should you write back saying, "Thanks for your letter, but we know what we're doing," then ignore the advice and proceed with production as usual? Or should you take the information very seriously? Contemplate this: When a customer takes the time to write to you with information about one of your products, that customer has field tested that product and is actually sending you a free test report. It would cost your company hundreds (maybe thousands) of dollars to hire someone to conduct an identical test, notice that same problem, and properly document it, therefore that letter is actually worth hundreds of dollars to you at minimum. If all you do is have some low level underpaid customer service representative speed read that letter once (without sending a copy to the engineering department for review), answer briefly, then throw it away, your company is essentially throwing away large amounts of free money.

All companies making RVs approved by me would be required to do the following upon receiving a "test report" letter: 1) Have the letter read by someone in the designing/engineering department [preferably by two people]. 2) Catalogue and file the letter according to its category, for easy retrieval. 3) Keep the letter on file for at least ten years. 4) Write back, thanking the person for the advice and observation. For Thomas Childs certification, companies would also need to publicly offer and pay money as a reward/tip for all important well documented advice and observations. Don't be afraid to let your customers be your eyes, ears and testers.

By the way, many big wigs at companies take the attitude that they "know what they are doing" and are somehow "too good" to take any advice from the people who buy and actually use their products. If you take such an attitude, you almost certainly do not know what you're doing and are not "too good" for anybody.

Warranties

Many RV components are sold with warranties only lasting one year. Such warranties are almost useless, as many people occupy their RVs only a couple weeks per year. (Thus they may not notice defects until warranty periods expire). Furthermore, those who install things themselves may take months to get around to doing the jobs. (Such delays further cut into usable warranty periods.) RVs have lots of components. If every individual RV component/appliance lasted only a year and a day, each RV user would need to spend a year of full time effort fixing his/her RV, just to camp in it for a day! RV component manufacturers need to stand by their products. All RV appliances must be warranted for a minimum of five years. Also, self-installers should have some warranty coverage too (provided the installer or installation doesn't damage items).

The RV World is Loaded With Problems

There is something perplexing about RVs. They seem to be built to a much lower standard than many other things. Electrical & appliance systems are particularly bad. For example, if RV designers built airplanes instead of RVs, those planes would be falling out of the sky every day. Actually, the planes wouldn't even take off. Parts would be falling off on the runways. People would stop flying and airports would close everywhere.

Why Are RVs Built to Lower Standards than Other Things?

Perhaps the low standards for RVs has something to do with the people who generally buy RVs: retirees. People who are advanced in age are a logical target for commercial predation due to design/manufacture-corner-cutting because people advanced in age are weak. Many have only about a decade of good health left. If victimized, they may not have the time, the energy nor the money to fight like hell, even though it may be the right thing to do. They may figure that they've already paid their dues to society, and choose to leave the long battles for the young and middle-aged (who often don't fight because they don't buy RVs that often, or don't have much time to use RVs if they do buy). Retirees may just give up, accept their losses and spend the rest of their precious years visiting their grandchildren and planting gardens.

Why are RV Electrical & Appliance Systems Particularly Bad?

Perhaps a reason electrical & appliance systems are bad is because problems with them go unnoticed by prospective new RV buyers. This may be because electricity isn't visible. People who visit an RV for sale can clearly see the cabinets with their decorative door handles, the flooring, the wall paneling, the upholstery, the table and counter, the general layout, the windows, the exterior siding, but they can't see what the electrons do in the wires and batteries without diving into the technical specifications in the manuals, and few take the time to do this. They (like me) then become aware of the problems only after they buy the RV, take it home, take it to the woods and actually start using it. By then it's too late. RVs may be designed primarily to look good in the showroom, not to work well in the field, because good looking RVs are more likely to sell. Good off-grid electrical & appliance system performance may just not be a good RV selling point.

Other Problems

The RV world is loaded with other problems that go beyond just electrical components. There are those crappy parts--- like that set of leveling jacks that I bought (one busted after only one year, the other busted after two), or the vinyl propane tank cover, which the wind blew to shreds within the first few hundred miles of its first trip (it was trash three weeks after I bought it). How about that RV with a table and refrigerator that weren't even bolted or fastened to the floor or walls? (Those designers should have been comedians instead.) Can't forget about those roofs that are designed to leak. (Not all roofs have this problem though. Some seem really well built.) Then there is the road safety issue: Motorhomes usually can't roll if they tip; they disintegrate. Shreds of shrapnel then fly everywhere, which can kill the people in them. (This happened in one case where someone took an exit ramp too quickly and tipped the RV. There was one fatality, even though the speed was less than 40 miles per hour and the motorhome didn't actually collide with anything. Metal roll cages imbedded within the structures of walls/roofs have been proposed by some to solve this problem. Like all improvements though, these would add cost to RV production.) And what about aerodynamics (or the lack thereof)? Many RVs look like boxes going down the road. Others have sharp angles instead of streamlined curves. I know very little about aerodynamics, but even I can tell some of these have lots of air drag. They could be unsafe in the wind. (RV designers shouldn't be inconvenienced by reading books or taking classes on this subject.) Then there is the climate of indifference. One attorney, who often sues RV manufacturers for faulty designs and construction, said that he is often willing to meet with the RV manufacturers that he sues, to discuss ways they can learn from their mistakes and reduce future problems, but he is usually met with resistance by people who just want to settle the lawsuits as quickly and for as little as possible, then get back to "business as usual." Finally, there is that nasty problem of rapid depreciation. New RVs can cost as much as houses, yet they lose their values as quickly as cars. Anyone who buys a new motorhome is either very dedicated (to the lifestyle), very foolish or has lots of extra money that they don't mind parting with. I could do more research on the internet, find more problems, and go on and on lamenting, but I need to stop writing sometime. I also need to limit the scope of this book.

Is There Hope?

There may be a bright side. RVing is a lifestyle embraced by many. Every reasonably populated area has at least one dealer of RVs and their parts. Sure, there are problems, but many of these parts do work at least reasonably well. The facts that manufacturers and stores design and sell things specifically for RV users, and there is a significant buyer's market for such things, speaks volumes as to the dedication to the RV lifestyle of a widespread group of people. Surely there exists a good potential for many RV enthusiasts of all varieties (makers, fixers and users) to unite and start cleaning up the problems that exist within this industry.

What Can We All Do?

What Can New RV Buyers Do?

If you have enough money and desire to have an RV custom made, you are in a great position. If you like my ideas, you could try to hire me to design an RV for you (though someone else would have to do the actual construction). If I am unavailable because I am too busy or dead, well, you could do it without my physical presence. You've got the blueprint for a really good RV in your hands or laptop right now (this book). If somebody denounces my ideas, calling me a heretic, lunatic, fanatic, etcetera, and tries to talk you into getting an RV of more traditional design, smile and politely tell that person that you are the one with the money, so you are the one who will choose the design. Then walk away without buying. If enough people do this, maybe those in the industry will start listening and there will be some improvements in RV design. Do not let any salespeople or manufacturers bully you into taking something home that isn't going to work well.

What Can Existing RV Owners Do?

If you already own an RV that you plan on keeping, start making improvements. You need to have either money or time to do this. If you have the money, hire a customizer. Make sure though that the person isn't too friendly with the RV industry, too uncreative or too experienced with RVs, as such a person might have been trained to follow a long standing accepted protocol which leads to dysfunctional designs. If you've got the time, do the work yourself. It doesn't take much money, but it does take much patience. Be prepared to do a lot a swearing. (I did.) You've got one great advantage over me: *you've got my book in your hands*. I had no such book when I began my odyssey of making RV improvements. I had to learn most of the stuff written here the hard way.

What Can RV Manufacturers Do?

Manufacturers of RVs and their parts need to care about the quality of the products they make. They need to design RVs as good practical road worthy housing, not fashion statements. They need to be willing to accept well deserved criticism with dignity, even if some of that criticism might be harsh. They need to communicate with RV users. They need to remember that although the initial point of sale may be the *end* of the RV relationship for the manufacturers, it is the *beginning* of the RV relationship for the users. Mostly, conscientious RV designers need to stand up to the colleagues in their field who are non-conscientious, and not let the idiots within their ranks set the standards and run the show.

Let's Not Allow Off-Grid Camping To Die

Is off-grid camping dying? It certainly seems to be on the decline. Over the past few decades, there has been a trend toward electrifying campsites all over the United States, especially in State Parks. Greater electricity availability encourages poor off-grid RV design (because people see less of a need for good off-grid RV functionality) and poor off-grid RV design encourages greater electricity availability (because more people want, request and eventually get campsite electrification). Where is this vicious self-feeding cycle heading? Possibly toward the demise of simple cheap camping in State Parks and having those parks become exclusive country clubs available only to a rich aristocracy, rather than to everyone. I find this "excluding the poor" trend repulsive, especially in light of the fact that many State Parks all over the U.S.A. were developed by the C.C.C. using Federal taxpayer dollars.

In order for wilderness areas to be protected (including State Parks), they must be loved. In order to be loved, they must be visited. In order to be visited, they must be *accessible* to everyone of reasonable health. This includes struggling students, unemployed people between jobs, the disabled, limited income seniors, as well as the rich and privileged. If you are rich, or even just financially comfortable, remember that poor people can be nice neighbors. Poor people can also make good contributions that can benefit everyone. (I wrote this book, which is an intellectual contribution. I have been below the poverty line all my adult life.) All persons should insist that *some* camping areas on all public lands (where camping is allowed) remain primitive and cheap so that camping can remain open to all.

Greater wilderness awareness, appreciation and protection would be fostered if more people camped off-grid. If more people camped off-grid, more people would care about these areas. Camping off-grid is not an impossible or unreasonable task, even in this 21st century. I've learned to do it. You can too.

Let's Communicate

If you are an RV user with horror stories, I want to hear from you. (Google my name to get my current business address/e-mail, if the ones in the back of this book aren't current.) Tell me your stories. I just can't get enough of them. In fact, we should form a club and share stories on a community bulletin board. If enough RV users share information, without adding too much sugar and perfume, and without having RV manufacturers infiltrate and dilute our experiences with propaganda, we could get a better picture of what's really going on in RV design and construction. This book should be considered only a starting point. Also, share your good experiences. We need to encourage RV designers to make stuff that actually works.

If you are a designer of RVs or their parts, I also want to hear from you. Please be honest about who you are. If you think I've been unfair, please tell me so and tell me why. I will consider your opinions, though I make no guarantee that I will convert to them. Also, please tell me about good products that you design and make. While I do a lot of complaining, I also must be willing to give praise for things done right.

Some Final Philosophical Thoughts

We humans tend to view our world in terms of what we know and do best. A trapeze artist or clown might see the world as being one great big circus. An internet system hardware designer might see the world as being one great big part-human-part-electronic brain, with the wire/radio pathways as its interconnecting neurons. A homicide investigator might see the world as being one great big crime scene--- put a huge yellow crime-scene tape into orbit around Earth (giving Earth a ring like Saturn), to declare our planet off-limits during the investigation. (The way humans behave sometimes, declaring Earth a crime scene may not be a bad idea.) I am no exception to this tendency to see the world in terms of what I know. As a person who has studied the RV and written a book about it, I tend to see the world's problems in terms of the RV. And as a person who has camped quite a bit, I tend to see the world as being one great big campground.

The RV as a Metaphor For Waste in America

The RV for me has become a metaphor for wastefulness in America (and to a lesser extent the rest of the world). We waste everything in America: oil, natural gas, coal, water, even human imagination (especially when schoolchildren who don't blindly obey authority get drugged). Many Americans pride themselves on waste. A person who buys and eats two lobster tails per day is seen as being far more successful than a person who buys and eats hamburgers, potatoes and salads, simply because lobster tails cost more--- whether or not that person likes lobster tails better is seen as irrelevant. But if a person buys fifty lobster tails per day, and is only able to eat two, and throws the other forty eight to the dogs or has the butler bury them in the garden as fertilizer, that person is held in the highest esteem possible--- especially if his actions push lobsters to the brink of extinction due to over-harvesting of the sea. Tell me, how many people buy lottery tickets, dreaming of that big win, so that they could buy and live in mansions instead of houses? They yearn to heat every room in their mansion in the dead of winter, and when it's 15° F outside, they'll leave the windows open and heat the outdoors--- this will show the neighbors how successful they are. He who destroys the most of the world's natural resources the fastest wins.

In the midst of all of this is the RV, both a symbol and a symptom of American wastefulness. The RV design trend for quite awhile has been: have more components, and make every component bigger, heavier, more electricity consumptive and more expensive than last year's models (but don't bother adding more batteries or chargers). This sounds eerily similar to the U.S.A.'s slash-and-burn energy policy, where we keep over-consuming fossil fuel, while stonewalling the development of clean renewable energy. The current RV design must have been borne in America. Where else could such a crate with a battery charging system efficiency well below 1% have originated, but in a nation that consumes 20%-25% of the world's oil with only 5% of the population? And don't forget, RVs not only waste electricity, because of component over-complexity and rapid depreciation, RVs also waste money.

Wastefulness makes the RV an irony. On the one hand, an RV is supposed to be a vessel that allows people to see, experience and appreciate the best of what nature has to offer, and therefore motivates people to love and preserve these places. On the other hand, the RV's over-consumptiveness makes it one of the most planet destroying contraptions that we humans make, therefore somewhat defeating its own purpose.

The Future of RVing
(Can RVs Go "Green?")

RVs consume a lot of fuel to move around, so if we humans don't develop clean renewable energy, when the fossil fuel runs low, RVs may be among the first human devices to die, and so too will the pastime of RVing die with them.

But maybe not. What if RVs could be made "green?" In this book, I have proposed ways of making RV systems of electricity and appliances both very efficient and effective, and therefore sustainable much further into the future. But there is still that nasty problem of RVs needing lots of fuel to move around, and there is no getting around this, as distances between desirable locations are formidable and RVs don't get many miles per gallon because they are big and must be so--- one can't be comfortable living in something the size of a compact car. But even this has a solution: *take longer trips*. It takes the same amount of motor fuel to move an RV to and back from New York to Colorado (for example), regardless of whether one stays for two weeks, four months or even a year. Thus, if one is able to take really long trips instead of short ones, although one doesn't cut down on the *actual* fuel consumed, one does drastically cut down on the *average* motor fuel consumption per day. Long trips could potentially make RVing still doable even in times of fuel shortage. And if some people shed their regular houses and live in RVs full time, RVing could actually become *less* fuel consumptive than regular house living. Why? It has everything to do with the need for furnace use. First, because RVs are much smaller than regular houses, RVs need a lot less fuel to heat in cold weather. Second, because RVs are movable, they could actually drive away from cold weather locations and into warmer regions, thus greatly reducing their need for heating. If the motor fuel needed to relocate an RV to a warmer climate is less than the furnace fuel needed to remain in a cold area, then relocating is actually the greener choice.

I believe that if both RV systems were engineered to be "green" and RV users adopted "green" habits, then RVing could have a healthy and long future--- though it would be a new and very different form of RVing than that which we are familiar with today.

RVers are in my opinion among the most creative people on the planet, so they might well be capable of finding solutions to their many challenges. If the fossil fuel dependency habit isn't kicked, big over-consumptive RVs will die over time, and so too will the luxury campgrounds at which they stay. Simple cheap off-grid camping might then make a comeback, because simple cheap off-grid camping will be the only type of camping available at such a time.

What Have I Learned From My RVing Days?

The experience of rebuilding a trailer and camping in it three full-time years has taught me plenty *technically*. But it has also taught me some things *philosophically*. Yes, there is the obvious--- the experience has taught me a profound love and appreciation for the geological and biological natural wonders of the United States. But the experience has taught me another somewhat unexpected lesson that might be even more important: *it has taught me that we humans actually only need a small amount of resources to be comfortable and happy*. I have learned to survive and be happy with only the electricity that a couple of batteries store. I have learned how to survive and be happy with only a few gallons of water that I sometimes transport miles into the wilderness (to wash dishes and take showers). And I have learned how to survive and be happy with only a small amount of money (most of which I got from SSI or Social Security Disability).

It bewilders me when someone makes calculations as to all the things one person supposedly "needs." A single person supposedly "needs" MegaWatt hours of electricity, tens of thousands of gallons of freshwater per year, a new car every three years, at least 800 channels of TV, internet at the house (when I graduated from high school, people didn't even have the internet), a six figure income to support all this, and at least a bachelor's degree (but at least a master's by age 30) to procure this six figure income. Collectively, we "need" to turn on a billion lights after-hours at night in buildings that are closed for business and unoccupied by humans. And when we die, each person "needs" a funeral costing at least $8,000. We need all this stuff? Horse feathers we do. We "need" because we waste. We waste because we seem to have developed some kind of collective psychiatric disorder that compels us to do so--- we as a nation are perhaps as addicted to waste as the average crack cocaine user is addicted to crack.

Camping as a Metaphor For Life

As a person who RV travels and camps, I have come to see the world as being one great big campground, and life itself as being one extended camping stay. When one camps, one arrives, sets up equipment, stays awhile, then packs up and leaves. When one lives, one is born, sets up shop, stays awhile, then sheds possessions and dies. I religiously follow a rule for camping, which is also a great piece of advice for others: *Leave your campsite cleaner than you found it.* In the grand scheme of life, I hope to follow the same rule (and advise it for all others): *Leave your "campsite" cleaner than you found it.*

In RV terms, a campsite is a small piece of land upon which you park and live for awhile. In the grander more philosophical terms of life, a "campsite" could be many things. Your "campsite" could be your small contribution to the collective intellectual knowledge that we humans possess. Your "campsite" could be the small piece of land on all the continents upon which you reside or do business. Your "campsite" could be your small share of Earth's natural resources and the trash you produce (which can either be discarded indiscriminately or disposed of responsibly). *Leave your "campsite" cleaner than you found it.* If you can do this, when you die, for sure your life will have been a success.

With the population of humans continuing to expand, if we are to continue to thrive (or even survive), we must each tread more lightly on the planet. The notion, that the way to happiness is through over-consumption of the Earth's resources, must die. Frugality is not the entire way to happiness, but frugality is a big piece of the puzzle that is the way to happiness.

index

A

AC (definition) 67-69
AC appliances run by power inverters 371-372
air compressors, compression 341-342
air conditioners (active)-- 304-313, more specifically:
ammonia absorption 306, 311-312
compression 305-306
evaporative 305, 307-311
solar powered fans 306, 312-313
thermoelectric 313
air conditioning (passive) (blocking unwanted heat) 314-330, more specifically:
chimney effect 328-329
greenhouse effect 323-326
heat storage within objects 327-328
Insulation 317-321
reflection 314-317
ventilation 322-323
air pressure-- units of measurement 339
alternating current (definition) 67-69
alternators 150-168
aluminum (electrical and corrosion properties) 378-380
aluminum foil 316-317
aluminum siding 315-316
Amish people 249, 295
ammonia 299
amp (definition) 56
amp hour (definition) 55-56
ant colony 271
antidepressant drugs 15
Apollo 13-- 197, 226
appliances-- all of Chapter 5, see individual specific appliances
artificial obsolescence 283
awnings (for windows) 325-326

B

batteries 75-82, 98-99
battery problems 78-82
battery placement in vehicles 151-158
swapping batteries 152, 162-167
biography (of author) 14-23
building codes (for RVs) 388

C

calculations-- see "calculations index" immediately following this index
California State Parks 48
camping rules (Federal lands) 42
camping terms 32
candoluminescence 240
capacitors 201-202, see also *supercapacitors*
capacity (synonym "charge" *(noun)*) (definition) 57
carbon dioxide 225-226, 269-271
carbon monoxide (poisoning) 222-225, 242, 270, 275
catalytic burning of propane
for comfort heat 263-264
for thermoelectric generator 188-189
charcoal grills 289
charge *(noun)* (definition) 55-57
chargers and **charging batteries** 82-85, 100-107, 125-135, all of Chapter 3, more specifically:
3 Amp charging rate 102-106
batteries for small devices (cell phones, etc.) 375
charging as you drive 150-168
comparing all RV battery charging strategies 209-211
from municipal outlets 101-102
ideal 120 Volt battery charger 134
supercapacitor aided 206-208
circuit boards (in refrigerators) 302
coal burning electric power plants 149
coal gas 250
coffee (cooking) 292
compressed air-- units of measurement 339
computer (monitoring RV electrical system) 142-145
computers (laptops) 374
connectors (electrical) 378-381
convection 265-266, 271, 328-329
cooking without electricity 292-293
cool white-- see "lights-- fluorescent & LED"
costs per kilowatt hour-- comparative table of RV electricity strategies 213-214
coulomb (definition) (synonym-- amp second) 55-56
current (definition) 55-57
current consumption table (of author's RV) 384-385
customer feedback 398

D

DC (definition) 69
defects (in RV electrical system designs) 98-113
direct current (definition) 69
disclaimers (for RV electrical systems) 51
dryers, drying cloths 369-370

E

Edison, Thomas 250-251
efficiency
of 3 Amp battery chargers 104-106
of alternator battery charging 161
of appliances 139-141, also:
of furnace heating 274-275
of RV lighting 259-260
of charging with multi battery high current generator systems 133
of thermoelectric generators (propane fueled) 192
table comparing efficiencies of all RV battery charging strategies 212
electrical terms 66
electricity-- the whole book
electricity vs. fire-- efficiency comparison:
for heat 234
for lights 252-258
electrification (putting electrical outlets in campgrounds) trends:
National Forests 50
National Parks 49
State Parks 47-49
electrolysis, reverse electrolysis 196
energy
definition 58-62
waste (in appliances) 109-110
energy density (definition) 195
of propane vs. RV batteries 234
evaporation (for cooling) 299, 305, 307-311

F

fiberglass insulation 319
filament (of light bulb) 236-237
flooring (both vinyl and rug) 394
Florida State Parks 47
fluorescence 237, see also "lights"
foam insulation (rigid sheets, liquid foam hardening in place) 320
formulas
electrical 66
energy storage in capacitor 204
heat flow (through insulation) 318
voltage change current relationship in capacitor 205
frugality 405-406
fuel cells 193-197
vs. batteries 193-194
furnace stories:
"I Just Can't Figure Out…" 277
"My Old Reliable Furnace" 277
"My Stove is Also My Furnace" 275
"Some Like it Cold" 276
furnaces-- 261-284, more specifically:
best overall heating plan 273-274
characteristics 261-262
types:
catalytic 263, 269-270
central radiant 262, 266
direct discharge 264, 270
electric 264, 271-272
forced-air 263, 266-268
electricity consumption 275
heat pumps 264, 272
hot water/steam 263, 268-269
passive convection 262, 265-266
furnaces that don't heat-- see "furnaces—forced-air"

G

gas grills 291
generators 75, 86-94, specifically:
effect of generator noise on people 90
generator noise 88-94
powering battery chargers 102-106, 198
rules for use 91
starting 108, 136-138
vs. solar panels 171-172
generaphiles, generaphobes 90-94
gensets-- see "generators"
glass (windows) 323-325
gold (plated connectors) 379
Grand Canyon National Park 35
gravity 300, 348-351
"green" (eco-friendly) RVs 404
grid dependent camping 44-46
grills, propane gas 291

H

harvesting electricity from nature 148-149
heat blocking capability of RVs-- see "air conditioning (passive)"
heat capacity 327
heat pumps 264,272, 305-306
history of artificial lighting 250-251
hot water heaters 354-359, more specifically:
combined with furnaces 358
D.S.I. (direct spark ignition-- gas) 355-356
electric 356
hot water recovery valve (suggested improvement) 358-359
MotorAid 357
pilot light (gas) 355
tankless 356-357
humidity (effect on evaporative air conditioners) 307-309
hydrogen 193, 196, 299

I

ice (frozen water) 294, 296-298
ice, dry (frozen carbon dioxide) 297
incandescence 237, see also "lights"
infrared radiation (absorption/reflection of) 315-316, 324
insulation 272-273, 317-321

J

joule (definition) (synonym-- watt second) 58

L

Lake Powell National Recreation Area 33
lamp-warming food 290-291
laptops 374
leprechauns 118, 280
lifestyle (RV lifestyle) 14
lightning (strikes & safety) 394
lights-- 236-260, more specifically:
best overall plan 259
electricity vs. fire (efficiency comparison) 252-258
fluorescent 237-238, 244-245
gas mantle 240-242, 247-249
incandescent 236-237, 243-244
LEDs 238-240, 245-247
portable 258

M

malfunctioning electrical system (20 things that can shut an RV system down) 112
master disconnect (switches) 377
microwave ovens 287-289
motorhome
that inspired this book 23
towing a car 155-156
municipal electricity 67-69
for battery charging 199-200
for plugging RVs into 44-52

N

natural gas 265
nuclear power plants 149

O

"octopus effect" (overcomplexity of electrical systems) 111-113, 142-145
off-grid camping 33-43, 402
reasons for 36-43
off-grid house air conditioners (ammonia absorption) 311-312
Ohm's Law 64-65
ovens 287-290
conventional 287
makeshift, improvised, alternatives to 290-291
microwave 287-289
battery drain from use of 288
oxygen 194
sensor 269

P

parallel battery configurations 121-124, 126-128, 153-155, 168
pilot lights 265-266, 354-355
popcorn (cooking) 293
power (definition) 58-59, 63
power converters 72
power inverters 72-74
preformed mantles 248
printers 375

propane-- all of chapter 4, more specifically:
- changeover 228-229
- detecting it 219-221
- energy density of propane vs. RV batteries 234
- properties 218-219
- tank recertification 227
- tanks (for motorhomes) 230-232
- tanks (for trailers) 227-229
- vs. electricity 218, 233-234

R

R value (thermal resistance) 272, 318-321
radiators (for hot water furnaces) 268-269
rain entering RV 306
rebuilding a trailer 16-21
reflectivity 314
refrigerators 294-303, specifically:
- best plan 303
- types:
  - ammonia (gas) absorption 295, 298-302
    - repair of above 301
  - compression (heat pump) 294, 298
  - ice boxes & coolers 294-298
  - thermoelectric 295, 303

repair of refrigerators 301
resistance (definition) 64-65
resistance (in wires and connectors) 100, 378-381
resistivity (thermal, not electrical) 318-321
reverse progress 283
RMS voltage (definition) 68
roof vents (& roof vent covers) 322-323
rubber roofs 316
rules for camping 42

S

safety (general overview) 3
saltwater 297, 310
screens (for windows) 322
separating heat from exhaust (from burned propane) 276
showering (without water pressure) 351
slang electrical terms 65
solar panels 169-175
- vs. generators 171-172
- vs. thermoelectric generators 188

soldering wires 381
sounding board effect (for generators in RVs) 88
specific energy (definition) 195
specific heat 328
standards (existing standards for RVs) 399-400
stereos 374
storm windows 272-273
stoves 285-286
"Stupercar" (a really stupid car design) 96-97
supercapacitors 201-208
- vs. batteries 202
- supercapacitor aided battery charging 206-208

swamp coolers-- see "air conditioners-- evaporative"
sway (of towed trailer) 328, 349
switches 377-378

T

Teflon 292-293
temperature control (in furnaces) 265
terms (electrical) 66
Tesla, Nikola 251
testing appliance currents 382-383
testing furnaces (99 night field test) 279-282
testing RVs (three month field test) 395-397
Texas State Parks 48
thermal mass 323-326
thermal resistance 318-321
thermoelectric coolers 187, 295, 303
thermoelectric generators 182-192
- in RVs (author's proposed use) 188-189
- thermoelectric effect 182-186
- vs. solar panels 188

Thomas Childs Certification 387-398
Thomas Childs RV 387-395
tie-on mantles 241, 248
timers 377-378
toast (cooking) 292-293
toilets 360-368, specifically:
- chemical (small portable) 367
- composting 364-365
- DryFlush 364
- folding toilets, bags & jugs 361-364
- water toilets for RVs 365-367
- wet vs. dry 360

tungsten 236, see also "lights-- incandescent"
TVs 372-374

U

ultraviolet radiation 237, 315-316

V

vacuum panel insulation 320
ventilation 269-270, 272-273
Venus (planet) 326
voltage, volt (definition) 53-54

W

warnings (safety) 3
warm white-- see "lights-- fluorescent & LED"
warranties 398
washing machines, washing cloths 369-370
wastefulness 403-404
water heaters-- see "hot water heaters"
water pumps, pumping 336-353, specifically:
  compressed air 339-346
  demand pumps (12 Volt) 337-338
  gravity 348-351
  hybrids (uses both demand pumps and compressed air) 347-348
  manual pumps (both hand and foot) 352-353
  municipal water 353
water systems 331-370, see specific water system components
water tanks
  accumulators 347-348
  freshwater tanks 332
  high pressure freshwater tanks 343-345
  pressurizing low pressure freshwater tanks 345-346
  wastewater tanks 333
watt (definition) 58
watt hour (definition) 59
Welsbach, Welsbach mantle 251, see also "lights-- gas mantle"
wind turbines 176-181
  finding wind 179-181
  vs. sun 178
winterization 334-335

Y

Yellowstone National Park 34

Calculations Index

air pressure in compressed air water tank 340
batteries vs. gasoline (equivalent energy stored-- by weight 79, by volume 80)
battery charging while driving (cost of fuel) 161
battery drain from 1200 Watt microwave oven 288
efficient lighting (energy savings) 260
forced-air furnace electricity consumption 275
gas mantle lights vs. electric lights (size, weight, costs of propane, gasoline and batteries) 252-258
gasoline containing 1 kiloWatt hour of energy (gallons and weight) 214
gasoline needed to charge a battery (at 100% efficiency) 160
generator powering battery charger (efficiency)
  3 Amps charger (sold standard with RVs) 105
  multi-battery high current charging 133
number of batteries a 3000 Watt generator could charge 129
off-grid vs. on-grid camping (money saved) 36
propane vs. batteries (weight needed to store equal energy) 234
refrigerator powered by 12 Volt battery (time span it can run) 301
slow vs. quick generator powered battery charging (cost of fuel) 131
supercapacitor aided battery charging (generator run time) 207
thermoelectric battery charging (efficiency) 192

Post Script:

I would like to hear from you if you have any pertinent information about RV electrical and appliance systems that is not included in this book or differs from my opinions. I might want to include such information in a future edition or supplement to this book.

Contact me:

-- if you have any relevant observations about RV equipment or its use,

-- if you have any information on RV or camping laws,

-- if you feel something in this book is inaccurate (be very specific as to why you think it is inaccurate),

-- if you feel any of my criticisms of RV or component designs are overly harsh or unfair (I can't guarantee that I will change my opinions, but I will consider opposing points of view).

Thomas Childs
campianocal@gmail.com